Extel One Ltd

20. 1. 1984

THE PHYSICAL METALLURGY OF CAST IRON

THE PHYSICAL METALLURGY OF CAST IRON

I. Minkoff
Professor of Metallurgy
Technion-Israel Institute of Technology
Haifa, Israel

A Wiley–Interscience Publication

JOHN WILEY AND SONS
Chichester · New York · Brisbane · Toronto · Singapore

Library of Congress Cataloging in Publication Data:

Minkoff, I.
The physical metallurgy of cast iron.

'A Wiley–Interscience publication.'
Includes bibliographies and index.
1. Cast-iron—Metallurgy. 2. Physical metallurgy.
I. Title.
TN693.I7M56 1983 669'.96141 82-21984
ISBN 0 471 90006 0

British Library Cataloguing in Publication Data:

Minkoff, I.
The Physical metallurgy of cast iron.
1. Cast-iron—Metallurgy
I. Title
620.1'7 TA474

ISBN 0 471 90006 0

Typeset by Preface Ltd, Salisbury, Wilts
Printed by Page Brothers, Norwich

For Rena, Yael, Gidon, Gila, Boaz, Michael, and Nadia

Acknowledgements

I am grateful to the following for making material available and for other assistance while writing this book. Dr. H. Morrogh, FRS Director, and Mr. I. C. H. Hughes, Deputy Director, BCIRA, Alvechurch, England; Professor G. Ohira, University of Sendai, Japan; Professor K. Engels, Verein Deutscher Giesserei Fachleute, Dusseldorf, Germany; Professor F. Neumann, Brown, Boveri and University of Aachen, Germany; Professor B. Lux, Technical University, Vienna, Austria; Professor M. Hillert and Professor H. Fredriksson, Royal Institute of Technology, Stockholm, Sweden; Professor W. Kurz, ETH, Lausanne, Switzerland; Dr. H. Mayer, Sulzer Bros., Winterthur, Switzerland; M. H. Georgi, Georgi Publications, St. Saphorin, Switzerland.

NOTE ON UNITS

SI units have been used throughout the text. Where I have thought it useful to quote published information in the original units, I have indicated these in parentheses. Some of the graphs taken from various sources show units of the original publications. I have adopted an accepted convention of temperature units using degrees Celsius, °C, or the degree kelvin, K. The following short guide may be useful:

Stress: 1,000 psi = $6.89476\ \text{N mm}^{-2}$
1 kgfmm^{-2} = $9.80665\ \text{N mm}^{-2}$
Angström unit: $1\ \text{Å} = 10^{-8}\ \text{cm} = 0.1\ \text{nm}$
Thermal conductivity: $1\ \text{cal (cm s °C)}^{-1} = 418.68\ \text{W (m K)}^{-1}$

Contents

Preface x

1 Solidification and Microstructure 1
1-1 Solidification of Metals and Alloys 1
1-2 Homogeneous and Heterogeneous Nucleation 2
1-3 Growth Rate of Phases 6
1-4 Phenomena of Instability in Growth 15
1-5 Eutectic Solidification 22

2 Phase Diagrams; Examination of Cast Iron Structure 28
2-1 The Binary Phase Diagram of the Iron–Carbon System . 28
2-2 The Iron–Carbon–Silicon System 32
2-3 The Iron–Carbon–Phosphorus System 34
2-4 The Iron–Carbon–Silicon–Phosphorus System 37
2-5 Instrumental Techniques in Cast Iron Structure Research . 40

3 Solidification and Cast Structure of Gray and White Cast Iron . . 54
3-1 Nucleation 54
3-2 The γ-Graphite (Gray Iron) Eutectic 63
3-3 The Ledeburite Eutectic 81
3-4 Gray or White Solidification of Cast Iron 81

4 Thermodynamics of the Iron–Carbon System 86
4-1 The Iron–Carbon System; Fundamental Thermodynamic Relationships 87
4-2 Multicomponent Systems 89
4-3 Structural Diagrams for Cast Iron 94
4-4 Thermodynamics and Structural Diagrams 97
4-5 The Liquid–Cementite or Liquid–Graphite Transformation . 98
4-6 The Transformation of Austenite to Pearlite or Ferrite in Cast Iron 98

5 Solidification of Spheroidal Graphite Cast Iron 102
5-1 The Original Papers by Morrogh and Williams 102
5-2 Spherulitic Morphology and Surface Energy 103

5-3 Surface Energy Models of Spherulitic Graphite Growth in Cast Iron 107
5-4 Growth Models of Spherulites 108
5-5 Undercooling and Spherulitic Growth 113
5-6 Spheroidal Graphite Growth Studied by Thermal Measurements 124
5-7 Reviews of Growth of Graphite Spherulites from the Liquid . 127
5-8 Thermal Measurements as a Means of Control of Spheroidal Graphite Cast Iron Melts 127
5-9 Melt Chemistry and the Periodic System 129

6 Liquid Metal Preparation 134
6-1 Oxygen Adsorption in Iron Melting; The Iron–Oxygen–Carbon–Silicon System 134
6-2 Nitrogen in Cast Iron 137
6-3 Study of Desulphurization Variables in Cupolas 143
6-4 Melting in the Induction Furnace 148
6-5 Solubility of Hydrogen in Iron Based Alloys 151
6-6 Effect of Dissolved Gases 155
6-7 Effect of Elements Dissolved in Cast Iron during the Melting Process 155
6-8 Liquid Metal Melting and Inoculation 156
6-9 Liquid Metal Melting and Refining 157

7 Production of Cast Iron with Intermediate Structures 159
7-1 Intermediate Structures 160
7-2 Obervations of Intermediate Structures in Iron–Carbon–Silicon Alloys 161
7-3 Cast Iron with Vermicular Graphite (Compacted Graphite) . 165
7-4 Experiments by Thury on the Form of Graphite 167
7-5 Combinations of Elements Producing Intermediate Graphite Structures 169
7-6 Review of Theory with Respect to Morphological Change . 171
7-7 Scheil Segregation Equation 172

8 Alloy Cast Iron Systems 175
8-1 Alloy White Cast Irons 175
8-2 Gray or White Solidification in Alloy Cast Iron 184
8-3 Nickel in Cast Iron 185
8-4 Nickel–Chromium Alloy Martensitic White Cast Iron (Ni-Hard Cast Irons) 187
8-5 Austenitic Cast Irons 188
8-6 Low Alloy Gray and Spheroidal Graphite Cast Irons . . 190
8-7 Bainitic and Martensitic Spheroidal Graphite Cast Irons . 193
8-8 Aluminium Alloyed Cast Iron 195

8-9 Silicon Cast Irons 199
8-10 Manganese (Austenitic) Cast Irons 200
8-11 Tin as an Alloying Element in Cast Iron 201
8-12 Copper in Cast Iron 201

9 Phase Transformations and the Heat Treatment of Cast Iron . . 203
9-1 Internal Stresses in Cast Iron 203
9-2 Phase Transformations in the Heat Treatment of Cast Iron . 208
9-3 Surface Hardening 220

10 Graphitizing Reactions in the Production of Malleable Cast Iron . 228
10-1 Division of Malleable Iron Types 228
10-2 Kinetics of Graphitization 229
10-3 Other Graphitization Models 233
10-4 The Influence of Alloying Elements on Graphitization Kinetics in Blackheart Malleable Iron 235
10-5 Stabilization of Carbides 237
10-6 Dissolution of Cementite; Influence of Alloying Elements . 241
10-7 Shape of the Graphite Phase in Malleable Iron 245

11 Strength and Fracture of Cast Iron 248
11-1 Orowan's Model of a Brittle Solid 248
11-2 Stress–Deformation Models of the Mechanical Behaviour of Cast Iron 249
11-3 Stress–Strain Behaviour of Cast Iron 255
11-4 Stress at the Periphery of a Graphite Flake 257
11-5 Modulus for Cast Iron 257
11-6 Tensile Failure of Cast Iron 263
11-7 Ductile–Brittle Failure of Spheroidal Graphite Cast Iron . 268
11-8 Thermal Fatigue 272

12 Testing and Inspection 278
12-1 Mechanical Testing 278
12-2 Fracture Toughness Testing 279
12-3 Fracture Toughness Testing of White Cast Iron 287
12-4 Sonic Methods of Testing 288

Index 299

Preface

The main purpose in writing this text on the physical metallurgy of cast iron is to present aspects which have developed since the major work on this subject by the late Professor E. Piwowarsky. His text, *Hochwertiges Gusseisen*, was a remarkable summary of information available up to approximately 1950. Based mainly on the physical metallurgical knowledge and experimental techniques of that time, it relied extensively on thermodynamics of phase equilibria, phase diagrams, mechanical testing, and metallography. The information presented has provided invaluable knowledge to all concerned with this field.

The final period covered by his text coincided approximately with papers presented by R. Mehl, C. Zener, A. Hultgren, and others on the kinetics of phase transformations in metallurgical systems. At the same time, the important results of A. Gagnebin in the United States and H. Morrogh and W. J. Williams in the United Kingdom on the change of graphite form by melt treatment were published.

Cast iron is a unique material in which the properties are largely determined by the mode of solidification, in which a metastable and stable phase may crystallize, and graphite is an important component in the structure. The period covered by the present text has seen important advances in solidification theory both with respect to single-phase and to multi-phase systems. Part of this theory is related to the growth of metallic phases and much progress has also been made in understanding faceted growth of materials, including graphite. It is thus possible to present at least the outline of an understanding of the physical metallurgical basis of how structure develops in these alloys.

I have devoted approximately one half of the text to outlining different aspects of this theory. The rest of the text is concerned with other subject matter of importance to understanding structure and properties. Some attention is paid to structural diagrams for cast iron as related to freezing and solid state transformation in the Fe–C–Si system. I have reviewed some of the theory on the malleablizing process and some current understanding of the tensile failure of cast iron. Finally, I have discussed instrumental techniques available both for structural control and research. One of the important advances made in the period discussed is in instrumentation. Both the scanning electron microscope and electron probe microanalyser have found extensive application. Combinations of electron optical instrument techniques with testing methods including those used in evaluating fracture toughness have given new results of much interest.

I must apologize for various omissions. In considering the voluminous literature on cast iron during this period, it has not been possible to include a large number of the studies made. I have selected the material in an attempt to keep the text within reasonable proportions. Part of the subject matter presented has been used for lectures to industry where I have found that engineers and metallurgists are prepared to understand the physical basis for the structure–property relationship of these cast alloys. I have indicated additional reading matter for some of the sections. The twelve chapters are roughly divided into the following subject matter:

Chapter 1 A short review of solidification and growth of phases to introduce the iron–carbon system.
Chapter 2 Some published corrections of the Fe–C phase diagram and the Fe–C–Si ternary diagram. Some developments in instrumental techniques applied to the structure of cast iron. Study of graphite.
Chapter 3 The gray and white iron eutectic systems. The mode of growth of the graphite and ledeburite eutectics. The influence of elements in solution on the mode of growth and the fineness of the graphite eutectic.
Chapter 4 Some recent thermodynamic analyses centred on the Fe–C–Si structural diagram, particularly the manner of representing the influence of alloy elements on the eutectic composition. Developments starting from the Maurer and related diagrams. Phase transformations on cooling.
Chapter 5 Spherulitic graphite, examined on the basis of undercooling at the graphite–melt interface. Unstable modes of growth. Relationship between melt chemistry and graphite undercooling. Definition of role of elements in growth.
Chapter 6 A brief review of some of the published papers on the melting process of cast iron and the equilibrium between Fe–C and the gases O, N, and H. The relationship between oxidation equilibria and other phenomena such as nucleation and desulphurization.
Chapter 7 Adjusting the melt analysis to give graphite structure intermediate between flake and spherulite. Additive action of elements in solution on graphite growth. Degree of reactivity of elements. Effect of segregation on growth forms. Problems of segregation affecting structure in thick-section castings.
Chapter 8 A short review of the metallurgy of some of the alloy cast irons.
Chapter 9 Some contributions to transformations in heat treatment.
Chapter 10 A review of solid state graphitization research, drawing on different sources for the analysis of kinetics of the transformation. Discussion of carbide dissolution and alloy element effects on graphite growth. Shape of the graphite phase.
Chapter 11 A review of mechanical behaviour of cast iron in tensile loading. Brittle fracture theory. Ductile extension of a crack.

Chapter 12 Some selected methods of testing, including ultrasonic techniques and fracture toughness.

I commenced collecting material for this text during a Sabbatical leave at the Department of Metallurgy and Science of Materials, Oxford University. I wish to thank Sir Peter Hirsch FRS for accommodation and Mrs Betty Hall for library information. Considerable assistance in obtaining reference material was given by Mrs Margalit Goldberg, librarian of the Department of Materials Engineering, Technion, Haifa. I also wish to thank Abigail Solomon and Mrs. Hadassah Nesher for typing, Gai Proaktor, Batia Abramovitz, and André Zisserman for illustrations, and Aharon Levi and my daughter Gila for preparing the graphic work.

Chapter 1
Solidification and Microstructure

In this chapter, an introduction is given to some of the aspects of solidification which are of importance for an understanding of the structure of cast iron.

Initially, some fundamentals of the theory of nucleation are presented. An outline is then given of the calculations involved in computing the diffusion controlled growth rate of metallic phase. The growth of a non-metallic faceted phase is then discussed with reference to graphite. At small undercoolings, according to growth theory such a faceted crystal should grow by a defect mechanism. The types of defect present in graphite, and the growth mode, are then described.

In the following sections growth instability theory for a metallic system is presented. The unstable growth behaviour of graphite, a non-metallic material, is described as an introduction to understanding some of the forms in which graphite crystallizes in cast iron.

Finally, a brief introduction to eutectic growth is given. Note is made of the differences both in appearance and in mode of growth between normal metallic two-phase eutectic structures and the γ-graphite eutectic in which one phase is metallic and the second phase is a non-metal.

1-1 Solidification of Metals and Alloys

Solidification in metallurgical systems is the process of transformation of a liquid metal or alloy to a solid. When a pure metal solidifies, growth of the solid phase involves the transfer of heat, while for an alloy, solidification involves both the transfer of heat and the diffusion of matter.

After solidification is complete, further changes may occur on cooling. The result of these processes is the cast microstructure. This consists of the microconstituents which might be (a) primary phases, directly crystallized from the liquid, (b) phases formed by a peritectic reaction (c) eutectic phases, (d) secondary phases resulting from transformation of a primary phase.

The study of solidification also involves aspects of crystal growth. The two fields, solidification and crystal growth, enable an understanding and a description to be made of cast microstructure. The common subject matter between the two studies comes under various headings such as nucleation, dendritic growth, instability in growth, growth kinetics, the nature of surfaces, growth morphology, impurity effects, segregation, fluid flow.

1-2 Homogeneous and Heterogeneous Nucleation

The initial stage of phase separation from the liquid takes place by nucleation and requires that the first solid to appear grows to critical dimensions. When this occurs with the aid of some already existing surface it is called heterogeneous nucleation. The direct aggregation of atoms in the liquid in crystallographic order to achieve critical dimensions is called homogeneous nucleation. Nucleation theory is discussed, developing first the equations for homogeneous nucleation and second for heterogenous nucleation.

Homogeneous Nucleation in Solidification

A homogeneous nucleus in a solidification transformation forms by the association of atoms directly from the melt. The formation of the nucleus creates a new surface between the solid and liquid phases. Homogeneous nucleation may be the initiating mechanism of graphite formation in melts of very high purity. The following analysis of nucleation stems from the theoretical research of Gibbs.[1]

The change in free energy for the liquid to solid transformation is given by $\Delta G = \Delta H - T\Delta S$, where ΔG is the free energy differences per mole between the solid phase formed and the initial liquid, $\Delta G = G_S - G_L$, ΔH is the enthalpy change, and ΔS is the entropy change.

At the liquidus temperature, the free energy of liquid and solid are equal so that $\Delta G = 0$. Gibbs took into account the surface created between the liquid and the solid, as a result of the transformation. If the nucleus is considered as a sphere of radius r, the total free energy change will be given by:

$$\Delta G = 4/3\pi r^3 \Delta G_v + 4\pi r^2 \sigma \qquad (1\text{-}1)$$

where ΔG is now composed of two free energy terms of which ΔG_v is a bulk free energy term for the volume of the sphere excluding its surface and σ is the free energy of the surface per unit area created between the liquid and solid.

For any temperature below the liquidus, ΔG_v will be negative while σ is a positive term. The resultant ΔG has a maximum as shown in Figure 1-1.

The value of the radius of the sphere at the maximum of ΔG is the critical radius for nucleation r^* and is given by maximizing Equation (1-1):

$$\frac{d\Delta G}{dr} = 4\pi r^2 \Delta G_v + 8\pi r \sigma = 0 \qquad (1\text{-}2)$$

$$r^* = -\frac{2\sigma}{\Delta G_v} \qquad (1\text{-}3)$$

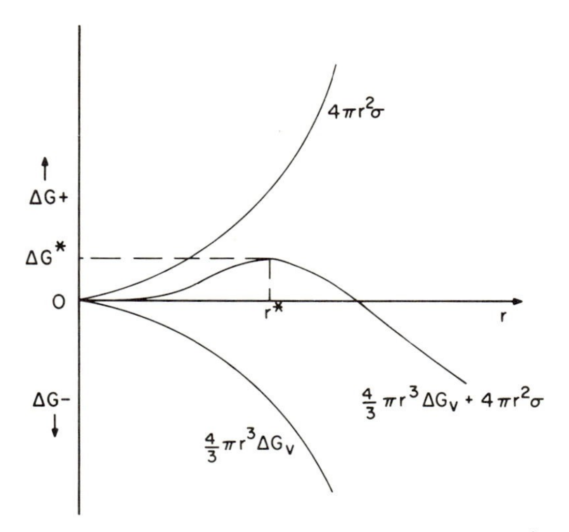

Figure 1-1 Change of free energy in homogeneous nucleation of a spherical phase having radius r

The value of ΔG at r^* is the critical free energy for nucleation:

$$\Delta G^* = \frac{16\pi\sigma^3}{3\Delta G_v^2} \qquad (1\text{-}4)$$

Radius r^* is directly dependent in value on σ and inversely dependent on ΔG_v. Since ΔG_v is proportional to the temperature of the systems below the liquidus, i.e. to the undercooling or ΔT, the value of r^* decreases as the undercooling ΔT increases. Hence nucleation becomes more probable at large undercoolings. The value of σ is considered independent of temperature.

Heterogeneous Nucleation

In practice, nucleation in liquids under normal conditions of solidification takes place heterogeneously. The new solid phase grows either at a mold surface or on a surface within the melt. This surface may be present as the result of different mechanisms. It can arise from a particle in the liquid originating at some extraneous location, by dissolution of parts of solid dendrites growing within the liquid, by reaction in the melt (e.g. sulphide formation, oxides, intermetallics, etc.), or by a prior transformation (e.g. peritectic).

Different phases may act as nuclei for graphite. Among those identified are duplex structures of sulphides and oxides. This will be discussed in Chapter 3.

A model for heterogeneous nucleation can be taken as a spherical segment on a flat surface[2] (see Figure 1-2). The phase β grows from the phase α on the

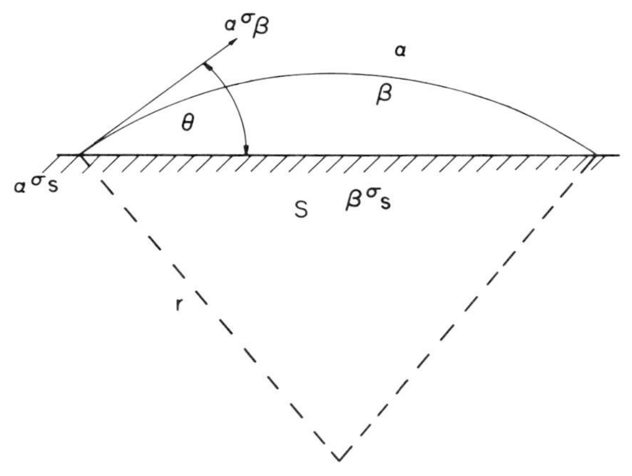

Figure 1-2 Model for heterogeneous nucleation of a spherical segment on a flat sub-strate

flat sub-strate S. The geometry of the β phase is a segment of a sphere of radius r, θ is the angle of contact, and the interfacial energies between the three phases, α, β, and S are ${}_{\alpha}\sigma_{\beta}$, ${}_{\alpha}\sigma_{S}$, and ${}_{\beta}\sigma_{S}$. The following relationship is obtained among the interfacial energies:

$$ {}_{\alpha}\sigma_{\beta} \cos\theta + {}_{\beta}\sigma_{S} = {}_{\alpha}\sigma_{S} \qquad (1\text{-}5) $$

Equation (1-1) now becomes, for the segment of Figure 1-2:

$$ \Delta G = \frac{\pi r^3}{3}\,\Delta G_v(2 - 3\cos\theta + \cos^3\theta) + {}_{\alpha}\sigma_{\beta}[2\pi r^2(1 - \cos\theta) - \pi r^2 \cos\theta(1 - \cos^2\theta)] \qquad (1\text{-}6) $$

Differentiating Equation (1-6) with respect to r and equating to zero, the value of the critical radius r^* is found to be:

$$ r^* = -\frac{2\,{}_{\alpha}\sigma_{\beta}\sin\theta}{\Delta G_v} \qquad (1\text{-}7) $$

The value of ΔG^* is given by:

$$ \Delta G^* = \frac{16\pi\,{}_{\alpha}\sigma_{\beta}^3[2 + \cos\theta)(1 - \cos\theta)^2/4]}{3\Delta G_v^2} \qquad (1\text{-}8) $$

When $\theta = 0$ in Equation (1-8), $\Delta G^* = 0$ and the only energy required for the formation of the nucleus is that necessary to form its periphery as the equilibrium liquidus temperature is passed, i.e. there may be little barrier to nucleation.

It can be shown that the value of θ tends to zero as the value of ${}_{\beta}\sigma_{S}$ tends to zero. The nucleation barrier for nucleation of a solid phase from a liquid by a solid surface is then related to the interfacial energy between the nucleated phase and the nucleating surface. This is discussed in Chapter 3.

The nucleation rate $\dot{N}$, the number of nuclei formed per unit of time, is given by the following expression:

$$\dot{N} = \frac{dN}{d} = K_v \exp[-(\Delta G_D + \Delta G^*)/kT] \tag{1-9}$$

where ΔG_D = activation energy for diffusion of the atoms across the interface to the nucleus
ΔG^* = activation energy for nucleation of β
$K_v = nkT/h$ in the Turnbull and Fisher theory[3]
n = number of atoms or molecules per unit volume of α
k = Boltzmann's constant
h = Planck's constant
T = temperature

Since ΔG_D is very much less than ΔG^* in liquids, the nucleation rate for heterogeneous nucleation is dependent on ΔG^*.

The value of the temperature at which $\dot{N}$ becomes appreciable is generally taken as that at which solidification commences (Figure 1-3). The variation of $\dot{N}$ with respect to the temperature below the liquidus, for heterogeneous nucleation, will depend on the value of the interfacial energy between the nucleated phase and the heterogeneous nucleus. Different values of the undercooling at which $\dot{N}$ becomes appreciable will correspond to different nucleating substrates and to the influence of contaminants. These varying values of ΔT are noted in graphite nucleation in cast iron.

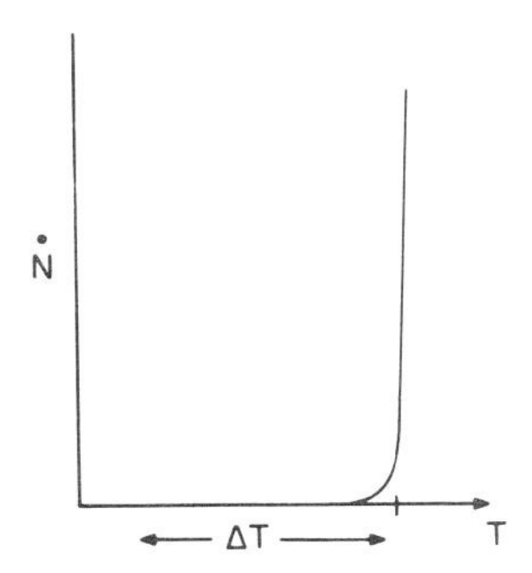

Fig. 1-3 The solidification temperature taken as that at which the nucleation rate becomes appreciable

1-3 Growth Rate of Phases

The growth rate of a phase is dependent on the mechanism which controls the rate of atom attachment to the surface. The controlling mechanisms can be diffusion, in which the flux of atoms controls the rate, all atoms reaching the surface being incorporated. Alternatively, the structure of the interface determines the rate at which atoms are accepted from the diffusion flux.

Two types of phase and their related mechanisms will be considered: (a) metallic phases and (b) non-metallic phases.

The metallic phase which crystallizes from the melt and is of importance in cast iron metallurgy is γ. Graphite is a typically non-metallic phase. It is a faceted crystal and is bounded by crystallographic planes of low index. At small undercoolings its growth is apparently determined by the presence of steps of a defect structure at the interface between the bounding crystal faces and the melt.

The γ phase, in contrast, must grow at a rate determined by solute diffusion in the liquid. In the development of the γ-graphite eutectic these features are of special interest since the structure is determined by the two phases growing by different mechanisms.

The comparative growth rates for the graphite and Fe_3C phases are of particular interest. Where these two phases may grow with equal probability at a particular temperature, their relative growth rates may determine which phase will dominate the structure. The kinetics of growth of the gray and white eutectics will be discussed in more detail in Chapter 3.

Diffusion Controlled Growth

The growth of phases by diffusion control and the subject of dendritic growth have been modified considerably by new research. A list of different books and papers covering these subjects, and including material directly applicable to castings, is given in the Bibliography at the end of this chapter.

A paper by Huang and Glicksman[4] discusses the theory and presents new experimental data. Huang and Glicksman divided dendritic growth into two parts. In one, the thermal diffusion process is solved neglecting sidebranch effects, and the steady state growth of a dendrite tip is calculated. In the second part, the non-steady state development of dendrite sidebranch structure is described. Steady state growth is described in terms of an effectively paraboloid shape. The problem of predicting a relationship between the growth velocity and the undercooling was treated and for this a criterion based on the stability of a dendrite tip was used.

This research requires a re-examination of calculations made involving dendritic growth in the Fe–C system. However, some of the published research is illustrative of the physical problems involved and will be briefly described here.

An early growth model of Zener[5] may be taken as a simple illustration of a

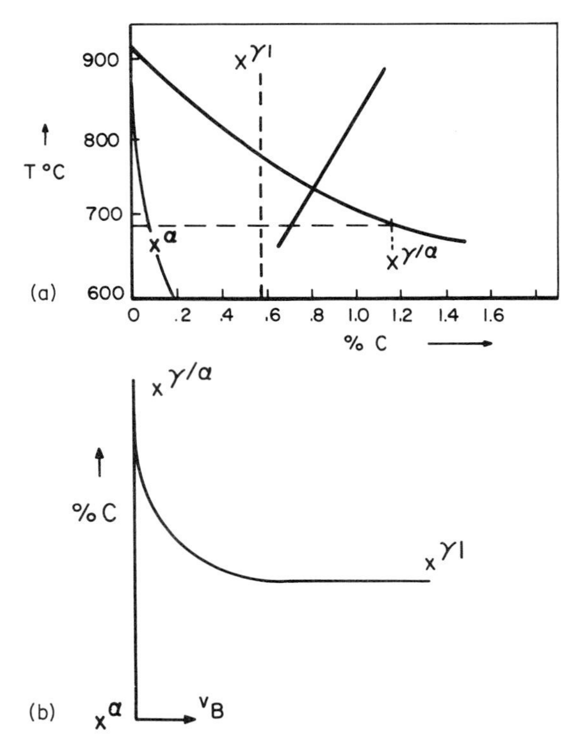

Figure 1-4 (a) Part of the Fe–C phase diagram with notation used for the growth of μα from γ.[6] (Reproduced by permission of The Metallurgical Society of American Institute of Mining Engineers.) (b) Distribution of solute ahead of growing α phase

treatment of the kinetics of growth. Zener calculated diffusion controlled growth for the case of the α phase in steel growing as a plate into the γ phase. For this geometry, growth was at a constant growth rate and could be measured. The phase diagram and solute distribution are shown in Figure 1-4(a) and (b).

Hillert[6] later modified Zener's equations. The nomenclature on the phase diagram follows that given by Hillert:[6]

x^{α} = composition of α phase
$x^{\gamma/\alpha}$ = composition at the interface between α and γ phases
$x^{\gamma 1}$ = initial composition of γ phase

The movement of the interface is given by

$$-D\left|\frac{\mathrm{d}c}{\mathrm{d}x}\right|_{\gamma/\alpha} = (x^{\gamma/\alpha} - x^{\alpha})v_{\mathrm{B}} \tag{1-10}$$

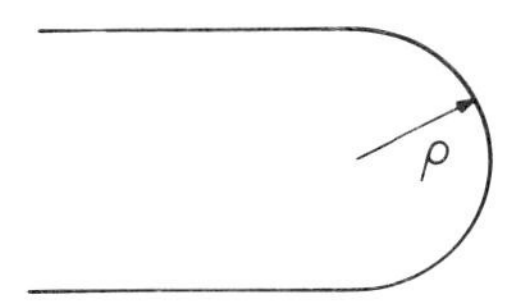

Figure 1-5 Tip of the growing phase in Zener's analysis[4]

where v_B is the velocity of the interface:

$$v_B = -\frac{D}{x^{\gamma/\alpha} - x^{\alpha}} \cdot \frac{\Delta C}{L} \tag{1-11}$$

In this equation, an average concentration gradient $\Delta C/L$ was employed. In Zener's analysis, L, the size of the solute layer, was taken equal to ρ, the radius of the tip of the growing phase. This is shown in Figure 1-5.

From this assumption, the smaller the tip radius ρ, the smaller L and the bigger will be v_B. Zener then set a limit to ρ which he put at ρ_c, the critical radius. At this radius, the growing phase will be in equilibrium with the matrix phase and hence no growth will occur. The maximum rate was derived to be at a radius equal to $2\rho_c$. The velocity v_B is then given as

$$v_B = \frac{2D}{\rho} \frac{x^{\gamma/\alpha} - x^{\gamma 1}}{x^{\gamma/\alpha} - x^{\alpha}}$$

This relationship was modified by Hillert[6] who gave the following expression

$$\frac{v_B \rho}{D} = \frac{1}{2} \frac{\Omega}{1 - \Omega} \quad \text{where} \quad \Omega = \frac{x^{\gamma/\alpha} - x^{\gamma 1}}{x^{\gamma/\alpha} - x^{\alpha}}$$

Thus

$$\frac{v_B \rho}{D} = \frac{1}{2} \frac{x^{\gamma/\alpha} - x^{\gamma 1}}{x^{\gamma 1} - x^{\alpha}}$$

Hillert's use of Zener's equation for estimating growth rates in an examination of eutectic growth in gray and white cast iron is given in Chapter 3 (Figures 3-17 and 3-18).

Interface Controlled Growth of the Graphite Phase

The graphite phase is a faceted crystal bounded by low index planes. Figure 1-6 shows some faces of a crystal observed in nature.[7] In a graphite flake crystallized from a liquid iron carbon alloy the normally observed bounding planes are (0001) and $\{10\bar{1}0\}$. Figure 1-7(a) shows a (0001) face of a graphite crystal bounded by planes of the $(10\bar{1}0)$ faces. The atomic arrangement within the hexagons is shown related to the bounding faces of the crystal. The

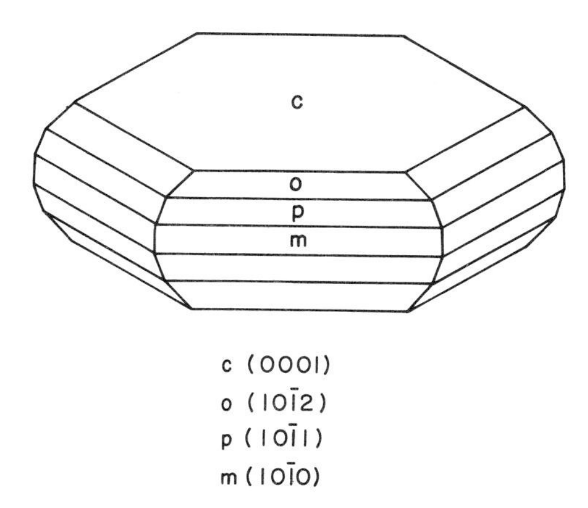

Figure 1-6 Facets on a natural crystal of graphite after Palache.[7] (Reproduced by permission of Mineralogical Society of America)

crystallographic structure of graphite is shown in Figure 1-7(b). In eutectic graphite the edges of the plate-like graphite crystals are not well defined. Unstable growth occurs on $\{10\bar{1}0\}$ so that these crystals may be bounded by faces of differing orientations.

Structure and Rate of Growth of a Faceted Graphite Crystal

An approximate division can be made between crystals which grow at a rate controlled by diffusion in which all atoms find their place in the growing solid and crystals which grow at a rate determined by the manner in which the structure incorporates them into the surface. Crystals of the first kind are typified by the metallic structures of which an example is the γ phase in

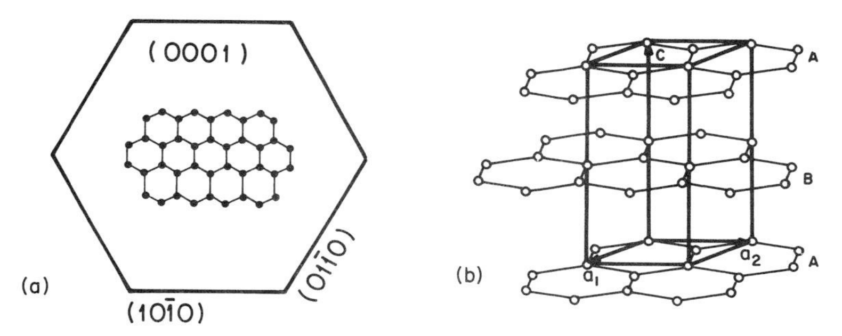

Figure 1-7 (a) Crystal of graphite bounded by (0001) and (10$\bar{1}$0) type faces. The hexagonal arrangement of the atoms within the (0001) plane is shown related to the bounding (10$\bar{1}$0) faces. (b) The hexagonal structure of graphite showing the unit cell (heavy lines)

Fe–C. Crystals of graphite should be of the second type, i.e. the growth probably controlled by the rate at which the different surface structures can incorporate atoms.

A calculation can be made of the degree of roughness to indicate whether the surface structure is controlling the rate of crystal growth[8]. It is shown that crystals can be divided by a structural parameter into two classes, faceted and non-faceted. This parameter is:

$$\alpha = \frac{L_0}{kT_E} \cdot \frac{\eta_1}{\nu} \tag{1-12}$$

where L_0 is the change in internal energy on freezing or condensation, k is Boltzmann's constant, and (η_1/ν) is equal to the number of nearest neighbours lying in a plane parallel to the crystal face being considered divided by the maximum number of nearest neighbours in the bulk solid. The calculation made by Jackson shows that for an α factor greater than 2, surfaces would be essentially smooth and correspond to a structure considered for a faceted crystal.

Growth of Faceted Crystals

Figure 1-8(a) shows the structure of different faces of a two-dimensional crystal as a function of the index of the face. The [01] face which is the lowest index face is structurally smooth while the [11] and [13] faces have a stepped structure.

The faces [11] and [13] can advance in directions normal to their planes by atoms adding on to the steps. This is shown in Figure 1-8(b). New atoms join at [1], followed by [2], followed by [3], and the face advances from a to b.

The processes can occur until the situation shown in Figure 1-8(c) takes place. In this, all the high index faces have grown out of the crystal, by virtue of their movement due to the attachment of atoms to their stepped surfaces. These provided the growth mechanisms until the surfaces have finally grown out of the crystal, which is then bounded by the low index faces.

Growth of the crystal can now occur by two mechanisms:

(a) Growth from steps due to defects (dislocations, boundaries, etc.).
(b) Growth by nucleation of new faces.

Movement of Steps over a Graphite Crystal Surface

A faceted crystal grows by the movement of steps over the crystal surface and the atoms attach to kinks on the steps (Figure 1-9a). For a solitary step on a low index plane reaching the edge of the crystal, further growth would require the nucleation of a new plane. However, defects normally present in crystals can provide steps for growth. A mechanism involving growth from the step of a screw dislocation was proposed by Burton, Cabrera, and Frank.[9] In this mechanism, the step of the screw dislocation shown in Figure 1-9(b) becomes,

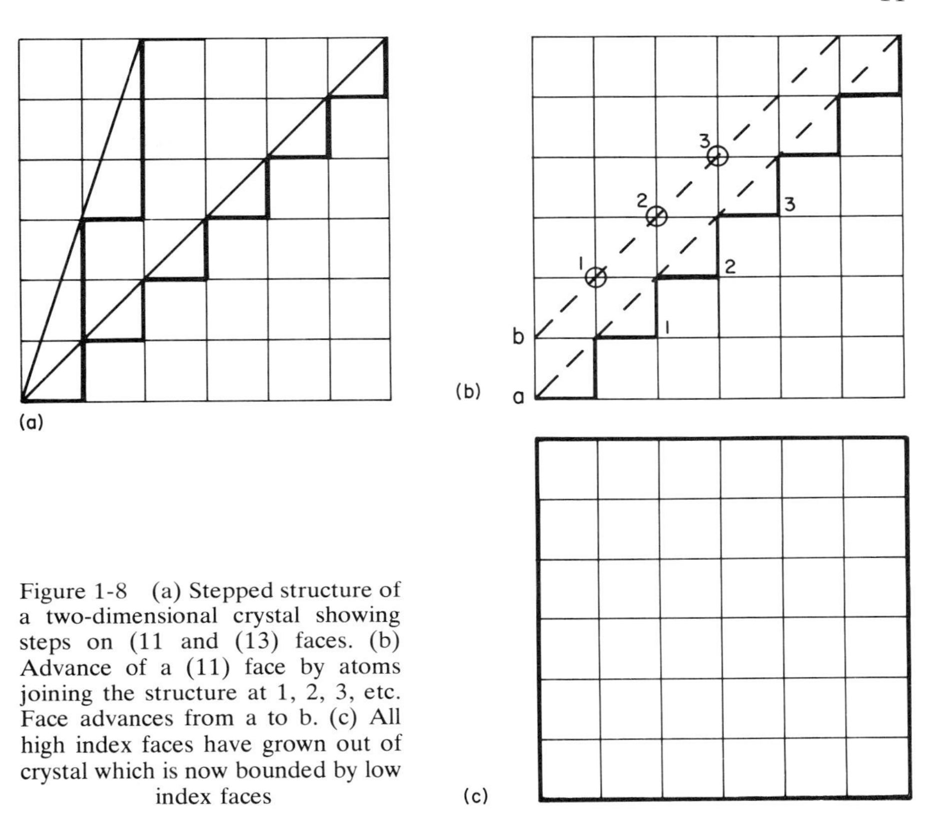

Figure 1-8 (a) Stepped structure of a two-dimensional crystal showing steps on (11 and (13) faces. (b) Advance of a (11) face by atoms joining the structure at 1, 2, 3, etc. Face advances from a to b. (c) All high index faces have grown out of crystal which is now bounded by low index faces

in growth, the spiral of Figure 1-9(c). Atoms attach to the step of the spiral and the rotation which results gives growth of one step height per revolution. Examination of graphite crystals[10,11] shows a pronounced dislocation density, the Burgers vectors of which have screw character. At small undercoolings, this is a suggested mechanism of growth of the (0001) face of graphite.[12,13]

Step Defect in Graphite Due to a Rotation Boundary

Another step defect in graphite is shown in Figure 1-10(a) and is in the form of a twist or rotation boundary.[14] It is a type of twin. The two parts of the crystal are in rotational symmetry by an angle ϕ. This angle, when measured for graphite, has fixed values which can be determined theoretically.[15] The growth of $\{10\bar{1}0\}$ faces at this type of step is shown schematically in Figure 1-10(b). Ordinarily, the $(10\bar{1}0)$ faces would require individual nucleation processes for each growth plane. The presence of the twin provides a step on which the new faces can be nucleated.

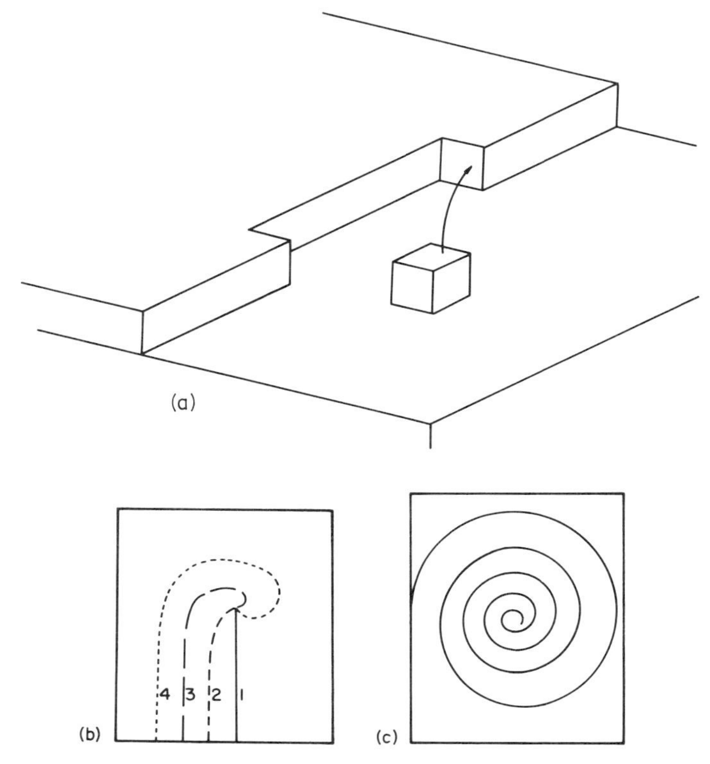

Figure 1-9 (a) Growth of a face on a faceted crystal due to atom attachment at kinks. (b) Initial movement of step of a screw dislocation. First position 1; succeeding positions 2, 3, and 4. (c) Growth spiral resulting from movement of step of screw dislocation

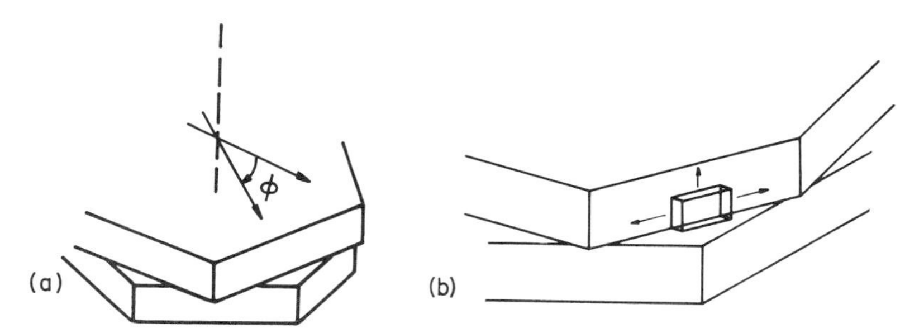

Figure 1-10 (a) Step defect in graphite in the form of twist or rotation boundary. (b) Nucleation of $(10\bar{1}0)$ face on step of rotation boundary

Rate of Growth of (0001) and (10$\bar{1}$0) Faces; Lamellar Form of Graphite

The rate of growth of the (0001) graphite surface at small undercoolings is suggested to have the temperature dependence shown in Figure 1-11(a). This is by correlation with the calculated temperature dependence of growth rate of a surface from screw dislocations.[9] At small values of ΔT, the growth rate dependence on ΔT is parabolic, i.e. $R \propto \Delta T^2$. At a higher undercooling, the growth rate dependence is linear, i.e. $R \propto \Delta T$.

The growth rate dependence of (10$\bar{1}$0) is suggested to be exponential with respect to undercooling. Figure 1-11(b) suggests that the graphite crystal may grow edgewise at different values of ΔT depending on the nucleation of (10$\bar{1}$0) planes at the step. Three possible curves, numbered 1,2, and 3, are shown. The greatest undercooling for growth, curve 1, would represent the case for growth if no defect boundary were present. Curve 2 represents the case for a lowering of ΔT when the rotation boundary step is present. Curve 3 is suggested as a minimum undercooling for growth when a contaminating impurity such as, for example, sulphur is present in the melt. This would provide the minimum critical free energy for nucleation at the step by reducing the edge energy of the nucleus (see Chapter 3).

The lamellar form of graphite is the result of the difference in rate of growth between the nucleation induced growth at a rotation boundary for (10$\bar{1}$0) and the growth of (0001) controlled by atom attachment at the steps of screw dislocations. The ratio of the two growth rates should provide the approximate dimensions of lamellar graphite which is close to 10 : 1 by measurement.

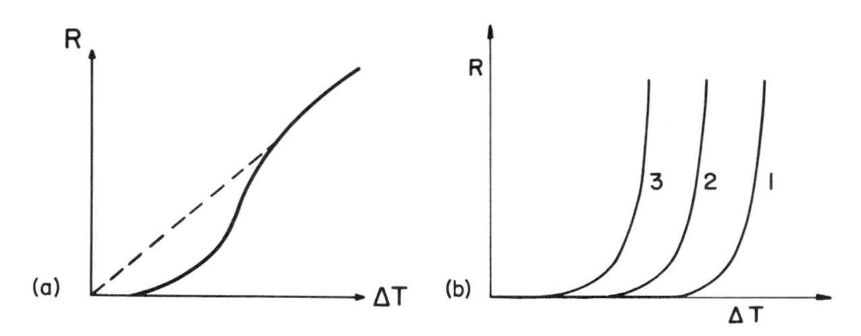

Figure 1-11 (a) Growth rate dependence of a surface from screw dislocations.[9] At small undercooling, growth rate dependence is parabolic. At higher undercooling, growth rate dependence is linear. (b) Growth rate of (10$\bar{1}$0) faces is suggested to be nucleation dependent and exponential. Different undercoolings corresponding to curves, 1, 2, and 3 relate to different conditions for nucleation

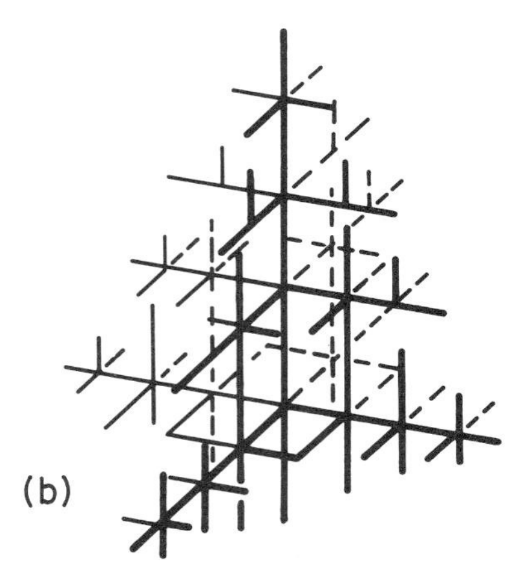

Figure 1-12 (a) Faceted crystals show different unstable morphologies than those normally shown by metals. Illustrated is a dendritic hopper crystal of Mg_2Si[16] (× 2,700). (Reproduced by permission of Chapman and Hall Ltd.) (b) Dendritic branching mode normally observed for a metallic crystal.[17] (Schematic)

1-4 Phenomena of Instability in Growth

Developments in the theory of instability are discussed in Chapters 3 and 5. Review papers on this subject, particularly those of Müller-Krumbhaar and Langer are quoted in the Bibliography at the end of this chapter.

In the observations of interface instability between a growing solid and liquid in metallic systems, a cellular interface first develops which may be followed later by dendritic branching. Dendrites are common growth forms in cast metallurgical systems. The γ phase in cast iron is an example.

Faceted crystals show different morphologies, such as the dendritic hopper crystal of Mg_2Si in Figure 1-12(a)[16]. This is to be compared with the dendritic branching mode for a metallic crystal generally observed in liquids (Figure 1-12b).[17]

Graphite has unusual instability behaviour. At small undercoolings, branched dendrites from $\{10\bar{1}0\}$ faces can be noted. This mode of growth establishes the lamellar radiating structure of a eutectic cell. At larger undercoolings, the instability manifests itself in rod forms and in spherulitic crystals.

Intermediate forms between branched dendrites and spherulites can be observed by varying the solute type and content of the alloys. Small impurity contents can have large effects on interface stability.

Two theories relating interface stability to undercooling, principally for metallurgical systems, will first be briefly described. Discussion will then be given of interface instability in a faceted system, of which graphite is an example.

Constitutional Supercooling Theory[18]

A schematic solute distribution ahead of a growing surface is illustrated in Figure 1-13(a). Different conditions exist in growth for which different solute distributions obtain in the liquid and in the solid.[17] The case considered here is that for limited diffusion in the liquid.

After an initial growth period of solid in an alloy, the solute concentration at the interface between solid and liquid reaches a steady-state value of C_0/k_0, where C_0 is the initial composition of the liquid and k_0 is the equilibrium distribution coefficient of solute between solid and liquid. The distribution of solute in the liquid is given by the following equation:

$$C_L = C_0\left[1 + \frac{1 - k_0}{k_0}\exp\left(-\frac{Rx}{D}\right)\right] \qquad (1\text{-}13)$$

where R is the rate of growth and D is the diffusion coefficient of solute in the liquid.

From the equilibrium diagram, Figure 1-13(b), the equilibrium temperature at any composition C_L in the liquid is given by $T = T_0 - mC_L$, where m is the gradient of the liquidus and is equal to $\Delta T/\Delta C$. Hence the liquidus tem-

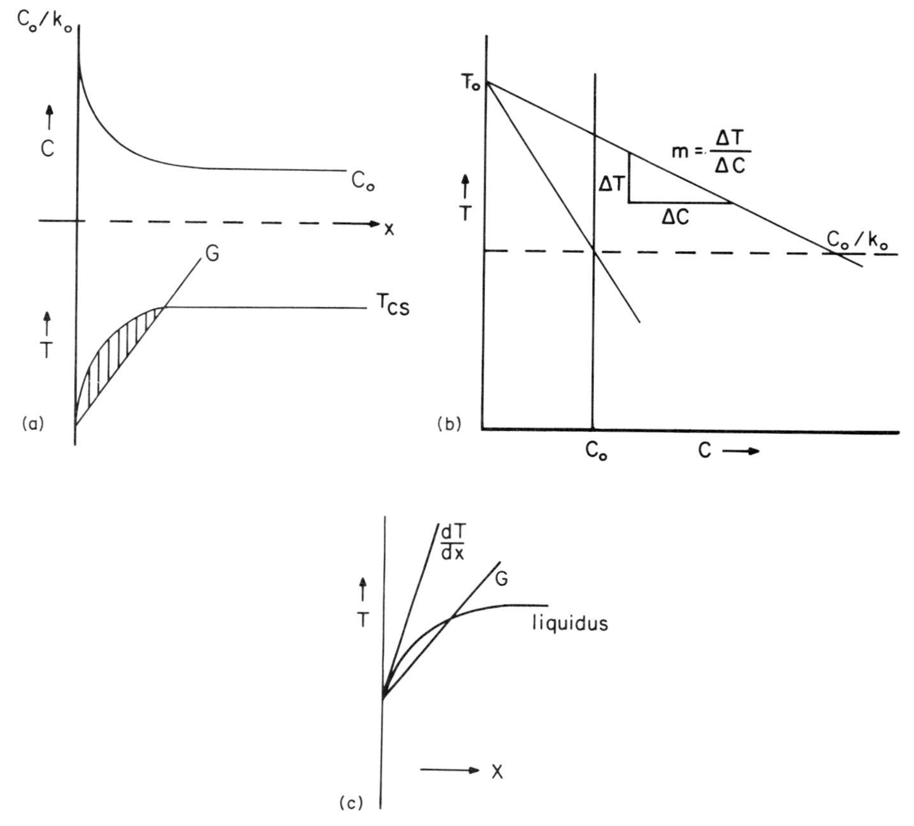

Figure 1-13 (a) Schematic distribution of solute ahead of a growing interface (upper curve). The lower curve shows the temperature distribution according to the liquidus of the solute profile T_{Cs}. Curve G is the temperature gradient in the liquid. (b) Schematic equilibrium diagram showing gradient of the liquidus(m). (c) The surface becomes unstable according to constitutional supercooling theory of dT/dx for the liquidus profile is greater than G, the existing temperature gradient

perature profile corresponding to the solute profile is given by:

$$T = T_0 - mC_0\left[1 + \frac{1 - k_0}{k_0}\exp\left(-\frac{Rx}{D}\right)\right] \qquad (1\text{-}14)$$

The difference between the liquidus profile and the actual temperature gradient existing in the liquid gives the supercooling of the liquid ahead of the interface. Any surface instability, according to this theory, should grow. From Figure 1-13(c), the limit of instability is when

$$| dT/dx |_{x=0} > G \qquad (1\text{-}15)$$

where $| dT/dx |_{x=0}$ is the temperature gradient of the liquidus profile at the interface and G is the actual temperature gradient. Writing $m = \mathrm{d}T/\mathrm{d}c$, Equation (1-15) can be written:

$$m\frac{\mathrm{d}c}{\mathrm{d}x} > G \qquad \text{for instability} \tag{1-16}$$

or

$$-G + mG_c > 0 \tag{1-17}$$

where $G_c = \mathrm{d}c/\mathrm{d}x$ is the gradient of the solute at the interface. In terms of the physical parameters, Equation (1-16) can be written:

$$mC_0\frac{R}{D}\left(\frac{1-k_0}{k_0}\right) > G \qquad \text{for instability} \tag{1-18}$$

or

$$\frac{G}{R} < \frac{mC_0}{D}\left(\frac{1-k_0}{k_0}\right) \tag{1-19}$$

For metallic systems, this expression has been tested and gives a representation of the conditions required to unstabilize a surface. No parameter related to the solid actually enters these expressions.

The Mullins and Sekerka Theory

In the Mullins and Sekerka theory[19] a small sinusoidal perturbation is imposed on the system and calculation is made of the dependence of the perturbation on time. If the perturbation grows without limit, the interface is unstable, while if the perturbation eventually decays to zero, the interface is stable. The theory is applicable to an isotropic surface and includes capillarity. Therefore it also takes into account surface energy. By including temperature gradients in the solid, it relates instability with velocity of growth of the interface.

A criterion obtained by Mullins and Sekerka is the following:

$$-\tfrac{1}{2}(\zeta' + \zeta) + mG_c > 0 \qquad \text{for instability} \tag{1-20}$$

where $\zeta' = \dfrac{K_s}{\bar{K}}G$

$\zeta = \dfrac{K_L}{\bar{K}}G$

and
G_c = constitutional gradient
m = slope of the liquidus as in the previous section
$\bar{K} = \tfrac{1}{2}(K_S + K_L)$
K_S = thermal conductivity of the solid
K_L = thermal conductivity of the liquid

Equation (1-20) resulting from the Mullins and Sekerka theory is similar to Equation (1-17) of the Chalmers theory but is not identical. The Mullins and

Sekerka theory involves the temperature gradients in the solid as well as in the liquid, the interface velocity, and the latent heat. In the Chalmers theory, an interface becomes unstable when it experiences a perturbation in the sense of an irregularity of growth, which can continue in the undercooled liquid.

In the Mullins and Sekerka theory, the perturbation for instability can be calculated. As has been pointed out, it is possible to have constitutional supercooling but no instability. The opposite is also true, i.e. an instability of the interface without constitutional supercooling.

Instability of a Faceted Surface; Chernov's Theory[20]

The constitutional supercooling and the Mullins and Sekerka theories are applicable to the case of materials having an isotropic surface energy and isotropic interface kinetics, which is the case for metals. For this type of material, the rate at which atoms attach to the surface is independent of crystallographic orientation.

For non-metallic materials like graphite, which grow in a faceted manner, faces close to the low index planes have a stepped structure and their growth rate is orientation dependent. As the orientation is changed, therefore, the growth rate of a face should change.

Chernov[20,21] has made analyses of instability in materials which have a prounounced growth rate anisotropy. The unstable forms of these materials look different from the normally observed metallic dendrites. Also their eutectics do not have the organized geometries noted in metallic systems. An example is the hopper type of crystal (Figure 1-12a) which has no morphological equivalent in metallic systems. It has been shown by Minkoff and Lux to be part of a growth requirement for the development of a spiral eutectic.[16] Chernov presents an analysis[21] for an hour-glass form of crystal which is close in morphology to a hopper form.

The fact that the manner of growth of metallic dendrites and unstable non-metallic crystals is different requires that this growth be approached analytically in a different manner. The metallic dendrite is considered to grow, when an interface becomes unstable in a sinusoidal manner. As growth proceeds, arms develop as side branches.

In faceted crystals, the unstable form may arise by a different process. A planar face of a faceted crystal has initially a homogeneous supersaturation over its surface. As the size of the crystal increases, the supersaturation varies, especially between centres and corners of edges of crystal. The limiting size of crystal in which the supersaturation is uniform is given by

$$\frac{\beta R}{D} \ll 1$$

where β = attachment coefficient, defined below with units of centimetres per second
R = growth rate
D = diffusion coefficient

Figure 1-14(a) shows the form of the supersaturation existing over a surface between the outer boundary and the face centre.

The kinetic coefficient β in the case of a rough surface having many steps is given by

$$\frac{D\,\delta c}{\delta n} = \beta(C - C_e)$$

where C = solution concentration
C_e = equilibrium concentration
D = diffusion coefficient
$\delta/\delta n$ = derivative along the normal to the surface

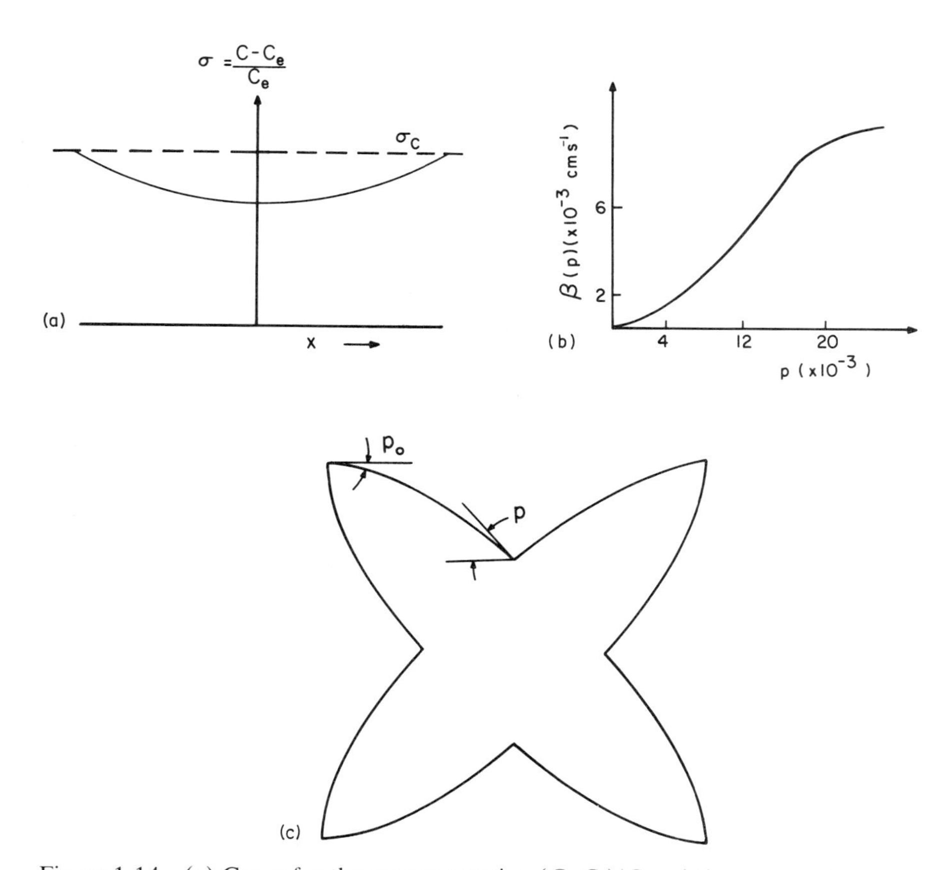

Figure 1-14 (a) Curve for the supersaturation $(C-C_e)/C_e$ existing over a planar surface between the centre and outer boundary σ_c is the critical supersaturation for generation of layers at crystal vertices. (b) Variation of the kinetic attachment coefficient β as a function of angle p made by a surface with a singular (faceted) surface of β-methylnaphthalene. (c) Hour glass form related to angle (p) made with faceted surface.[21] (Reproduced by permission of North-Holland Publishing Company)

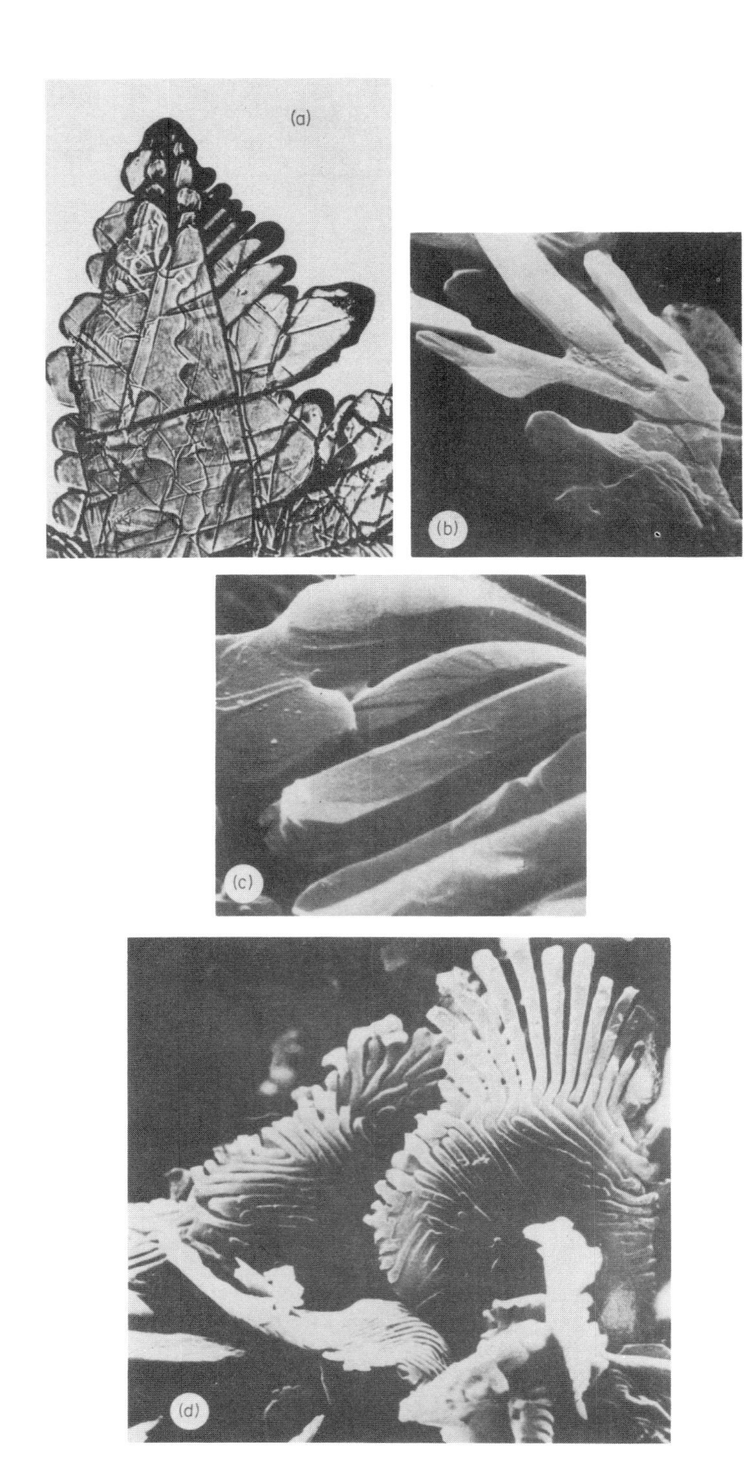
(a)
(b)
(c)
(d)

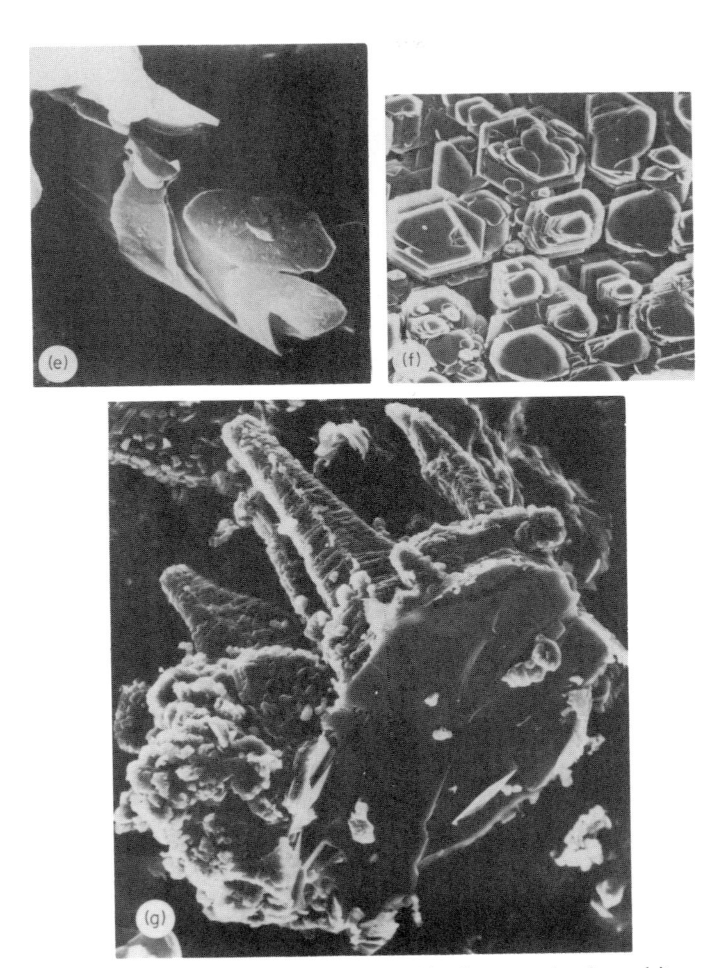

Figure 1-15 Different types of instability noted in the growth of graphite crystals. (a) Dendritic branching of primary graphite crystal extracted from Bi–C alloy[14] (× 40). (Reproduced by permission of Taylor and Francis Ltd.) (b) Eutectic graphite crystal branching[22] (× 2,280). (Reproduced by permission of American Foundrymen's Society.) (c) Steps on a graphite crystal surface which became unstable and formed elongated ledges[23] (× 4,600). (Reprinted by permission from *Nature*, Vol. 225, p. 541, Fig. 2. Copyright © 1970, Macmillan Journals Limited.) (d) The ledges formed by steps becoming unstable on crystal surfaces can leave the crystal surface and grow into the liquid[24] (× 600). (Reprinted with permission from *Micron*, Vol. 2, p. 286, Fig. 3, by I. Minkoff and B. Lux, Copyright © 1971, Pergamon Press, Ltd.) (e) Ledges can grow round crystals[25] (2,800). (Reprinted by permission from Georgi Publ. Co., Switzerland.) (f) Pyramidal growth on (0001)[26] (× 1,200). (g) Elongated pyramidal growth bounded by $(10\bar{1}1)$[27] (× 4,000)

Chernov shows the anisotropy of the kinetic coefficient, an example being given in Figure 1.14(b). The variation is shown of an effective coefficient as a function of tangent of the angle (p) for faces vicinal to a singular (faceted) surface in β-methylnaphthalene. The value of β increases with p.

An hour-glass form results from the change of geometry of the surface to give increasing values of p towards the face centre (Figure 1-14c). Then the kinetic coefficient β is greater at the centre where the supersaturation is the smallest so that the hour-glass geometry grows while remaining parallel to itself, satisfying the growth rate relationships. Near the edge of the crystal, the saturation is high and the surface adjusts to a smaller slope p. This is a shape preserving mode.

Chernov also discusses the instability of plane faces by forming hillocks, the angle of the hillock increasing with the supersaturation. Chernov's papers are of interest here because they deal with the instability of faceted surfaces. In his analyses, changes of orientation are accounted for by changes in the kinetic coefficient. This is different from the case of metals whose surfaces have isotropic growth rates.

The theory for faceted crystal instability must still be developed, to take into account the quite different growth forms which are observed as compared with metallic dendrites.

Types of Instability Observed in Graphite Growth

Different types of instability which occur in graphite are summarized as follows:

(a) Primary crystals branch dendritically from $(10\bar{1}0)$ faces (Figure 1-15a).[14]
(b) Eutectic crystals branch out of the edge of the lamellae (Figure 1-15b).[22]
(c) Steps on graphite crystals become unstable and form elongated ledges. This can be seen on plate, cylindrical, or spherical crystals (Figure 1-15c).[23]
(d) The ledges formed on surfaces by steps becoming unstable can leave the crystal surface and grow into the liquid (Figure 1-15d).[24]
(e) The ledges can grow round the crystals (Figure 1-15e).[25]
(f) Pyramidal growth can occur on (0001) (Figure 1-150f).[26]
(g) Elongated pyramidal crystals grow bounded by $(10\bar{1}1)$ (Figure 1-15g).[27]

These instabilities of growth can be related to the undercooling as calculated on the basis of impurity effect,[26] and the morphologies of graphite observed in cast iron, at least in part, can be related to the cases of instability studied. This represents some of the subject matter dealt with in Chapters 5 and 7.

1-5 Eutectic Solidification

Eutectic solidification will be considered here for two cases:

(a) The growth of two non-faceted phases as, for example, in metallic systems.

(b) Growth of a two-phase structure having a non-facted and a faceted phase as in the γ-graphite eutectic.

Growth theory applied to eutectics in metallurgical systems employs analysis related to diffusion kinetics. The theory for eutectics having one faceted phase should also take into account interface kinetics.

Eutectics in metallurgical systems are ordinarily regular in their geometry and consist either of plates or of rods. Eutectics in systems having one faceted phase are generally irregular in their geometry as, for example, in freely solidifying gray iron eutectic.

Growth of Eutectic Having Two Metallic Phases

The theory for the diffusion controlled growth of a two-phase structure was first presented by Zener[4] for the pearlite transformation. The equations of Zener were modified by Hillert[6] and applied to the eutectic transformation. Further analyses have been given by Tiller[28] and by Hunt and Jackson.[29] These researches have provided theories giving the relationship between the growth velocity (R) or the undercooling (ΔT) and the lamellar or rod spacing (λ). It is also possible to evaluate the interface shape. A summary of some of the theoretical work of growth of two-phase structures was given by Hillert.[6]

In Zener's analysis, the velocity of growth of the two-phase mixture was calculated from a diffusion analysis. An effective diffusion distance L was taken, related to the interlamellar spacing S. Zener wrote $L = \alpha S$ where α is a numerical coefficient.

In the equations of Zener, different values of the spacing S could be assigned for a particular undercooling, allowing different values of the growth velocity. Only one value of the lamellar spacing is observed for any one value of ΔT and Zener assumed that the structure which gave the maximum velocity for a given undercooling would be that which developed. Only one value of S is then given, calculated to be equal to $2\ S_c$. The value of S_c is a critical lamellar spacing where no concentration difference exists and the growth rate would be equal to zero.

The Jackson–Hunt analysis[29] presented the steady-state solution for the diffusion equation for a lamellar eutectic growing with a plane interface. A similar solution is also found for phases having rod morphology. Theoretical analysis was made of the solute distribution and diffusion at an α/β interface of plates or rods in contact with liquid. The interface undercooling required for the transformation was divided into several quantities which included the following:

ΔT_D = undercooling required to drive the diffusion
ΔT_λ = undercooling required for the formation of phase boundaries
ΔT_K = kinetic term for the undercooling required for solute attachment kinetics
ΔT_σ = capillary term due to a non-planar interface

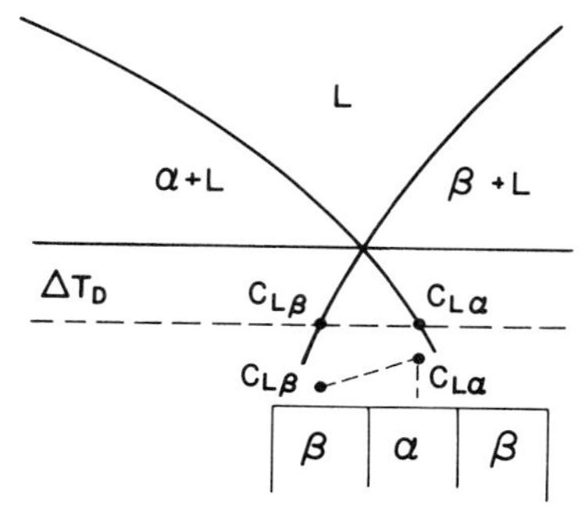

Figure 1-16 The composition of the two phases α and β in contact with liquid at a eutectic interface for an undercooling ΔT_D. The composition difference $C_{L\alpha} - C_{L\beta}$ drives the solute diffusion

The ΔT_K term, important for the graphite eutectic and discussed later in this text, is comparatively unimportant in the growth of metallic phases. The ΔT_D term allows the two phases α and β to be in contact with different compositions of liquid. This is shown in Figure 1-16. The composition difference ($C_{L\alpha}$ – $C_{L\beta}$) drives the solute diffusion.

The Jackson–Hunt analysis gave the following equation:

$$\frac{\Delta T}{m} = v\lambda Q^L + \frac{a^L}{\lambda} \tag{1-21}$$

where ΔT = undercooling term comprising the separate terms above
v = interface velocity
λ = eutectic spacing
m, Q^L, and a^L = constants given by:

$$\frac{1}{m} = \frac{1}{m_\alpha} + \frac{1}{m_\beta}$$

where m_α = gradient of the liquidus of the α phase
m_β = gradient of the liquidus of the β phase

$$Q^L = \frac{P(1 + \xi)^2 C_0}{\xi D}$$

where P = a term dependent on S_α/S_β, tabulated by Hunt and Jackson[29]
$\xi = S_\beta/S_\alpha$ where S_β and S_α are the widths of the α and β phases
D = diffusion coefficient

$$a^L = 2(1 + \xi)\,\frac{a_\alpha^L}{m_\alpha} + \frac{a_\beta^L}{m_\beta}$$

where a_α^L and a_β^L are derived from the Gibbs–Thomson equation and the superscript L refers to the case for lamellar growth.

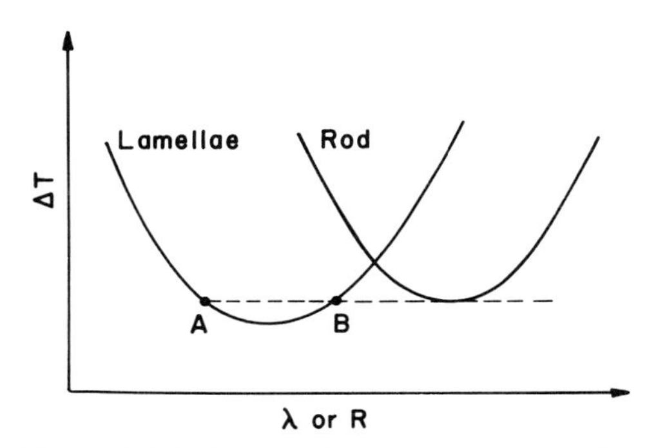

Figure 1-17 Schematic representation of the equation $\Delta T/m = v\lambda Q^{L} + a^{L}/\lambda$ for a given velocity. The lamellar structure should grow to the right (point B) of the minimum undercooling.[29] (Reproduced by permission of The Metallurgical Society of American Institute of Mining Engineers)

Figure 1-17 shows the relationships between ΔT and λ or R, Equation (1-21), where R is the rod spacing. It can be shown that for a lamellar eutectic, the spacing λ is that equivalent to point B. The equation has a minimum value of ΔT for λ or R at a constant growth velocity given by:

$$\lambda^2 v = \frac{a^{L}}{Q^{L}} \tag{1-22}$$

or

$$\frac{\Delta T^2}{v} = 4m^2 a^{L} Q^{L} \tag{1-23}$$

or

$$\Delta T \lambda = 2ma^{L} \tag{1-24}$$

The relationship λ = constant $v^{-0.5}$ has been tested for a number of systems and found to be applicable (see Section 3-2).

Observed Differences between Non-faceted and Faceted Eutectics

Observations of the microstructure in eutectic systems with a faceted phase indicate some of the differences between this type of eutectic and those in non-faceted systems. Some of the different features observed are:

(a) The faceted phase, e.g. graphite, in coarse eutectic structures appears at a discontinuous growth front.
(b) Irregular morphologies grow, e.g. the fanning branched structure of a eutectic graphite cell.
(c) Organized morphologies may appear other than lamellar or rod. For instance, a spiral eutectic in Al–Mg_2Si.
(d) Marked impurity effects occur, influencing the eutectic development, e.g. Na in Al–Si and Mg in γ-graphite.

Degree of Cooperation in Eutectic Growth

Cooperative growth at a joint interface of two phases can take place in metallic eutectic systems. It will occur if the velocity of growth of the two phases growing cooperatively is faster than the velocity of growth of either of the two phases developing independently. Alternatively stated, it may occur if the two-phase structure grows at a smaller undercooling than that for the two constituent phases.

In non-metallic systems, it may be difficult to ascertain the degree to which a phase grows by diffusion or by interface kinetics. Hillert[6] discussed the definition of a driving force for a reaction as the decrease in the free energy for the transformation per unit volume, $\Delta G/V_m$. If the interface is subject to mixed control, one should write:

$$\Delta G_m = \Delta G_m^D + \Delta G_m^B$$

where D represents diffusion and B represents the boundary or interface. In graphite eutectic growth, it is evident that interface kinetics plays a role in the growth of the graphite phase. Cooperation in the sense discussed for the metallic systems may not be a requirement under all conditions. This may lead to a degree of independence in growth of the faceted phase, in distinction to the total cooperation required of the phases in a metallic system.

This plays a marked role not only in the varying modes of development of the lamellar graphite eutectic under the different composition and temperature conditions existing in a melt, but also in intermediate morphologies in graphite eutectics. In the extreme, in nodular graphite cast iron, independent growth of the graphite phase leads to divorced eutectic growth of the spherulitic and γ phases (see Chapters 3 and 5).

References

1. Gibbs, J. W.: *The Scientific Papers*, Vol. 1, *Thermodynamics*, Dover Publications Inc., New York, 1961.
2. Hollomon, J. H.: *Heterogeneous Nucleation. Thermodynamics in Physical Metallurgy*, Am. Soc. Met., 1950.
3. Turnbull, D., and J. C. Fisher: *J. Chem. Phys.*, **17**, 71, 1949.
4. Huang, S. C., and M. E. Glicksman: Acta Metall., **29**, 701, 1981.
5. Zener, C.: *Trans. Am. Inst. Min. Eng.*, **167**, 550, 1946.
6. Hillert, M.: *Metall. Trans.*, **6A**, 5, 1975.
7. Palache, C.: *Am. Mineral*, **26**, 709, 1941.
8. Jackson, K. A.: *Liquid Metals and Solidification*, p. 174, Am. Soc. Met., 1958.
9. Burton, W. K., N. Cabrera, and F. C. Frank: *Proc. R. Soc. London, Ser. A.*, **243**, 299, 1951.
10. Roscoe, C, C., and J. M. Thomas: *Carbon*, **4**, 383, 1966.
11. Horn, F. H.: *Nature, London*, **170**, 581, 1952.
12. Minkoff, I.: *Acta Metall.*, **14**, 551, 1966.
13. Hillert, M., and Y. Lindbloom: *J. Iron Steel Inst. London*, **176**, 388, 1959.
14. Oron, M., and I. Minkoff: *Philos. Mag.*, **9**, 1059, 1964.
15. Minkoff, I., and S. Myron: *Philos. Mag.*, **19**, 379, 1969.
16. Minkoff, I., and B. Lux: *J. Mater. Sc.*, **9**, 1365, 1974.

17. Chalmers, B.: *Principles of Solidification*, p. 93, John Wiley, New York, 1964.
18. Tiller, W. A., K. A. Jackson, J. W. Rutter, and B. Chalmers: *Acta Metall.*, **7**, 428, 1953.
19. Mullins, W. W., and R. F. Sekerka: *J. Appl. Phys.*, **35**, 444, 1964.
20. Chernov, A. A.: *Sov. Phys. Usp.*, **4**, 129, 1961
21. Chernov, A. A.: *J. Cryst. Growth*, **25/25**, 11, 1974.
22. Minkoff, I., and B. Lux: *Cast Metals Res. J.*, **6**, 181, 1970.
23. Minkoff, I., and B. Lux: *Nature*, **225**, 540, 1970.
24. Minkoff, I., and B. Lux: *Micron*, **2**, 282, 1971.
25. Lux, B., I. Minkoff, and F. Mollard: *The Metallurgy of Cast Iron*, Cover illustration, Georgi Publ. Co. Switzerland, 1975.
26. Munitz, A., and I. Minkoff: Int. Foundry Congress, Budapest Congress, Papers 45th, Paper 32, 1978.
27. Munitz, A. D.Sc. Thesis. Technion. Haifa 1977.
28. Tiller, W. A.: *Liquid Metals and Solidification*, p. 276, Am. Soc. Met., 1958.
29. Hunt, J. D., and K. A. Jackson: *Trans. Metal Soc. AIME*, **236**, 843, 1966.

Bibliography

Chadwick, G. A.: *Eutectic Alloy Solidification*, Vol. 12 (2), Progress in Materials Science, Pergamon Press, Oxford, 1964.

Chalmers, B.: *Principles of Solidification*, John Wiley. New York, 1964.

Davies, G. J.: *Solidification and Casting*, Applied Science Publishers, London, 1973.

Flemings, M. C.: *Solidification Processing*, McGraw-Hill, New York, 1974.

Heine, R. W., Loper, C. R., and Rosenthal, P. C.: *Principles of Metal Casting*, 2nd ed., McGraw-Hill, New York, 1967.

Kondic, V.: *Metallurgical Principles of Founding*, Edward Arnold, London, 1968.

Langer, J. S.: Instabilities and pattern formation in crystal growth, *Rev. Mod. Physics*, **52**, 1, 1980.

Müller-Krumbhaar, H.: Kinetics of crystal growth, Chapter 1 in *Current Topics in Materials Science* (Ed. E. Kaldis), Vol. 1, North-Holland, Amsterdam, 1978.

Winegard, W. C.: *An Introduction to the Solidification of Metals*, Monograph and Report Series No. 29, Institute of Metals, London, 1964.

Woodruff, D. P.: *The Solid Liquid Interface*, Cambridge Solid State Science Series, Cambridge University Press, 1973.

Chapter 2
Phase Diagrams; Examination of Cast Iron Structure

This chapter introduces some of the phase diagrams of interest in understanding the Fe–C system. Diagrams important for alloy cast irons are given in Chapter 8. References should be made to calculated phase diagrams which may revise published systems and appear in journals such as *Calphad* (published by Pergamon Press).

Initially, a description is given of the Fe–C system as related to the stable (graphite) and metastable (Fe_3C) transformations. An introduction to the thermodynamics of these systems is presented by free energy diagrams and a calculated Fe–C diagram is described.

A discussion is then given of the Fe–C–Si and Fe–C–P ternary systems. Of these, the Fe–C–Si diagram is the most generally interesting both with regard to transformations in the liquid and also in the solid. Pseudo-binary diagrams based on the ternary Fe–C–Si system are briefly dealt with. For the Fe–C–P system, some of the possible transformations involving Fe_3P, Fe_3C, and graphite phases are described.

In the remainder of the chapter, some techniques for the examination of different aspects of the cast iron structure are described, with some details of their application. The techniques include X-ray diffraction, X-ray topography, and electron microscopy. The use of the scanning electron microscope in the study of cast iron structure is described and applications of electron microprobe techniques. Finally, some recent employment of Auger spectroscopy is presented.

2-1 The Binary Phase Diagram of the Iron–Carbon System

The phase diagram for the binary Fe–C system is characterized by the stable (graphite) and the metastable (Fe_3C) phase equilibria. Several contributions to the determination of diagrams for this system have been made and description will be given of the diagram published by Benz and Elliott[1] and that based on thermodynamic calculations by Hillert.[2]

The Iron–Carbon Diagram of Benz and Elliott

Benz and Elliott[1] redetermined the austenite solidus of the Fe–C system and published the complete diagram of Figure 2-1, basing this construction on a review of the published data for this system and suggesting revisions. The new austenite solidus was determined from diffusion couples of austenite and melt saturated with graphite. The main features of their diagram are as follows. The temperature horizontal of the peritectic reaction was selected as that which intersected their newly determined austenite solidus at 0.16 per cent carbon. This is the composition of γ-iron at the peritectic (1772 K). The γ solidus was that determined in their own investigation. The γ liquidus was corrected for the newly suggested peritectic horizontal. The γ-graphite eutectic composition was shown at 4.26 per cent carbon and the γ-Fe_3C eutectic composition was located at 4.30 per cent carbon. The respective eutectic temperatures were 1426 and 1420 K respectively.

For the graphite solubility in liquid iron, the following relationship was

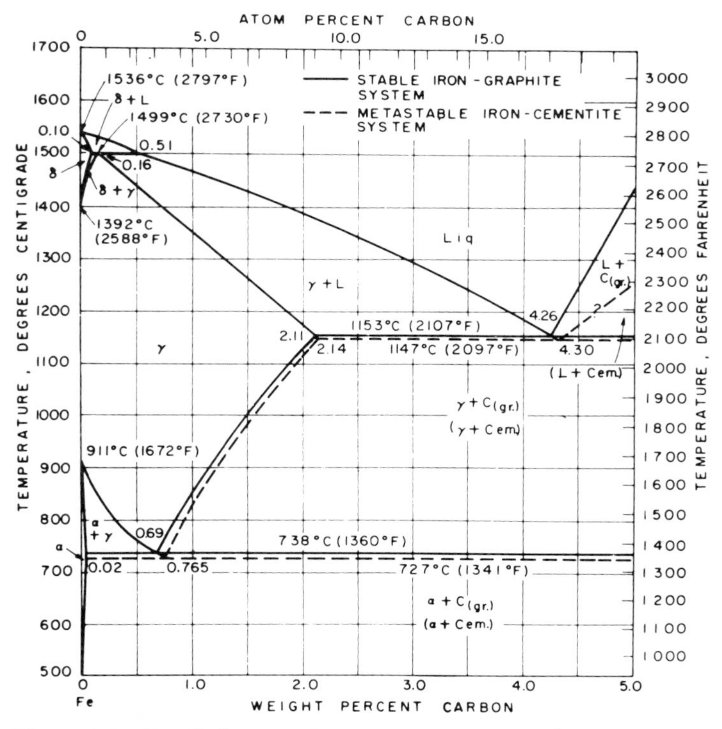

Figure 2-1 Fe–C diagram due to Benz and Elliott.[1] The austenite solidus was redetermined. The rest of the diagram was based on a review of published data. (Reproduced by permission of The Metallurgical Society of American Institute of Mining Engineers)

selected from the research of Neumann and Schenk:[3]

$$\text{Weight per cent of carbon} = 1.30 + 2.57 \times 10^{-3}\, T\ (^\circ\text{C})$$

A probable cementite liquidus was indicated as shown in Figure 2-1 with a question mark, as being unknown.

The two curves for the solubility of graphite and cementite in austenite were based on the following. The graphite solubility line was based on research of Wells,[4] Gurry,[5] and Smith.[6] This curve was extrapolated to meet the graphite eutectic horizontal at the intersection of the determined austenite solidus and the graphite eutectic horizontal. The curve of Fe_3C solubility in γ was based on Smith.[7]. This again was extrapolated to meet the cementite eutectic horizontal at the intersection of the determined γ solidus with the cementite eutectic horizontal. The two curves are approximately parallel.

The Iron–Carbon Diagram Due to Hillert

Hillert[2] calculated the equilibria of interest in the Fe–C diagram for the solidification of cast iron. The diagram is shown in Figure 2-2. Both the stable (graphite) and unstable (Fe_3C) equilibria were calculated. Comparison of data for the graphite equilibrium with other determinations appeared to be good (see, for example, Figure 2-1).

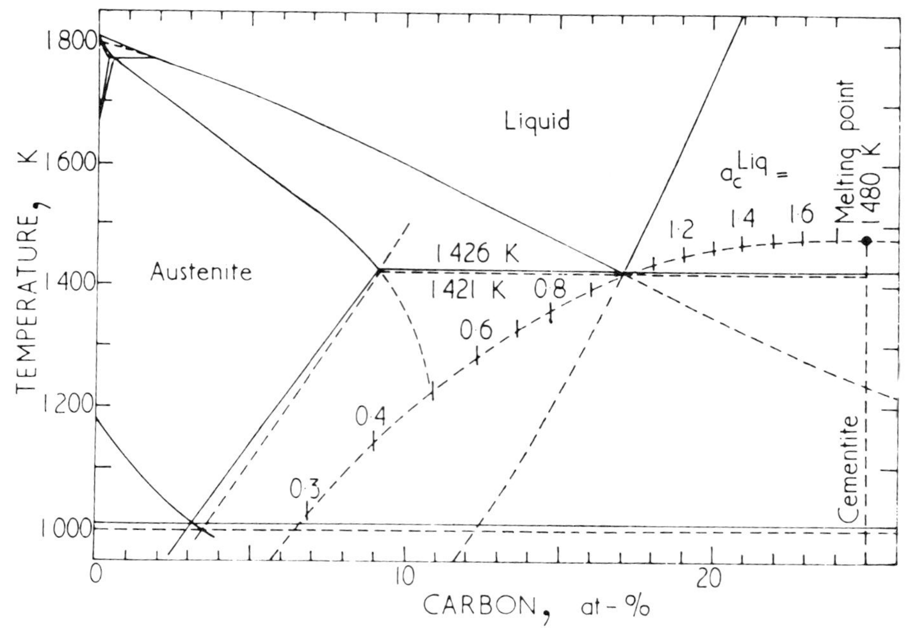

Figure 2-2 Calculated Fe–C diagram due to Hillert.[2] The melting point of the Fe_3C phase was located at 1480 K. (Reproduced by permission of Gordon and Breach Science Publishers Inc.)

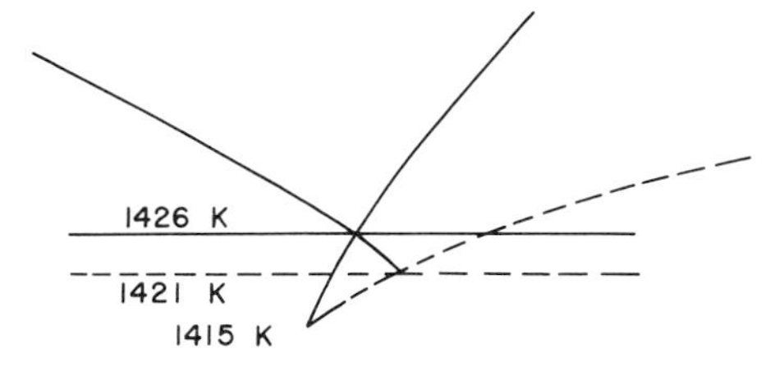

Figure 2-3 Eutectic region from Hillert's data showing temperatures for the stable (full line) and metastable (dashed line) reactions. The Fe_3C liquidus meets the graphite liquidus at 11 K below the graphite eutectic temperature

Hillert pointed out the lack of data for the Fe_3C equilibria but suggested that since the agreement was good for the stable equilibria, the metastable equilibria might be safely accepted. Note should be made of the position of the Fe_3C liquidus which predicted a melting point of cementite of only 1480 K, while tabulated values gave 2100 K.

The Eutectic Region Figure 2-3 shows details of Hillert's published eutectic reactions and their relative locations. The full line gives the phase equilibria for the graphite liquidus and the eutectic while the dashed line gives the Fe_3C equilibria. The temperature for the two reactions are 1426 and 1421 K. This leaves a temperature interval of 5 K between the two eutectic arrests in which the graphite eutectic alone can nucleate. Below 1421 K both eutectics can nucleate and grow. The slope of the Fe_3C liquidus which meets the graphite liquidus at 11 K below the graphite eutectic line indicates the more rapid supersaturation of the melt with respect to the Fe_3C phase.

The Iron–Carbon Binary System Represented by Free Energy Diagrams The different equilibria between liquid, γ, Fe_3C, and graphite phases are represented schematically by free energy diagrams. Hillert[8] showed the series of free energy diagrams of Figures 2-4 as representing the solidification of cast iron. A phase diagram with the temperatures marked and relating to the given sketches T_1 to T_5 is shown in Figure 2-4 (f).

At T_4, the γ, liquid and Fe_3C phases are shown by the common tangent construction to be in equilibrium, and this is the γ–Fe_3C eutectic temperature. Hillert use the common tangent to the liquid and Fe_3C phases and its intersection with the 100 per cent carbon axis to represent the carbon activity of cementite in equilibrium with liquid. As the temperature decreases from T_1 to T_4, the carbon activity is seen to approach that of graphite. At T_5, it is less than that of graphite and the melt is more supersaturated with respect to cementite than to graphite.

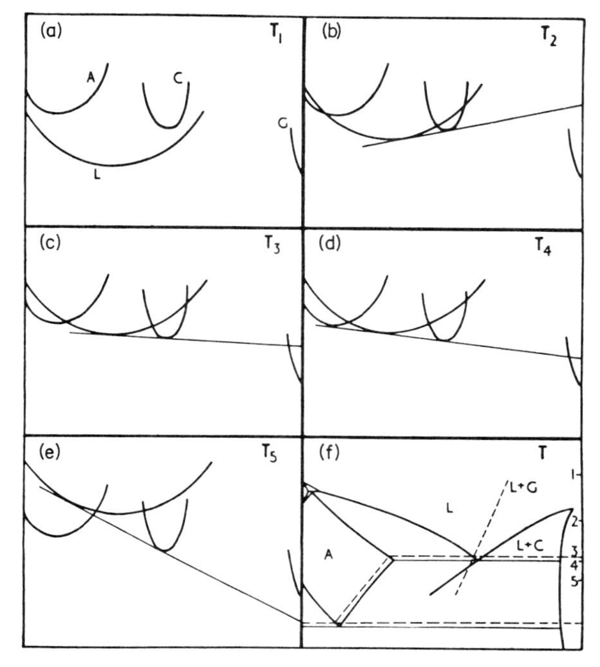

Figure 2-4 Representation by free energy diagrams of solidification of cast iron.[8] The phase diagram is shown at (f) with indicated temperatures T_1, T_2, T_3, T_4, and T_5 A common tangent construction to the liquid and Fe_3C phases cutting the 100 per cent carbon axis is used to represent the carbon activity of cementite in equilibrium with liquid. (Reproduced by permission of The Metals Society)

2-2 The Iron–Carbon–Silicon System

A study by Hilliard and Owen[9] was directed principally to the liquidus–solidus region of the Fe–C–Si system within the range of interest of cast iron metallurgy. The invariant four-phase plane for the equilibrium between liquid, γ, α, and carbide phases as suggested by Scheil[10] is shown in Figure 2-5. Hilliard and Owen placed the liquid point L on the opposite side of the plane. The resulting vertical, constant silicon, sections are shown in Figure 2-6.

An important result of the study of this system made by Hilliard and Owen was to resolve the problem of composition of the carbon-rich phase in the metastable system. This was suggested as being Fe_3C with negligible silicon content.

It is noted that there is an increase in temperature of the eutectic transformation from 1403 K in Fe–C to 1434 K at 5 per cent silicon. After this the metastable eutectic temperature once more decreases.

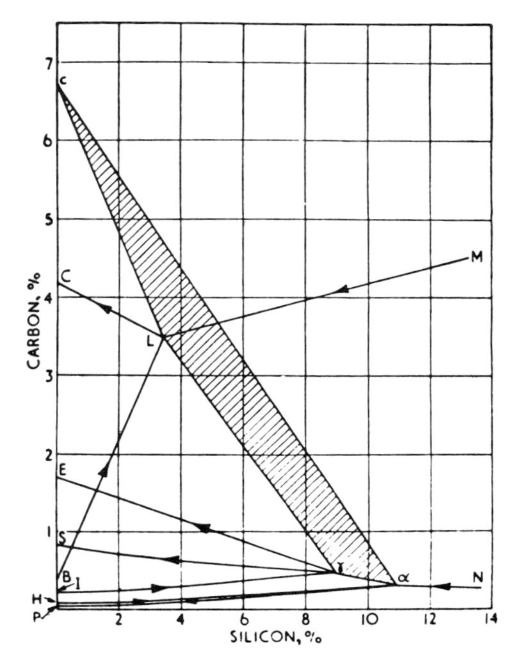

Figure 2-5 The invariant four-phase plane for the equilibrium between liquid, γ, α, and carbide phases as suggested by Scheil.[10] Hilliard and Owen in their research[9] placed the liquid point L on the opposite side of the plane. (Reproduced by permission of The Metals Society)

The diagrams show a progressive lowering of the γ–liquidus temperature with increasing silicon content and a progressive diminution of the γ phase region. The γ–Fe_3C eutectic composition is progressively displaced to lower carbon contents.

The Iron–Carbon–Silicon System; The Eutectoid Area of the Ternary Diagram

Diagrams which include the eutectoid region of Fe–C–Si are given in Figure 2-7.[11] These are vertical sections at 2.4, 4.8, and 6.0 per cent silicon, and show the progressive rise in temperature of the eutectoid transformation with increasing silicon content and the displacement of the eutectoid γ composition to lower carbon contents. At the same time a large temperature increase occurs in the γ–α–C equilibrium region. Figure 2-8 shows a horizontal section 1053 K[11] which is just below the start of γ eutectoid transformation for a cast iron with 2.4 per cent silicon.

Hillert[12] has published the horizontal section of the Fe–C–Si system at

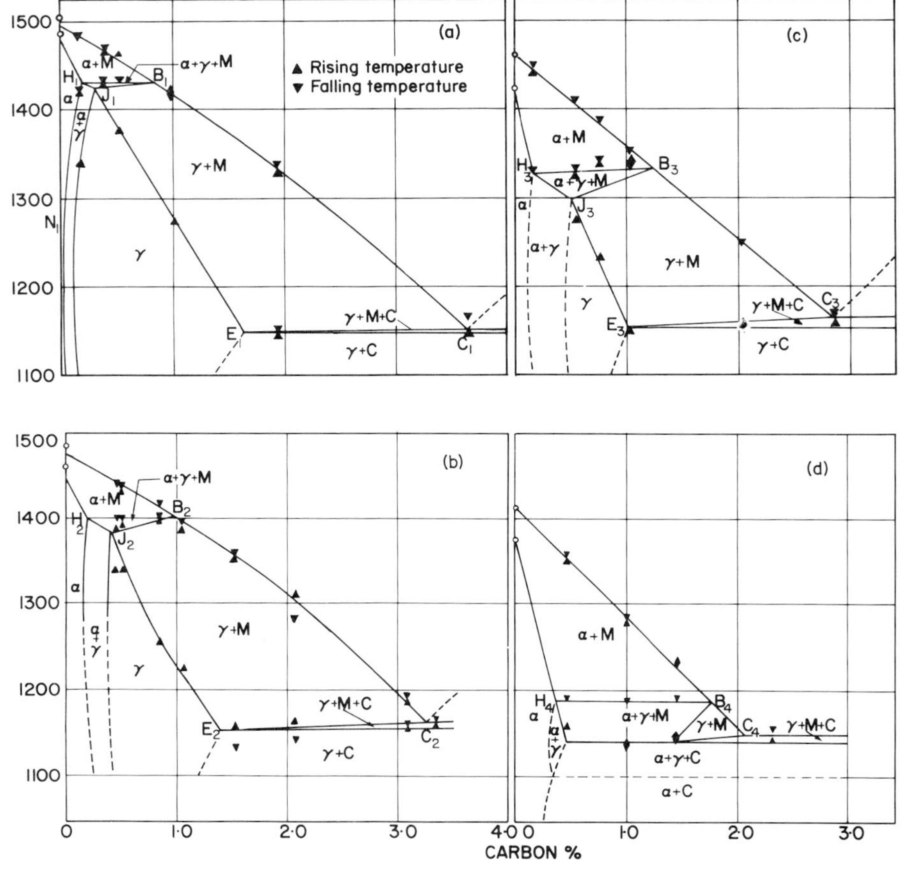

Figure 2-6. Vertical sections of the Fe–C–Si diagram in the neighbourhood of the γ liquidus, due to Hilliard and Owen.[9] (a) 2.3 per cent SI, (b) 3.5 per cent Si, (c) 5.2 per cent Si, and (d) 7.9 per cent Si. (Reproduced by permission of The Metals Society)

1053 K, shown in Figure 2-9. This shows calculated extensions of the γ–α and γ–Fe_3C equilibria, used by Hillert to estimate the compositions of γ in equilibrium with the α and Fe_3C phases in the ternary system during eutectoid decomposition.

2-3 The Iron–Carbon–Phosphorus System

The Fe–C–P system can be studied by first examining the character of the Fe–P diagram. This diagram, due to Hansen,[13] is shown in Figure 2-10. According to Hansen, Fe–P melts are characterized by a large susceptibility to undercooling. Where the phosphorus content becomes greater than 10 per cent by weight and in particular for a phosphorus content in excess of 15 per

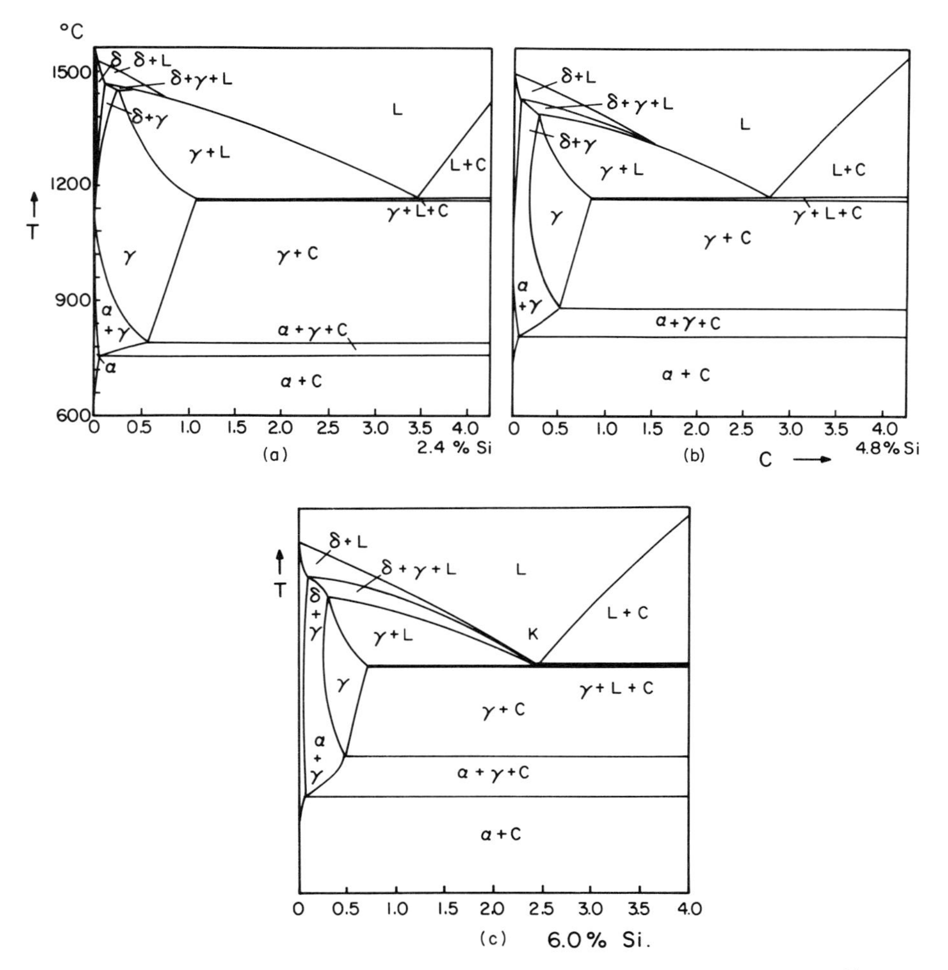

Figure 2-7 Vertical sections of the Fe–C–Si diagram from Piwowarsky.[11] The $\delta(\alpha) + \gamma + L$ region should be compared with Figure 2-6. Note the eutectoid transformation region and progressive increase in $\gamma + \alpha + C$ equilibrium range. (a) 2.4 per cent Si, (b) 4.8 per cent Si, and (c) 6.0 per cent Si. (Reproduced by permission of Springer-Verlag)

cent by weight, there is a tendency for a metastable mode in this binary system. This is shown by the dashed line. The Fe_3P phase is suppressed and an unstable eutectic between α and Fe_2P tends to form at 1218 K. This tendency is enhanced by rapid solidification or no inoculation.

The ternary system based on the Fe–Fe_3P and Fe–Fe_3C binary systems was given by Vogel[14] and reproduced in the text of Piwowarsky.[11] Figure 2-11 is a schematic basal projection of the Fe–C–P system after Morrogh and Tütsch.[15] The numerals indicate the areas of primary crystallization. In this

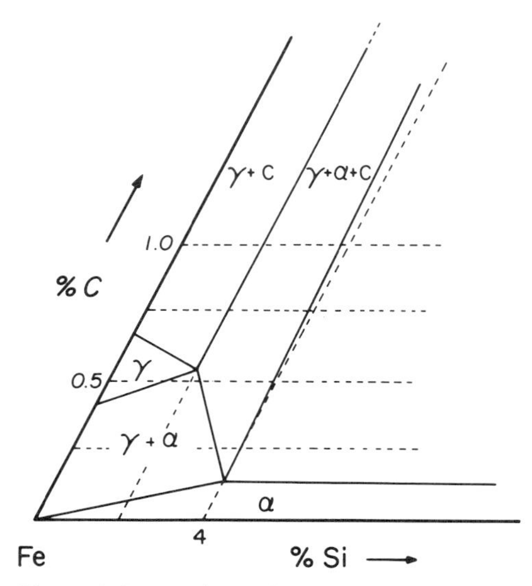

Figure 2-8 Horizontal section at 1053 K of Fe–C–Si ternary system from Piwowarsky.[11] This is just below the start of eutectoid transformation for a cast iron with 2.4 per cent Si. (Reproduced by permission of Springer-Verlag.) A detailed section calculated by Hillert is shown in Figure 2-9

diagram, I conforms to the γ primary phase crystallization region, II is for the Fe_3C phase, and III is for Fe_3P. Region Ia is for the phase area $\alpha + \delta$ and AU is for the peritectic $L + \alpha \rightarrow \gamma$ transformation.

Normally, for gray cast iron, the phosphorus contents are small and silicon is present. Therefore and phosphide eutectic may have the stable graphite phase as a constituent in addition to γ, Fe_3P, and Fe_3C. This is discussed in the following section. Figure 2-11 shows that in the Fe–Fe_3C–Fe_3P ternary system, different binary eutectics may exist between the γ, Fe_3C, and Fe_3P phases, and they can crystallize before the ternary eutectic.

Morrogh and Tütsch[15] performed an interesting series of experiments on alloys of this system to study the structures. They remarked on the tendency to primary crystallization, in particular of the γ and Fe_3C phases where two-phase or three-phase structures would be predicted. In the Fe–C–Si system, this is a noted effect (see Section 3-7). Either primary γ or primary graphite precedes eutectic growth, or in the metastable system the Fe_3C phase precedes the γ–Fe_3C eutectic. Morrogh and Tütsch suggested that the consistent primary phase growth observed in Fe–C–P would be related to the presence of phosphorus, presuming an effect of the undercooling related to this element in the system.

Ternary eutectic growth is complex and irregular structures may be

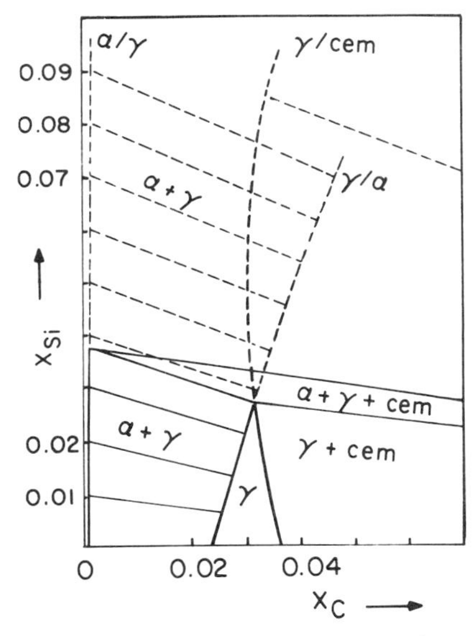

Figure 2-9 Calculated isothermal section by Hillert[12] of Fe–C–Si diagram at 1033 K, showing extensions of γ–cementite and γ–α boundaries. The data obtained from the extensions were used to calculate pearlite growth in the ternary system. (Reproduced by permission of the American Society for Metals)

obtained involving three phases. Flemings[16] has discussed the microstructures and the conditions under which plane front solidification can be obtained to give a lamellar or rod morphology. In the ternary eutectic region of the Fe–C–P system, the graphite or Fe_3C phases lead to random structures dominated by one of these phases. Morrogh and Tütsch referred to experiments by Williams[17] involving the lederburite eutectic, where in the presence of high sulphur concentrations growth of acicular carbides was noted.

2-4 The Iron–Carbon–Silicon–Phosphorus System

The system was discussed by Hillert and Söderholm.[18] The presence of silicon allows the possibility of solidification according to the gray mode. Phosphorus segregates into the liquid during solidification and the melt will finally be saturated with respect to Fe_3P. Two four-phase equilibria are then possible and the ternary eutectic may solidify according to the following:

$L \rightarrow \gamma + \text{graphite} + Fe_3P$, yielding a gray structure

$L \rightarrow \gamma + Fe_3C + Fe_3P$, yielding a white structure

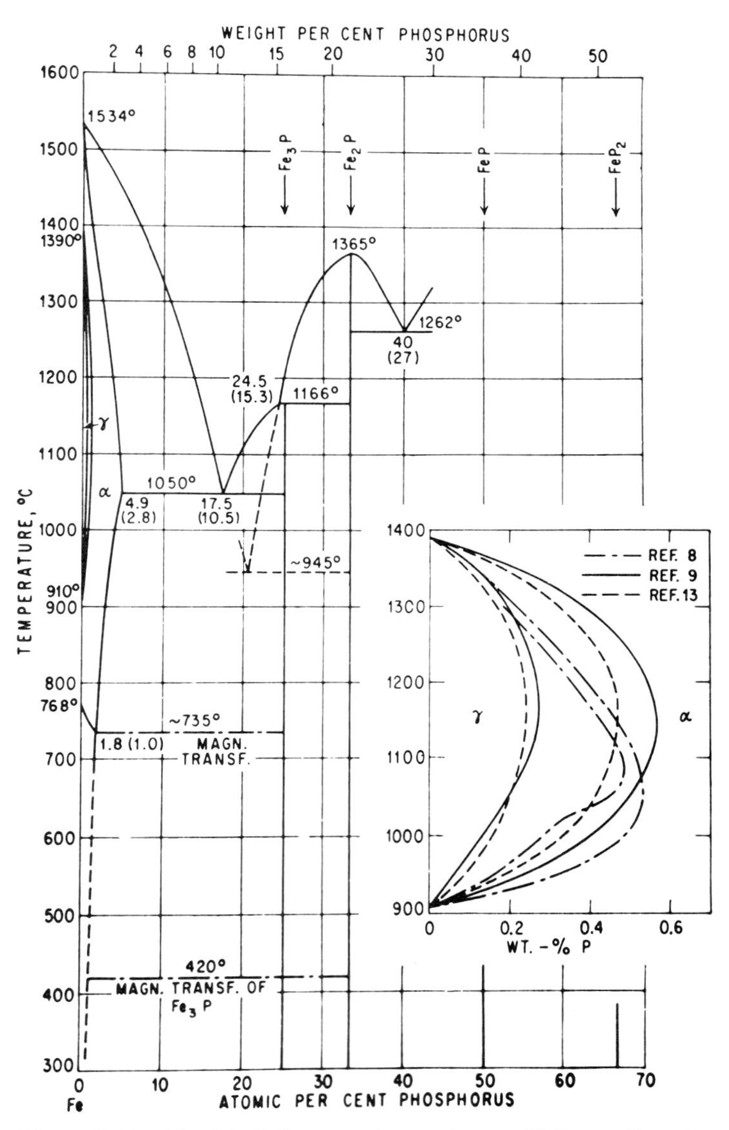

Figure 2-10 The Fe–P diagram due to Hansen.[13] (From *Constitution of Binary Alloys* by M. Hansen. Copyright © 1958 McGraw-Hill Book Company. Used with the permission of McGraw-Hill Book Company)

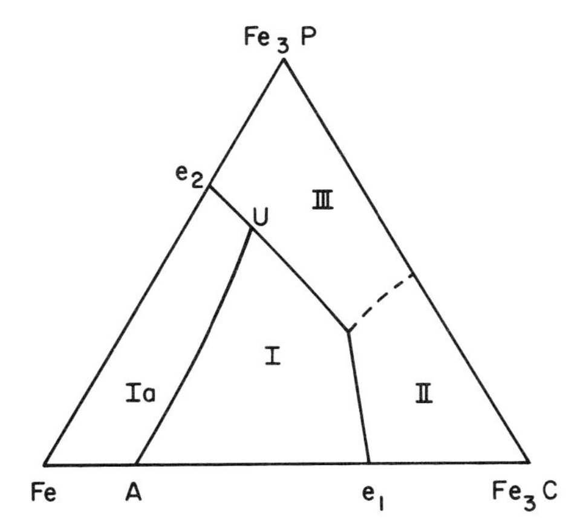

Figure 2-11 Schematic basal projection of the Fe–C–P liquidus surface due to Morrogh and Tütsch.[15] (Reproduced by permission of The Metals Society)

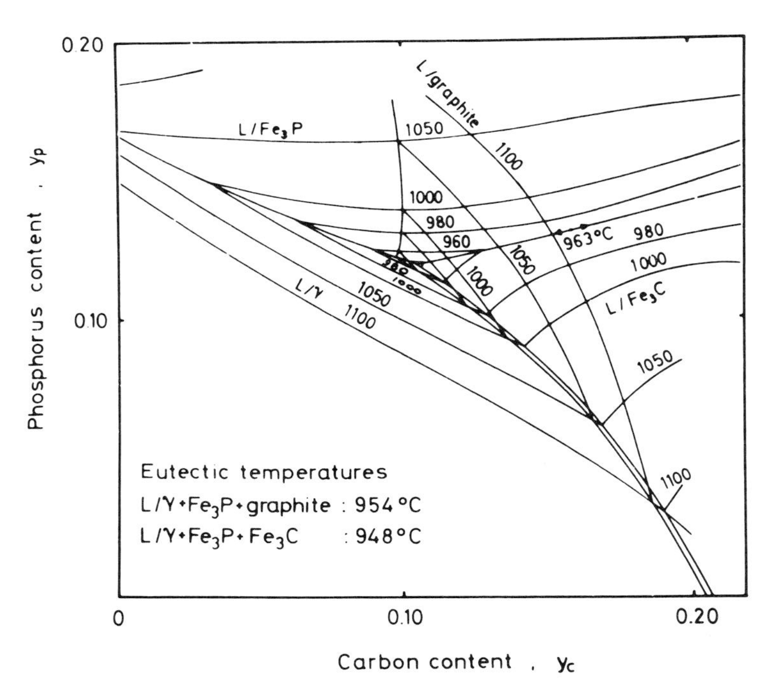

Figure 2-12 Calculated liquidus surfaces in Fe–C–P system after Hillert and Söderholm.[18] (Reproduced by permission of Georgi Publishing Co.)

The white structure in the ternary eutectic is normally obtained. Figure 2-12 shows calculated liquidus surfaces in the Fe–C–P system for the γ, Fe_3P, Fe_3C, and graphite phases. The eutectic lines formed by their intersections are noted. The coordinates are:

$$y_P = \frac{x_P}{1 - x_C}$$

$$y_C = \frac{x_C}{1 - x_C}$$

where x_P = weight fraction of phosphorus
x_C = weight fraction of carbon

The positions of the two ternary eutectic points were given by the calculations as follows:

Gray solidification, 1127 K: $y_C = 0.099$ $y_P = 0.123$
White solidification, 1121 K: $y_C = 0.106$ $y_P = 0.121$

The maximum point on the binary eutectic line $L/Fe_3P + Fe_3C$ was calculated to be 1136 K or 15 K higher than the point for the white ternary eutectic.

2-5 Instrumental Techniques in Cast Iron Structure Research

Some instrumental techniques in addition to optical microscopy are available in materials research for examining structure. They are discussed here in their applications to understanding different aspects of the physical metallurgy of cast iron. Some of the techniques can be used for control of a process, e.g. electron probe microanalysis for standardizing spheroidal graphite cast iron production.

The scanning electron microscope has provided a new insight into the development of morphology of graphite in cast iron. It is also a useful metallographic tool, e.g. in examining austenite transformation products in the cast iron matrix following heat treatment.

Some instrumental techniques have been used specifically for looking at the graphite structure, in particular X-ray diffraction and transmission electron microscopy. These have given insight into the role of defects in graphite in growth and fracture.

Techniques of more recent application which show promise in understanding the role of alloying elements in solid state transformation include the field ion microscope atom probe mass spectrometer. A brief summary of some of the instruments, some results obtained, and some applications are now given.

X-ray Diffraction Techniques; Structure and Growth of Graphite

The use of X-ray diffraction techniques to study graphite crystals extracted from cast alloys has been directed to understanding graphite dendrite growth.

Figure 2-13(a) shows a sketch of the graphite dendrite illustrated in Figure 1-15(a). The crystallographic directions are indicated. These were determined by Minkoff and Oron[19] from X-ray observations using both standard Laue techniques and also with greater perfection, using a conical beam camera.[20]

From measurement of the angles between the reflections obtained, the dendrites were observed to be bi-crystals in which the members were rotated

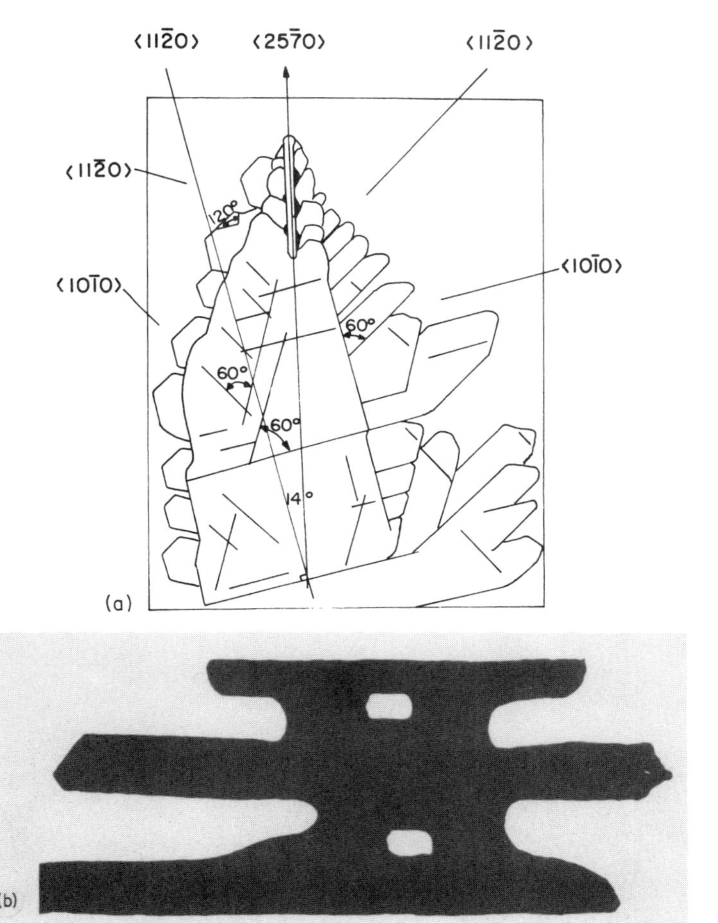

Figure 2-13 (a) Sketch of dendritic crystal of graphite, illustrated in Figure 1-15(a), showing crystallographic and growth directions determined from X-ray data by Minkoff and Oron.[19] The dendrite was determined to be a bi-crystal. (b) Metallographic cross-section of type of dendritic crystal examined in (a) shows upper and lower section and two asymmetric holes left in structure due to growth process.[19] (× 200). (Reproduced by permission of Taylor and Francis, Ltd.)

by a specific angle about the ⟨0001⟩ axis. The angles determined were 21.8° and 27.8°. This asymmetrical rotation of the two crystal parts of the dendrite leads to an asymmetrical growth direction, determined to be ⟨25$\bar{7}$0⟩ for a rotation of 27.8° (Figure 2-13a) and ⟨12$\bar{3}$0⟩ for a rotation of 21.8°. Figure 2-13(b) shows a cross-section of a crystal examined,[19] demonstrating an upper and lower section and two holes left in the structure due to the growth process. The two holes are asymmetrical in conformity with the rotation boundary at the crystal centre.

Minkoff and Myron[21] used X-ray diffraction techniques to study graphite crystals occurring in nature. In this research, a rotating crystal X-ray diffraction technique was employed for the measurement of rotation boundary angles. For this type of crystal, it was noted that there was no tendency to repeat values of rotation boundary angles. Coincidence boundary theory was employed and it was shown that in dendrite growth there was a conformity with coincidence, while in ordinary graphite crystal growth a wide spread of measured values was observed about the coincidence relationships expected.

This was in contrast to the recurring values obtained in dendritic graphite crystals. Therefore Minkoff and Myron[21] suggested that the maximum in the growth rate for dendrites has a specific structure requirement, which can be expressed from coincidence boundary theory as a maximum in the density of coincidence sites at a boundary. Double and Hellawell,[22] using transmission electron microscopy (see next section) to study the growth of graphite in the eutectic, determined rotation boundary angles conforming to theoretical values of maximum coincidence, thus indicating the dendritic nature of graphite growth in the eutectic.

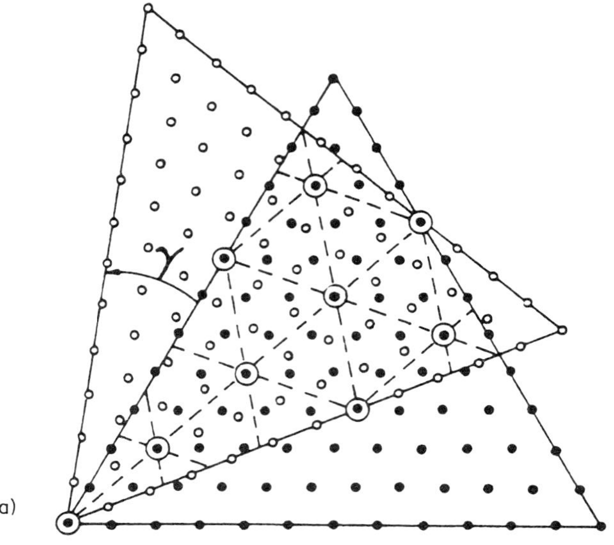

Figure 2-14 (a) Coincidence lattice created by a rotation of (a) $\gamma°$ in a crystal, for a

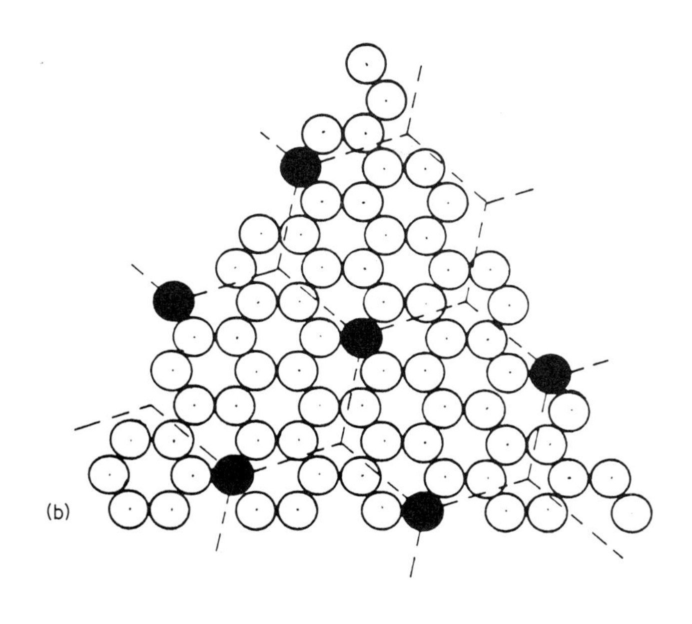

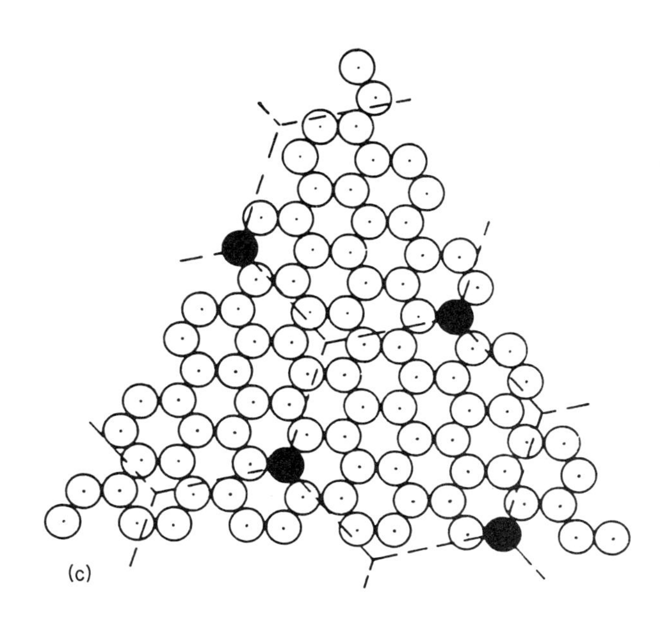

rotation of (b) 21.8° in graphite and (c) 27.8° in graphite.[21] (Reproduced by permission of Taylor and Francis, Ltd.)

Figure 2-14(a) shows a coincidence lattice created by a rotation of $\gamma°$ in a crystal. Figure 2-14(b) and (c) shows the two coincidence lattices for rotations of 21.8° and 27.8° in graphite. These are incomplete crystallographic rotations of the lattice, bringing the crystal into a condition of coincidence where a new lattice of matching planes is obtained. The new cell is a multiple of this primitive cell. The rotation boundaries create steps on the bounding $\{10\bar{1}0\}$ type faces of graphite which is considered theoretically to be a requirement of the growth process.

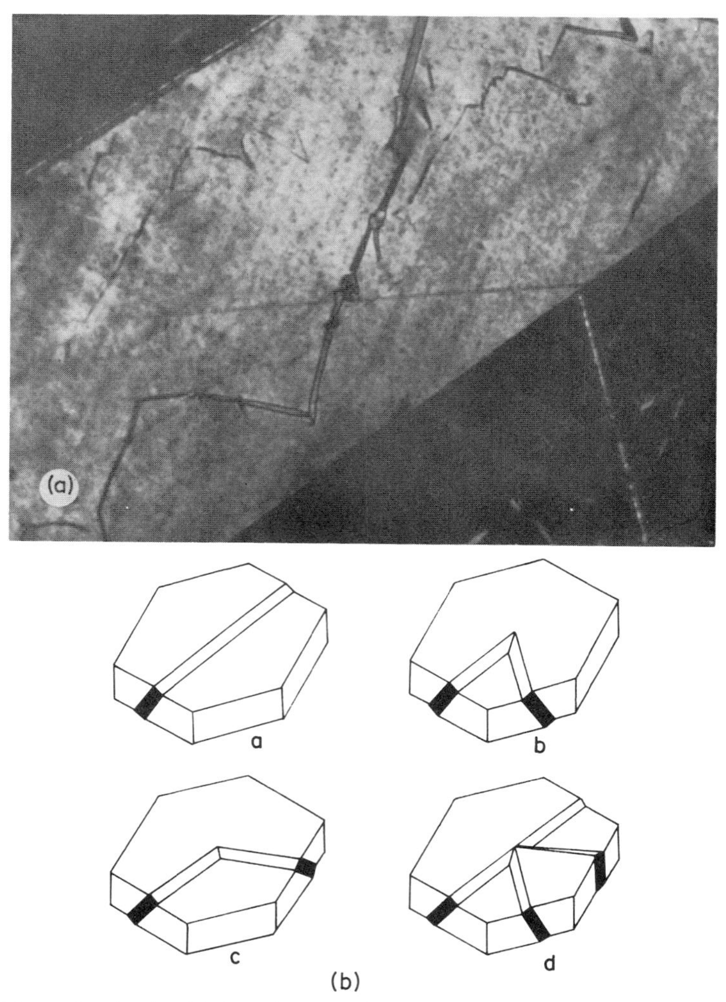

Figure 2-15 (a) Twinned area of graphite flake crystal viewed in polarized light. (× 325) (b) Schematic arrangements of twins in graphite crystals, after Freise[25] (Reproduced by permission of E. J. Freise)

X-ray Diffraction Study of Twinning in Graphite

Polished microsections of gray cast iron when observed in polarized light show the presence of misoriented regions of structure in the graphite crystals. The appearance is illustrated in Figure 2-15(a). These are twinned areas and were referred to by Morrogh and Williams[23] when discussing 'zig-zag' patterns in flakes. The crystallographic structure was studied by Freise and Kelly[24] using a combination of techniques which included Laue transmission methods and a rotating crystal camera. An X-ray microbeam camera with a beam of 100 μm diameter was used to examine some of the narrower markings. Figure 2-15(b) shows schematic arrangements after Freise[25] of various types of twin intersection in ideal hexagonal graphite crystals. This type of twin invariably occurs in flake graphite crystals and is a major imperfection of the structure. The most common twinning mode observed is $(11\bar{2}1)$—with habit plane $\{11\bar{2}1\}$—occurring as thin lamellae transversing the crystal.

A twinned area may be considered as formed by two symmetrical tilt boundaries, which are walls of discontinuouity in the graphite crystal. Tilt boundaries form a substructure in the crystal. Figure 2-16 shows one boundary of a $(11\bar{2}1)$ twin.[24] This is composed of partial dislocations with Burgers vector $a/3[10\bar{1}0]$, there being one partial dislocation on every layer plane, having b_e (edge) and b_s (screw) components.

For a discussion of dislocations in graphite, reference should be made to Delavignette and Amelinckx.[26] Freise and Kelly[24] suggested that a twin is initiated from partial dislocations which separate stacking faults in the graphite lattice. The partials which lie in the basal plane line up to form a short dislocation wall, normal to the basal plane. The twinned area extends

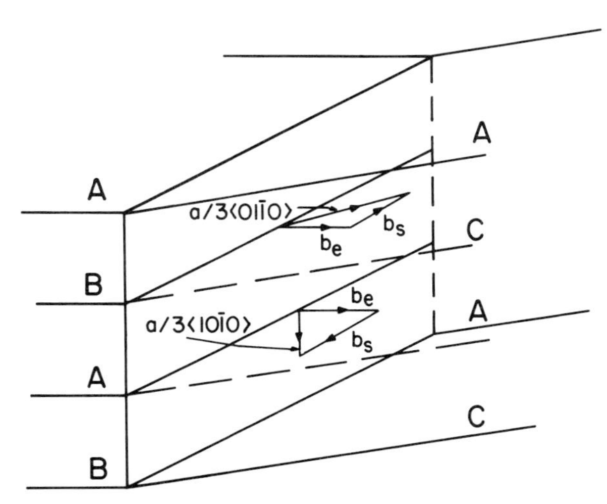

Figure 2-16 $(11\bar{2}1)$ twin in graphite.[24] (Reproduced by permission of The Royal Society)

under the influence of stresses which are associated with this wall. These stresses are relieved by further slip in the basal plane.

It is important to note that graphite fracture initiates at twin boundaries.[24] The application of stress to flake graphite cast iron promotes fracture of the flake. This gives the fracture appearance at twin boundaries, noted by Glover and Pollard[27] in their research on cast iron failure and described in further detail in Chapter 11.

X-ray Topography of Graphite Crystals and Presence of Tilt Boundaries

Tilt boundaries in graphite crystals differ from twins in that they can have any angle. They are essentially walls of discontinuity forming a substructure in the crystal. This substructure can be observed by X-ray diffraction topographic techniques, as shown by Austerman, Myron, and Wagner[28] using the Berg–Barrett method. Models of tilt boundaries in graphite crystals were described by Bollmann and Lux.[29]

Transmission Electron Microscopy

The transmission electron microscope has been used to study defects in graphite and the relationship between the type of defect and the mode of growth of the graphite crystal. Replica techniques have been employed to study etched sections of graphite crystals.

A basic study of edge dislocations and stacking faults in graphite crystals was made by Delavignette and Amelinckx.[26] The Burgers vectors of unit dislocations in the graphite structure are shown in Figure 2-17(a) where AB, AC, and AD are perfect dislocations. A perfect dislocation can dissociate into partials according to the following reaction, shown in Figure 2-17(b):

$$AB \rightarrow A\sigma + \sigma B$$

The partials separate leaving a ribbon of stacking fault. This has rhombohedral stacking, ababcaca, instead of the hexogonal stacking of graphite, abababab. Examination of graphite crystals shows dislocation nets in the c plane. These result from the intersection of two families of ribbons with a different Burgers vector, situated on the same lattice plane.

Tzuzuku[30] used transmission electron microscopy to study crystals grown by graphitization of carbon black in graphite crucibles at 2773 K. Two types of crystal were noted. Figure 2-18(a) is a pyramid plate type of crystal 1–5 μm in diameter and several nanometres in height. Tzuzuku considered that these grew by a spiral growth mechanism. Figure 2-18(b) shows crystals of conical habit. These were circular in outline and larger in dimension than the pyramidal plates, reaching a size of 3–7 μm. Tzuzuku suggested that the circular crystals also formed by a spiral growth mechanism under a vapour pressure somewhat higher than the case of the hexagonal crystals. The electron diffraction patterns from these crystals showed spotty Debye–Scherrer rings.

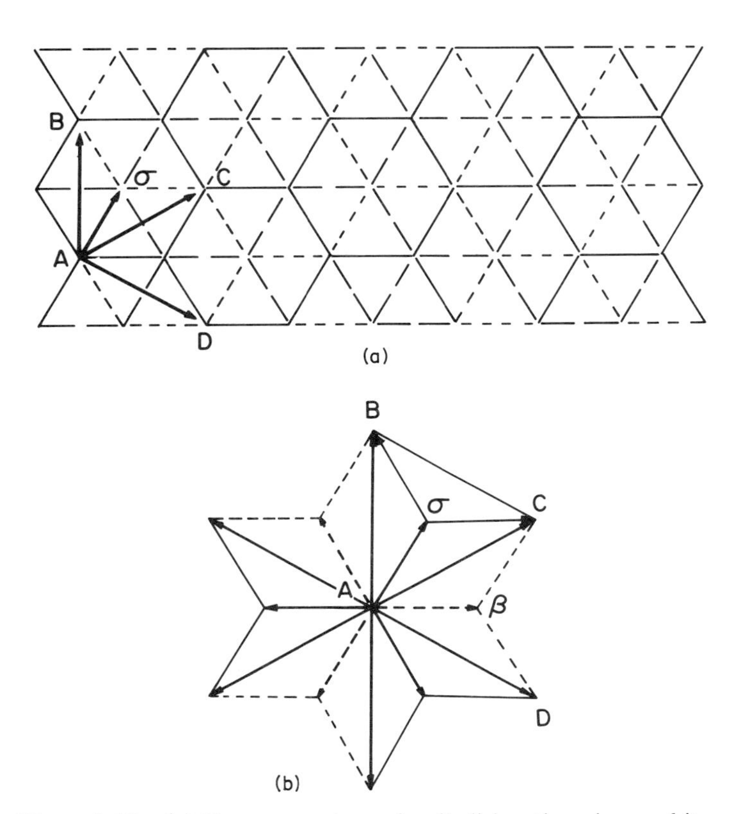

Figure 2-17 (a) Burgers vectors of unit dislocations in graphite, AB, AC, AD. (b) Dissociation of perfect dislocation into partials according to the scheme AB → Aσ + σB.[26] (Reproduced by permission of North-Holland Publishing Company)

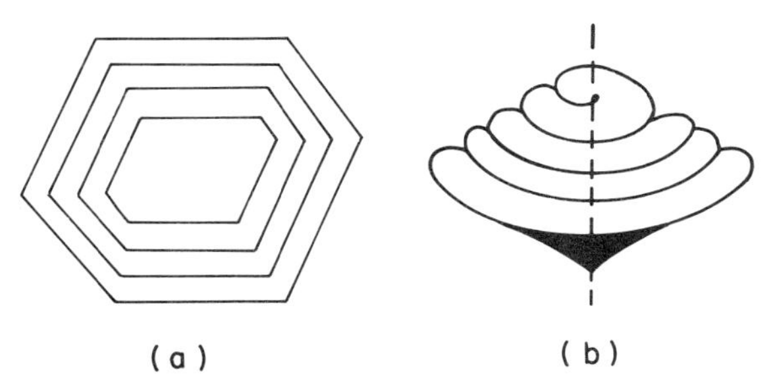

Figure 2-18 Two types of graphite crystal noted by Tzuzuku.[30] (a) pyramidal plates and (b) crystals of conical habit. (Reproduced by permission of Pergamon Press Ltd.)

Tzuzuku suggested that while the pyramidal plate crystal grew by an ordinary spiral mechanism, producing single crystal diffraction patterns, a special mechanism existed for the conical crystal at the initiation of growth. This could arise by rotational slip followed by mechanical buckling which produced a central screw dislocation. Further development of the crystal resulted in the conical spiral structure.

Techniques of Double and Hellawell[22]

A combination of techniques were employed, which included transmission electron microscopy, to study the structure of flake graphite in a nickel–carbon alloy. The crystals were eutectic, grown at approximately 3×10^{-3} cm s^{-1}. They were 300 nm thick and transparent to 100 kV electrons. Rotational faults of the same character as shown by X-ray diffraction methods in graphite crystals[20,21] were shown to exist.

Twin and kink bands were also studied. Double and Hellawell suggested that these types of defects allow crystallographic freedom for a flake to bend and convolute into different complex shapes. They showed a model of bending of a flake allowing growth into an incoherent spherical form from which a spherulitic form could then develop.

Electron Microscope Study by Bolotov[31] *of Solid State Spherulitic and Flake Growth*

The study was directed to an *in situ* investigation of graphite growth during annealing processes. Use was made of stereophotography and electron diffraction. A Ni–C alloy was studied having a composition of 0.6 per cent carbon. The specimen was quenched from 1523 K and annealed at 773, 823, and 923 K in either hydrogen or in a vacuum. In the hydrogen atmosphere, spherulites grew, while in a vacuum, Bolotov reported that flakes were produced.

An extraction carbon replica technique was employed to obtain specimens for the microscope. Bolotov showed three-dimensional models of hexagonal crystals in which $\{10\bar{1}0\}$ faces gradually became overgrown by (0001). This research is referred to again in relation to malleable cast iron in Chapter 10.

Electron Microscope Examination of Replicas of Etched Spherulitic Crystal Sections by Hunter and Chadwick[32]

The electron microscope was used to study diametral sections of spherulites after polishing and thermal etching. This treatment consisted of oxidation by heating the polished specimen in air to about 750–800 K for about 1 minute. A plastic replica was made, shadowed with gold at an angle of 30°, and onto this a carbon film was deposited from a source vertically above the replica. Finally, the polymer was dissolved in acetone and the resultant carbon replica

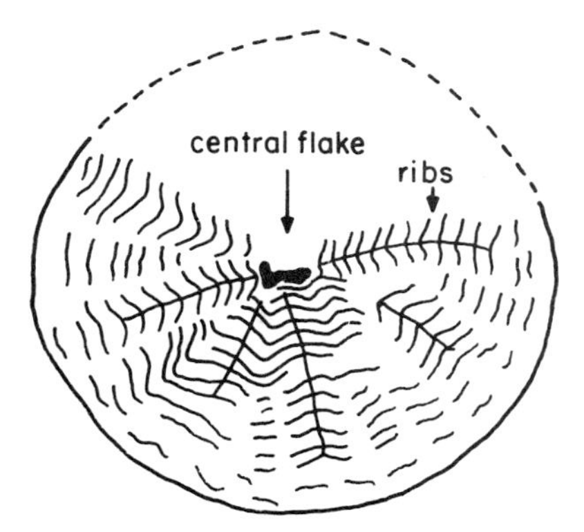

Figure 2-19 Schematic sketch from transmission electron micrograph of replica of etched spherulitic graphite crystal showing a central flake section which has become unstable. A zig-zag ribbed structure is noted. (From Hunter and Chadwick,[32] reproduced by permission of The Metals Society)

was examined. Figure 2-19 shows a schematic representation of the main features observed, in which a central flake section becomes unstable and a spherulite develops by crystal multiplication.

Scanning Electron Microscopy

Employment of the scanning electron microscope in cast iron metallurgy has been extensive since the first publication on its application in this field by Minkoff and Nixon.[33] A review of its potential application in materials science was published in 1967.[34] Applications of the instrument in understanding the fracture of cast iron are given in Chapter 11.

Different techniques have been employed for the study of the growth of graphite in cast iron. In the original applications to cast iron by Minkoff and Nixon, the graphite crystals were extracted from the material by dissolution in acid, and were then washed carefully, dried, and examined on the stage of the microscope.

Most of the later papers have made use of deep etching techniques so that the graphite crystals are observed *in situ* as they appear in the matrix. While this technique is more in keeping with the methods of metallography, there is some advantage to be obtained by observing separated crystals. The morphology can be observed more readily. An example is given of the crystal shown in Figure 2-20(a), taken from the research of Minkoff and Lux.[35] This shows the development of morphology in a primary crystal of graphite extracted from a Ni–C–B alloy. Two forms, a sphere and a cylinder, have grown jointly under conditions where steps on the surface become unstable.

Both forms may grow under these conditions. This type of spherulite is characterized by a surface appearance which differs from the spherulite of a magnesium-treated cast iron melt.

The latter is shown in Figure 2-20(b)[33]. The appearance of the hillock instabilities depends on local growth conditions.

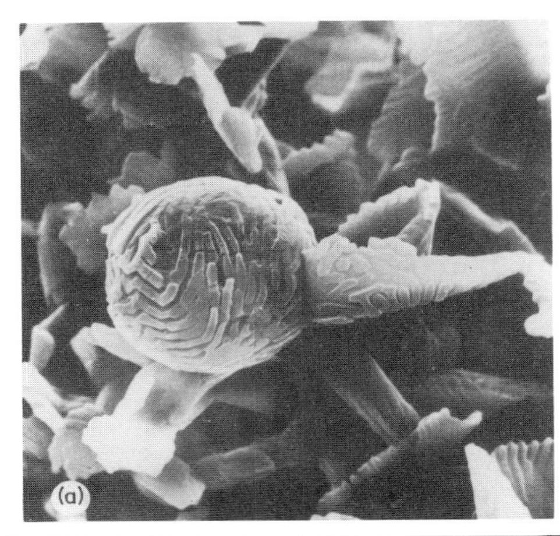

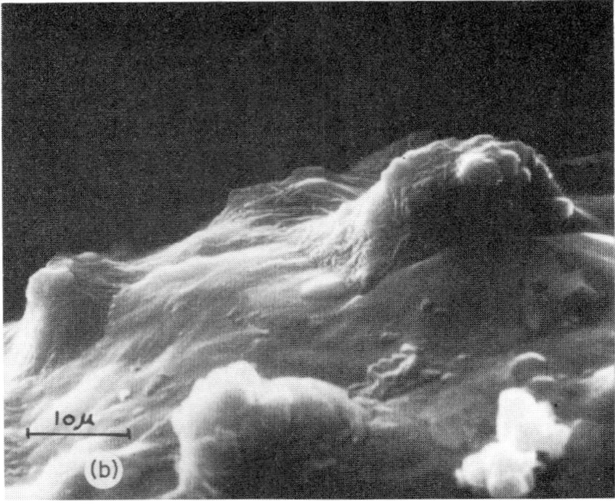

Figure 2-20 Scanning electron microscopy of individual graphite crystals. (a) Detail of a crystal of graphite extracted from a Ni–C–B alloy. Development of morphology showing sphere and conical crystal growing jointly under conditions where steps on the surface become unstable (× 1,100). (From Minkoff and Lux)[35] (Reprinted by permission from *Nature* Vol. 225, p. 541, Figure 3. Copyright © 1970, Macmillan Journals Ltd.) (b) Detail of the surface of a spherulitic crystal of graphite extracted from a commercial alloy showing instabilities on the surface in the form of growth hillocks. (From Minkoff and Nixon[33]. Reproduced by permission of American Institute of Physics)

Additional Techniques; Review by Menter[37]

A review of techniques available for studying problems of heterogeneity, in principle for steel, was made by Menter. This review included the electron probe microanalyser, Auger spectroscopy, electron microscope microprobe analyser (EMMA), and the field ion microscope atom probe. Reports have been made for some of these techniques applied to cast iron.

Problems of resolution, when analysing elements with an electron probe, arise because of the scattering of electrons into the volume of the material. This has been resolved by using thin film techniques in the transmission electron microscope together with crystal spectrometers. The electron microscope microprobe analyser (EMMA) allows a better defined element distribution to be obtained.

Auger spectroscopy uses a monochromatic incident electron beam on a surface. The atoms present at the surface can be examined from the energy spectrum of the ejected electrons.

The field ion microscope atom probe mass spectrometer has been used with much interest in studying phase transformations, e.g. pearlite formation from austenite[38] in which the role of alloying elements in controlling the transformation has been examined (see Chapter 4). In this method, a sufficiently strong electric field is applied to the specimen surface to cause surface atoms to evaporate and move as charged ions towards a screen. The ions pass through a small hole in the screen into a time-of-flight mass spectrometer. This establishes the mass and hence the atomic identity.

Studies of Graphite Growth in Cast Iron with Scanning Auger Spectroscopy

Johnson and Smart[39] used a scanning Auger microprobe (SAM) to determine the interfacial compositions between graphite flakes and spherulites, and the metal matrix. The SAM allowed a spatial resolution of 5 μm and the surface composition represented the top two or three atomic layers. The technique of examination consisted of fracturing the cast iron specimens in an ultra-high vacuum system at pressure 0.6 nN m^{-2}. A tensile device was used for fracturing and the fracture surfaces were analysed with the SAM. Inert ion sputter etching was also used to determine the depth of the chemical variations on the surface. The concentrations of sulphur and oxygen could be determined both for flakes and for spherulites, showing the variation of composition of these two elements at the surface when magnesium treatments were made with liquid alloys. This was discussed in relation to the stability of graphite surfaces.

Ion Microprobe Mass Analyser

Fidos[40] has used a combination of techniques, including the ion microprobe mass analyser, to study element distribution and spherulitic graphite growth processes in cast iron. The samples were selected to show spherulitic crystals with a ferrite halo. A high-energy beam of nitrogen ions was used to sputter ions from the surface and these were analysed for their mass to charge ratio in the spectrometer. Scanning ion micrographs were shown for several elements.

It was possible to observe element distribution within, the spherulitic crystal and also in the surrounding halo. The element boron tended to encapsulate the graphite spherulite and also formed a structure inside. Magnesium was found both within the spherulite and in the halo.

References

1. Benz, M. G., and J. F. Elliott: *Trans Metall. Soc. AIME*, **221**, 323, 1961.
2. Hillert, M.: *Recent Research on Cast Iron* (Ed. M. Merchant), p. 101, Gordon and Breach, New York, 1968.
3. Neuman, F., and H. Schenk: *Arch. Eisenhuettenwes.* **30**, 477, 1959.
4. Wells, C.: *Trans. Am. Soc. Met.*, **26**, 289, 1938.
5. Gurry, R. W.: *Trans. Am. Inst. Min. Eng.*, **150**, 147, 1942.
6. Smith, R. P.: *J. Am. Chem. Soc.*, **68**, 1163, 1946.
7. Smith, R. P.: *Trans. Metall. Soc. AIME*, **215**, 954, 1959.
8. Hillert, M.: *J. Iron Steel Inst. London*, **178**, 158, 1954.
9. Hilliard, J. E., and W. S. Owen: *J. Iron Steel Inst. London*, **172**, 268, 1952.
10. Scheil, E.: *Stahl Eisen*, **50**, 1725, 1930.
11. Piwowarsky, E.: *Hochwertiges Gusseisen*, 2nd ed., Springer-Verlag, 1958.
12. Hillert, M.: *Phase Transformations*, p. 181, Am. Soc. Met., 1970.
13. Hansen, M.: *Constitution of Binary Alloys*, McGraw-Hill, New York, 1958.
14. Vogel, R.: *Arch. Eisenhuettenw*, **3**, 369, 1929.
15. Morrogh, H., and P. H. Tütsch: *J. Iron Steel Inst. London*, **176**, 382, 1954.
16. Flemings, M. C.: *Solidification Processing*, McGraw-Hill, New York, 1974.
17. Williams, W. J.: *J. Iron Steel Inst. London*, **164**, 407, 1950.
18. Hillert, M., and P. O. Söderholm: *The Metallurgy of Cast Iron* (Eds. B. Lux, F. Mollard, and I. Minkoff), Georgi Publ. Co., Switzrland, 1975.
19. Minkoff, I., and M. Oron: *Philos. Mag.*, **9**, 1059, 1964.
20. Oron, M. and I. Minkoff: *J. Sc. Instrum.*, **42**, 337, 1965.
21. Minkoff, I., and S. Myron.: *Philos. Mag.*, **19**, 379, 1969.
22. Double, D., and A. Hellawell: *Acta Metall.*, **17**, 1071, 1969.
23. Morrogh, H., and W. J. Williams: *J. Iron Steel Inst. London*, **155**, 321, 1947.
24. Freise, E. J., and A. Kelly: *Proc. R. Soc. London, Ser A*, **264**, 269, 1961.
25. Freise, E. J.: Ph.D. Dissertation, Univ. of Cambridge, March, 1962.
26. Delavignette, P., and S. Amelinckx: *J. Nucl. Mater.*, **5**, 17, 1962.
27. Glover, A. S., and G. Pollard: *Fracture Toughness of High Strength Materials*, Iron Steel Inst. Publ. 120, 1970.
28. Austerman, S. B., S. M. Myron and J. W. Wagner: *Carbon*, **5**, 549, 1967.
29. Bollman, W., and B. Lux: *The Metallurgy of Cast Iron* (Eds. B. Lux, F. Mollard, and I. Minkoff), Georgi Publ. Co., Switzerland, 1975.
30. Tzuzuku, T.: *Proc Third Carbon Conf.*, p. 433, Pergamon, 1957.
31. Bolotov, I. Ye.: *Phys. Met. Metallogr. (Engl. Transl.)*, Vol. 20(2), p. 86, Pergamon, 1967.
32. Hunter, M. J., and G. A. Chadwick: *J. Iron Steel Inst. London*, **210**, 117, 1972.
33. Minkoff, I., and W. C. Nixon: *J. Appl. Phys.*, **37**, 4844, 1966.
34. Minkoff, I.: *J. Mater. Sc.*, **2**, 388, 1967.
35. Minkoff, I., and B. Lux: *Nature* **225**, 540, 1970.
36. Minkoff, I.: *Acta Metall.*, **14**, 551, 1966.

37. Menter, J. W.: *J. Iron Steel Inst. London*, **209**, 249, 1971.
38. Miller, M. K., and G. D. W. Smith: *Met. Sc.*, **11**, 249, 1977.
39. Johnson, W. C., and H. B. Smart: *Metall. Trans.*, **8A**, 553, 1977.
40. Fidos, H.: Presented to S. African Inst. Foundrymen, 1973.

Bibliography

Bulletin of Alloy Phase Diagrams, ASM, Metals Park, Ohio.
Bunshah, R. F. (Ed.): *Techniques of Metals Research*, Vols. I–VII, John Wiley, 1976.
Herman, H. (Ed.): *Experimental Methods of Materials Research*, John Wiley, 1967.
Masing, G.: *Ternary Systems*, Dover, 1944.
Minkoff, I.: *Graphite Crystallisation. Preparation and Properties of Solid State Materials* (Ed. W. R. Wilcox), Vol. 4, Marcel Dekker, 1979.
Murr, L. E.: *Electron Optical Applications in Material Science*, McGraw-Hill, 1970.
Rhines, F. N.: *Phase Diagrams in Metallurgy; Their Development and Application*, McGraw-Hill, 1956.

Chapter 3
Solidification and Cast Structure of Gray and White Cast Iron

This chapter extends the treatment of the solidification of cast iron commenced in Chapter 1. It deals with the structure of unalloyed cast iron. The principal subjects reviewed are:

1. The nucleation of graphite in cast iron and and the related problem of inoculants.
2. Theory of growth of the graphite eutectic.
3. Observations of growth of the ledeburite eutectic.
4. Competitive growth of the gray and white eutectics.
5. Mottled structures.

3-1 Nucleation

Nucleation in a melt has been described as either homogeneous or heterogeneous. In general, for commercial melt conditions, nucleation is heterogeneous and for cast iron this will be the case for all types of alloy, i.e. flake, spherulitic, or intermediate graphitic form.

However, using pure materials, and in particular with Ni–C alloys, where impurity free metal may be used for experimentation, it may be possible to achieve homogeneous nucleation. This may be the case when vacuum melting techniques are employed.[1]

Heterogeneous Nucleation of Graphite with Disc-Shaped Nucleus

In Chapter 1, the theory of homogeneous nucleation was given with a model employing a spherical nucleus. Since graphite is anisotropic and has different values of the interfacial energy for the different faces, it is convenient to discuss a disc-shaped nucleus. This has height h and radius r (Figure 3-1). The edge energy of the disc is uniform and has value ε while the flat faces of radius r have a surface energy σ. The free energy change[2] in formation of such a

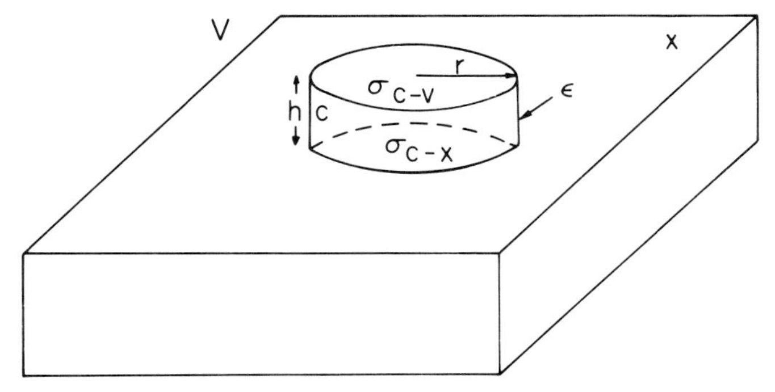

Figure 3-1 Disc-shaped nucleus of Phase C on a substrate of phase X growing from vapour V. The edge energy of the disc is ε^2

nucleus is:

$$\Delta G = 2\pi r\varepsilon + \pi r^2\sigma + \pi r^2 h \Delta G_v \tag{3-1}$$

In this equation, the value of σ is given by:

$$\sigma = \sigma_{C-V} + \sigma_{C-X} - \sigma_{X-V}$$

The critical free energy for formation of the nucleus is:

$$\Delta G^* = -\frac{\pi\varepsilon^2}{h(\Delta G_v + \sigma/h)}$$

The different components of σ are indicated in Figure 3-1. Hillert[3] has assumed values of the surface energy for the faces of graphite crystals and using calculated values of ΔG_v has estimated h both for graphite and Fe_3C as a function of composition and temperature.

Heterogeneous Nucleation of Graphite

Heterogeneous nucleation of graphite is an important study in cast iron metallurgy. It must be examined in relation to the materials added to the liquid alloy and also to the chemistry of the melt.

Some of the structural effects related to heterogeneous graphite nucleation are:

(a) The promotion of gray (graphite) crystallization rather than the alternative white (carbide) mode.
(b) Avoidance of undesirable undercooled structures.
(c) In gray cast iron, an increase of the eutectic cell count (frequency of eutectic nucleation).
(d) In spherulitic cast iron, an increase in the number of spherulites.

The Interface between Graphite and the Heterogeneous Nucleus

The interfacial free energy σ between the nucleated phase and the heterogeneous nucleus has an important influence on the critical free energy for nucleation. This is seen from Figure 3-2. When the interfacial energy is equal to zero, the value of ΔG^* will be zero and nucleation should theoretically commence on cooling as soon as the equilibrium temperature has been passed.

The value of σ is dependent on the structure of the heterogeneous nucleus and involves effects like coherency and semi-coherency.

Coherent Interface Figure 3-3 shows a coherent interface, in which matching occurs between the planes of atoms constituting the interface.

The matching of planes my involve strains on both sides. These strains are indicated in Figure 3-3 in which the nucleating substrate is dilated and the nucleating phase is contracted. If the two strains are ε_1 and ε_2, the total interface strain $\varepsilon = \varepsilon_1 + \varepsilon_2$.

Semi-coherent Interface A semi-coherent interface is formed by regions of perfect matching alternating with regions of mismatch. This can be rep-

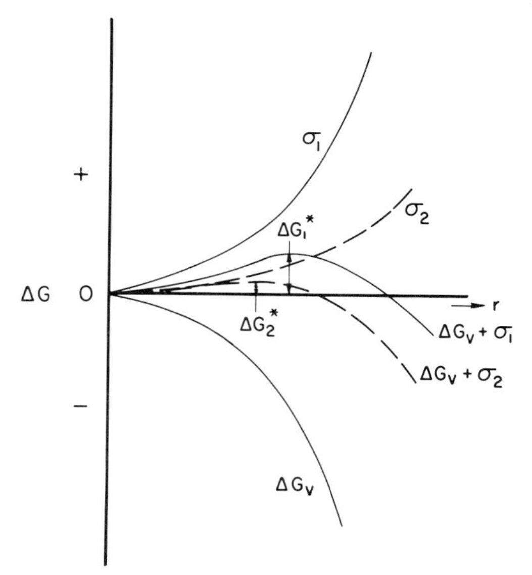

Figure 3-2 Free energy change during heterogeneous nucleation on a substrate. As the interfacial energy σ tends to zero, the critical free energy for nucleation tends to zero. σ_1 and σ_2 are different values of the interfacial energy; ΔG_1^* and ΔG_2^* are corresponding values of the critical nucleation energy

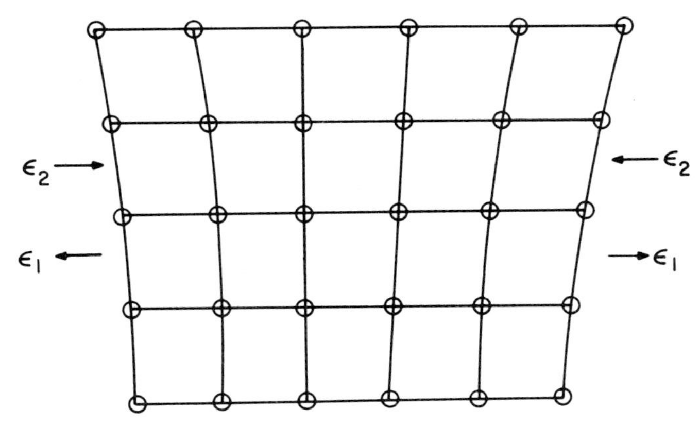

Figure 3-3 Model of a coherent interface between two phases. Matching of planes may involve strains on both sides. If the two strains are ε_1 and ε_2, the total interface strain $\varepsilon = \varepsilon_1 + \varepsilon_2$

resented by a model of a coherent interface with interspaced dislocations (Figure 3-4).

The interfacial energy is made up of the energy of the matching region σ_ε and the energy of the misfit region σ_S, where the latter may be considered due to dislocations:

$$\sigma = \sigma_\varepsilon + \sigma_S$$

This interface model is due to Van der Merwe.[4] A typical example proposed was the nucleation of a (0001) hexagonal crystal face on a (111) cube face. On these crystallographic planes, the structure is identical. It is applicable to the nucleation of the hexagonal graphite structure by cubic phases.

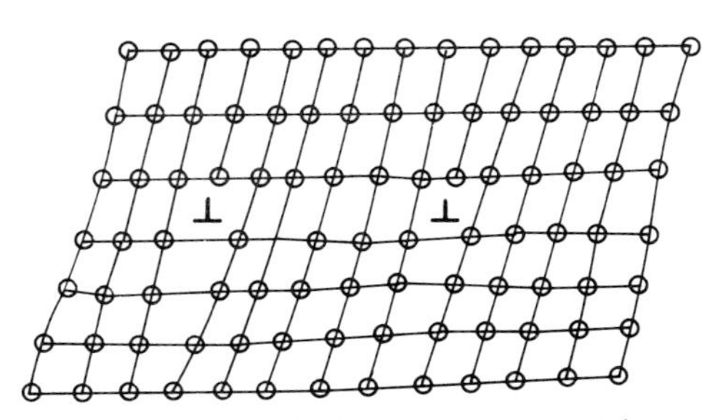

Figure 3-4 Model of a semi-coherent interface, showing regions of perfect matching alternating with regions of mismatch. (After Christian,[5] reproduced by permission of Pergamon Press Ltd.)

For the two lattices, there is a disregistry δ which can be calculated as follows:

$$\delta = \frac{a_\alpha - a_\beta}{a_\beta}$$

where a_α is the lattice spacing for hexagonal (0001) and a_β is the lattice spacing for cubic (111). The value of a_α can be obtained for graphite and a_β is applicable to the different phases suggested as nuclei for graphite. The disregistry is partly taken up by strain and partly by dislocations.

Incoherent Interface When the disregistry δ is large, the dislocation spacing is small. The boundary then becomes what is generally termed incoherent.

Free Energy for Nucleation with Coherent and Semi-Coherent Interfaces

For coherent nucleation, the strain energy may be large and the surface energy small. In this case the free energy term will be given as:[5]

$$\Delta G^* = \frac{\mu V^\beta \varepsilon^2}{1 - \nu} \tag{3-2}$$

where μ = shear modulus
ε = disregistry of atoms along one direction
ν = Poisson's ratio
V^β = specific volume of atoms in the β phase

As the disregistry δ increases, the contributions to ΔG^* by strain energy and dislocations change. This is shown in Figure 3-5.[5] For a small disregistry δ, the strain energy term is energetically more favourable so that coherency is preferred. As δ increases, it becomes energetically more favourable to take up the misfit with dislocations.

Inoculants

The main constituents of most of the commercial inoculants used in cast iron practice are Fe and Si. Small quantities of other elements may be present, e.g. Ca, Al, Zr, Ba, Sr, and Ti. In the Meehanite process[6] small quantities of alkaline earth metals, e.g. Ca, Mg, and Ba, are introduced.

Cast irons are mostly hypo-eutectic in composition and ordinarily would be characterized by primary growth of the γ phase. Inoculation effectively prevents structures dominated by this phase. The interdendritic late freezing of the graphite eutectic would lead to undesirable structures, and in thin sections the ledeburite eutectic could dominate growth.

The eutectic cell number is influenced by inoculation and is related to mechanical strength. The chemical influences on inoculation, i.e. the dependence of cell number on the concentration of silicon, sulphur, and minor

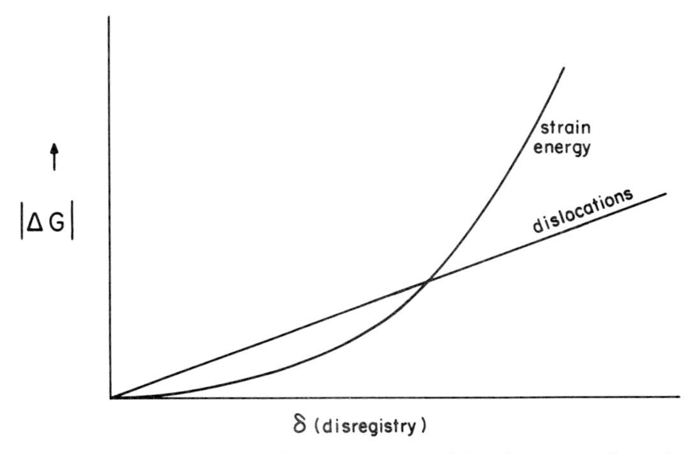

Figure 3-5 Free energy for nucleation with coherent and semi-coherent interfaces. As the disregistry δ increases, the contributions to ΔG^* by strain energy and dislocations change. For small δ, coherency is preferred. As δ increases, it becomes energetically more favourable to take up the misfit with dislocations. (From M. E. Fine, *Introduction to Phase Transformations in Condensed Systems*, Macmillan, 1965)

elements, indicate that the nucleus for graphite in a cast iron melt is very complex.

While different models of the inoculation mode of cast iron have been presented, the analysis of available experimental data indicates that nucleation of graphite is, in the main, by oxides. The oxide SiO_2 must form the basis and other elements are present in the lattice. These elements probably form important additions because of their affinity with oxygen, although it does not seem that they affect the lattice parameter of the SiO_2 phase. That nucleation is complex becomes apparent from experimental study.[7] One proposal to be described is that nucleation should be taken as a demonstration of a two-stage process, in which the formation of the oxide phase is preceded by the formation of a sulphide.

Notwithstanding, some other suggestions are presented, in which, for example, carbides are quoted as constituting the nucleus.[8]

Recording of Nucleation and Eutectic Crystallization Temperatures

Figure 3-6 shows the general form of a typical cooling curve for a gray iron of eutectic composition. T_N is the temperature recorded for undercooling of the melt. The eutectic freezes over a range of temperatures commencing before the maximum in the undercooling. T_E is related to the rate of heat removal and to the composition. Both these temperatures can be displaced with variation of these conditions.

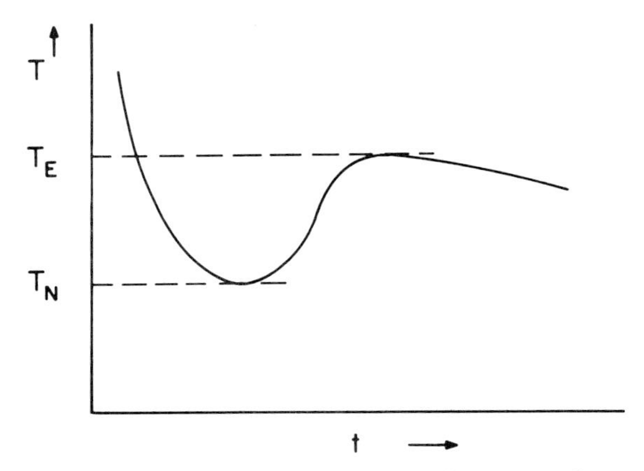

Figure 3-6 Form of a typical cooling curve for a gray iron of eutectic composition. T_N is the maximum undercooling. The eutectic freezes over a range of temperature commencing before T_N. T_E is related to the rate of heat removal and to the composition

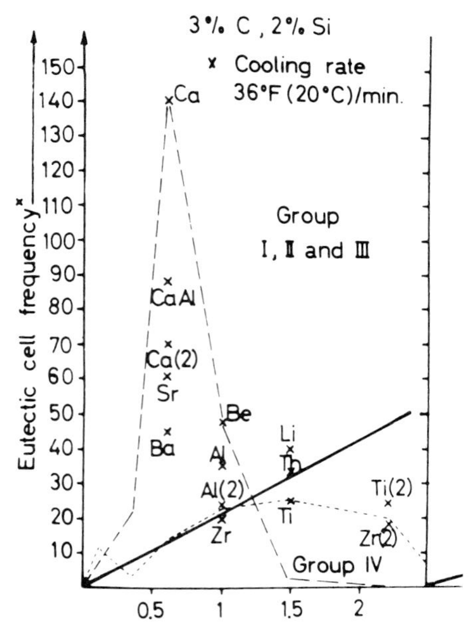

Figure 3-7 Experimental results due to Lux[8] for eutectic cell frequency in inoculation of cast iron by elements of groups I, II, and III of the periodic system. Eutectic cell frequency versus cell diameter (mm). (Reproduced by permission of American Foundrymen's Society)

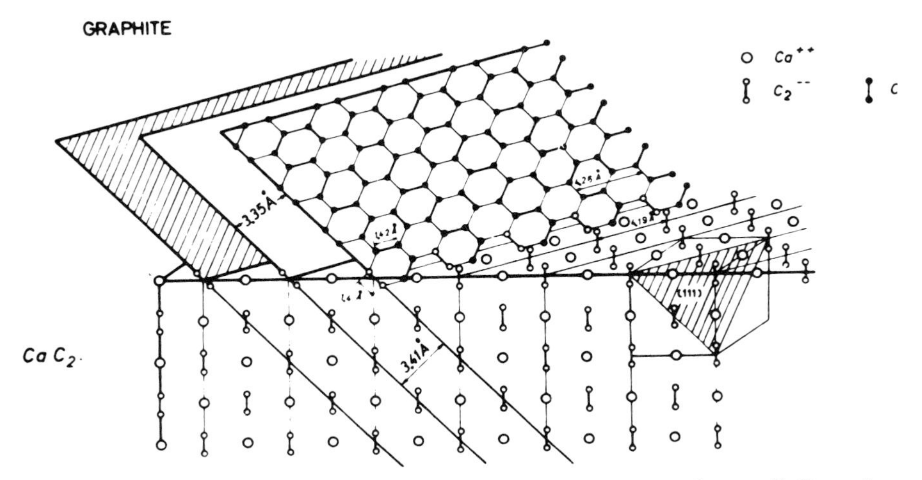

Figure 3-8 Model by Lux[8] for the growth of a graphite crystal on a CaC_2 nucleus. (Reproduced by permission of American Foundrymen's Society)

Model of Inoculation by Carbides

Lux[8] investigated the inoculation of a cast iron melt by element of groups I, II, and III of the periodic system. The effectiveness of nucleation was measured by determining undercooling during eutectic solidification and the number of eutectic cells per unit volume. Figure 3-7 shows the experimental results for the eutectic cell frequency related to the element added.

Lux considered from this that the nucleus could be a salt-like carbide and suggested that the compound CaC_2 was the most effective. He suggested that these could be the effective nuclei formed from the impurities in technical inoculants.

Figure 3-8 shows the model presented for the nucleus. Lux suggested that the carbon layers of the carbide represent pre-formed intermediate carbon aggregates for graphite crystals. The figure shows a CaC_2 crystal bounded by (100) faces. Graphite is growing parallel to (111) of the CaC_2 lattice. The distance between (111) planes of CaC_2 is 0.341 nm, while between the (0001) graphite layers it is 0.335 nm.

Model of Inoculation by Duplex Sulphide Oxide

Inclusions in graphite extracted from cast iron have been investigated by different techniques to determine the possible identity of heterogeneous nuclei. These techniques include electron microscopy and electron diffraction[9,10,11] and electron microscopy with X-ray microanalysis.[12]

An investigation by Jacobs *et al.*[12] was directed to determining the nature of nuclei and changes in their chemical composition and structure, during treatment of iron with magnesium ferrosilicon. Post-inoculation treatment included the alloy Superseed. Different variations of the treatment were made to examine the effects of elements like Al and Sr on nucleus identity. The results are interesting for cast iron in general since the examination showed a duplex structure for nuclei consisting of a sulphide core surrounded by an oxide shell. It was also possible to examine lattice strains in epitaxial graphite by observations of streaking in the X-ray diffraction patterns.

In the particular analyses of cast iron investigated, and depending on the type of inoculant used, the nucleus had the following composition:

Core (Ca, Mg) sulphide or (Sr, Ca, Mg) sulphide
Outer shell (Mg, Al, Si, Ti) oxide

The orientation relationships in the initially growing complex nucleus and between the oxide phase and graphite were established as follows:

Nucleus (110) sulphide $\parallel$ (111) oxide
$[1\bar{1}0]$ sulphide $\parallel$ $(2\bar{1}\bar{1})$ oxide

Nucleus/graphite (111) oxide $\parallel$ (0001) graphite
$[1\bar{1}0]$ oxide $\parallel$ $[10\bar{1}0]$ graphite

The research was performed investigating the particles by an extraction replica technique. These particles were examined both in the graphite nodules and as a free phase extracted from the iron matrix. The latter particles contained the same elements as the particles at the nodule centres. The particles at the centre of the graphite nodules were approximately 1 μm in diameter.

X-ray Diffraction Data of Epitaxy

The X-ray diffraction data for the oxide layer of the particles examined showed no appreciable change in the lattice parameter for relative changes in Mg, Al, Si, or Ti. This parameter was determined to be 0.770 ± 0.005 nm.

The data for the central (sulphide) region in part of the experiments showed an f.c.c. lattice with a parameter between that of CaS (0.5695 nm) and that of MgS (0.5200 nm). It was concluded that in these experiments the centre was a mixed sulphide (Ca, Mg)S.

Streaked X-ray reflections were obtained. The geometry of the observed streaks on the diffraction patterns were explained by strain in the graphite lattice. It was suggested that the first few graphite layers, adjacent to (111) oxide, had a dilated lattice, i.e. 0.264 nm in place of 0.246 nm. For the graphite layers further from the oxide surface, it was suggested that the spacing progressively decreased until the unconstrained spacing was attained. Dislocations were observed frequently in the matrix particles and it was believed that these were generated to partly relieve the elastic strain in the graphite layers adjacent to the oxide. This follows the theoretical concepts described in heterogeneous nucleation in the previous section. The possible nucleation order, i.e. the primary nucleation of a sulphide phase, followed by the nucleation on the sulphide of an oxide phase, was discussed. Further observations were made in this research of the prior nucleation of TiO before the growth of the mixed spinel, and the influence of S in contaminating nuclei.

3-2 The γ-Graphite (Gray Iron) Eutectic

The three-dimensional arrangement of graphite within a eutectic cell was described by Hultgren, Lindblom, and Rudberg,[13] Morrogh and Oldfield,[14] Brunin, Malinochka, and Fedorova,[15] and Kusakawa and Nakata.[16]

Figure 3-9(a) shows a model due to Morrogh and Oldfield[14] demonstrating the interconnected character of graphite growth within one eutectic cell. The idea presented was a continuous skeleton of graphite which branched with a frequency dependent upon the radial rate of growth of the eutectic cell.

Figure 3-9(b) shows a series of illustrations from Minkoff and Lux[17] as a further contribution to understanding the mode of development of a graphite eutectic cell. The series of sketches (i), (ii), and (iii) were made from observations of eutectic growth in a Ni–C alloy shown in (iv). The graphite crystal extends as a thin plate-like form bounded by (0001) surfaces and the second phase of the eutectic grows over it. Branching of the graphite then occurs from the side and the branch curves and grows over the parent crystal. Continued branching of the growing members and curvature extends the eutectic skeleton in three dimensions. The graphite behaviour is reminiscent of dendritic branching. it is also independent in growth with no apparent cooperation involving the second phase. The illustration (v) is a view of a graphite crystal dissolved from a cast iron alloy. This shows a branching mechanism basically similar to that of graphite in Ni–C. The sketch (vi) is a representation of this branching behaviour. In the cast iron eutectic, the graphite volume is greater than in the Ni–C alloy, which tends to obscure the mode of branching from the crystal edge.

Minkoff and Lux[17] ascribed this specific behaviour of graphite in the eutectic to the stability of the (0001) graphite face and the presence of a rotation boundary defect on $(10\bar{1}0)$. The branching from one face only imposed a constraint on eutectic development and was a primary factor in determining the coarseness of the two-phase structure.

The Interface between the Gray Iron Eutectic and the Melt

In observations of the structure of coarse graphite eutectic, the interface appears non-planar (Figure 3-10a). This is apparently because the graphite phase grows by an interface mechanism and requires a higher undercooling for growth than the γ phase. This results in a non-isothermal and hence non-planar interface. As the eutectic becomes finer, this departure from planarity is reduced. Figure 3-10(b) shows how the interface adjusts itself to the temperature conditions in directional growth experiments with an imposed

(a)

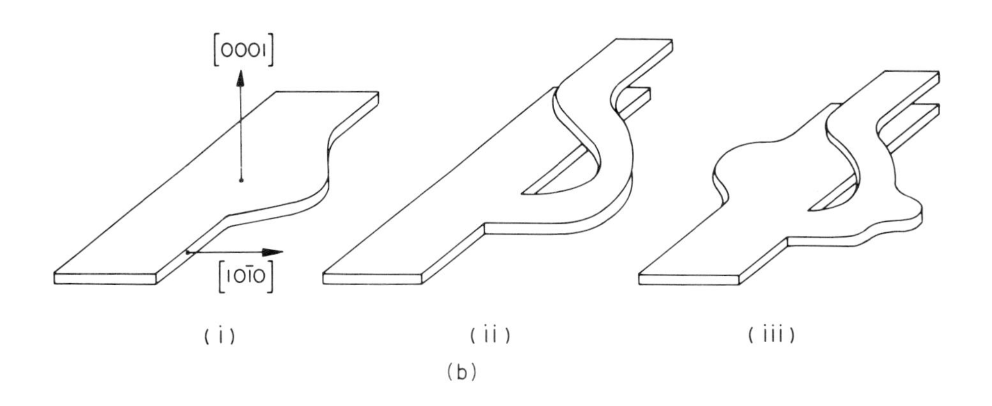
[0001]
[10$\bar{1}$0]
(i)
(ii)
(iii)
(b)

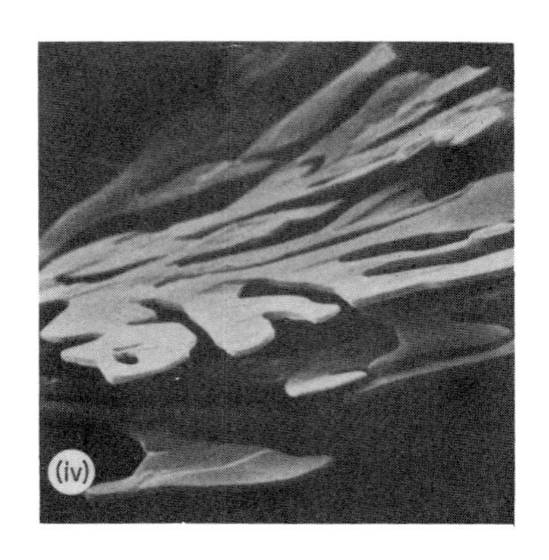
(iv)

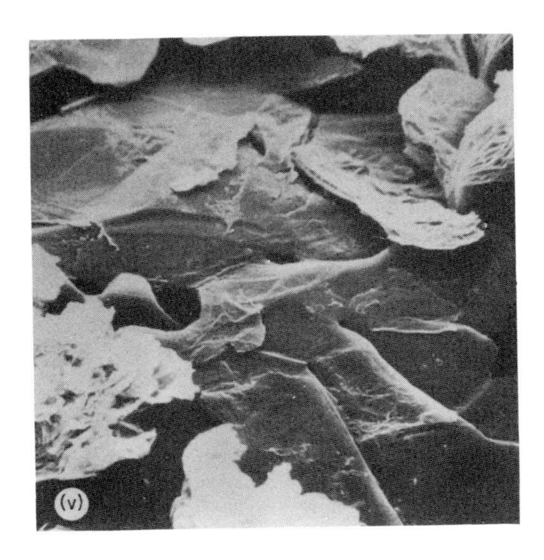

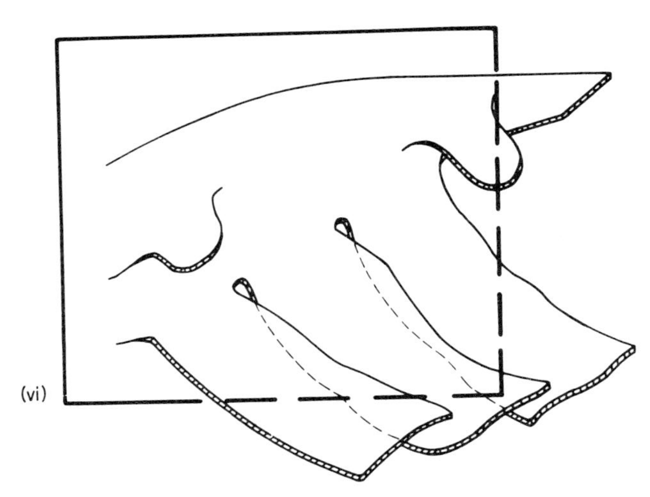

Figure 3-9 (a) Model of a graphite eutectic cell in cast iron due to Morrogh and Oldfield.[14] (Reproduced by permission of IPC Business Press Ltd.) (b) Details of branching mode of graphite in eutectic cell due to Minkoff and Lux.[17] (Reproduced by permission of American Foundrymen's Society.) (i) (ii) (iii) Stages of growth and instability of graphite phase in eutectic noted for Ni–C alloy. The graphite crystal branches out of the side. Extension of the eutectic in three dimensions is by the branch curling over the parent crystal and further sideways branching.[17] (iv) Extracted graphite crystal from eutectic in Ni–C alloy[17] (× 1,650). (v) Extracted graphite eutectic crystal from commercial cast iron alloy (× 3,850). (vi) Schematic representation of the branching mode observed in (v) (× 5,500)

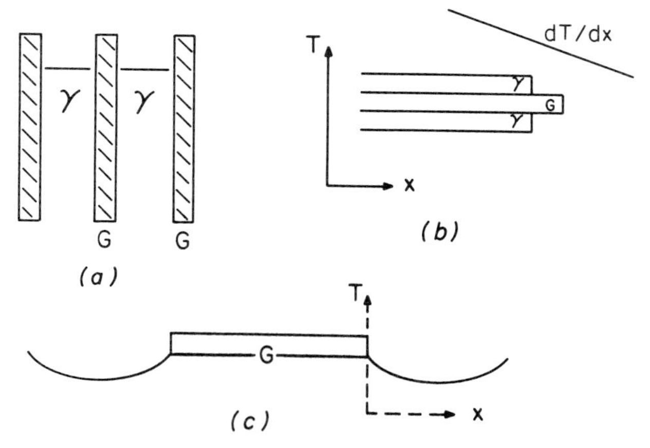

Figure 3-10 (a) Non-planar interface of graphite lamellae, G, and γ in eutectic growth. (b) Growth of graphite eutectic in a negative temperature gradient. (c) Temperature conditions ahead of a eutectic cell under conditions prevailing in a commercial melt of cast iron

negative temperature gradient. Figure 3-10(c) shows the temperature conditions ahead of a eutectic cell under conditions prevailing in a commercial melt of cast iron. The temperature gradient ahead of the eutectic cell is also negative. This originates because a liquid alloy of eutectic composition undercools by at least 4 °C, possibly before the graphite nucleates to initiate the eutectic. Primary growth of the eutectic with release of latent heat raises the temperature in the neighbourhood of the growing solid. The heat flow is now down the temperature gradient to the liquid and the graphite grows ahead of the interface into the gradient of the undercooling.

In hypo-eutectic cast iron, the γ phase grows first and solid separates from liquid until the latter has reached a composition and temperature where graphite can nucleate. The interdendritic liquid is undercooled and eutectic growth proceeds as above to fill the interstices between the γ dendrites.

In controlled growth experiments, the interface can be made to approximate planarity by considerably increasing the imposed temperature gradient. the fine structures resulting, having smooth interfaces, have been suggested to grow in a cooperative manner.[18]

Coarseness of the Graphite Eutectic; λ–R Relationships

The theoretical relationship $\lambda^2 R$ = constant, due to Jackson and Hunt,[19] is not obeyed in the growth of the α–graphite eutectic, and λ–R relationships in cast iron compared with the theoretical curve of Jackson and Hunt are shown in Figure 3-11. This departure from the theoretical value is indicative of a deviation from cooperative growth, all such deviations being in the direction of coarser structures.

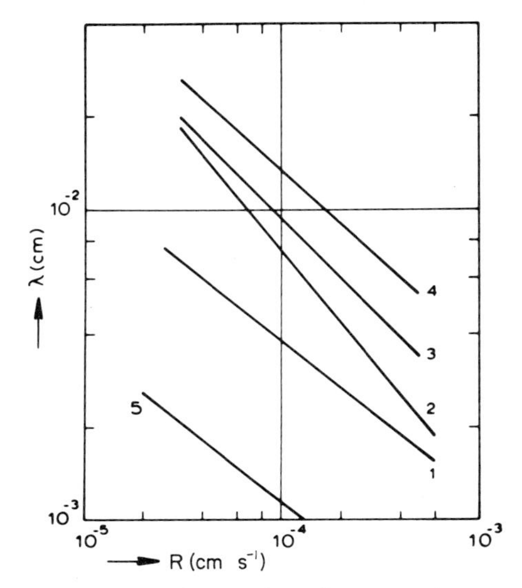

Figure 3-11 Comparison between theoretical (curve 5) and experimental λ–R values in gray eutectic solidification of cast iron. Curve 1 is due to Lakeland and Hogan.[20] Curves 2, 3, and 4 are due to Nieswaag and Zuithoff.[21] (Reproduced by permission of Georgi Publishing Co.)

The relationships shown are as follows:

1. Lakeland:[20] $\lambda = 3.8 \times 10^{-5} R^{-0.5}$ cm
2. Nieswaag and Zuithoff:[21] $\lambda = 0.56 \times 10^{-5} R^{-0.78}$ cm
3. Nieswaag and Zuithoff:[21] $\lambda = 3.1 \times 10^{-5} R^{-0.62}$ cm
4. Nieswaag and Zuithoff:[21] $\lambda = 7.1 \times 10^{-5} R^{-0.57}$ cm
5. Jackson and Hunt:[19] $\lambda = 1.15 \times 10^{-5} R^{-0.5}$ cm

Curves 2 and 4 are for varying sulphur content. Curve 3 is for a cast iron containing phosphorus. Curve 2 is for a cast iron containing 0.003–0.004 per cent sulphur while curve 4 is for a cast iron containing >0.02 per cent sulphur.

Influence of Sulphur on Coarseness of the Graphite Eutectic

Nieswaag and Zuithoff[21] summarized accumulated test results for the λ–R relationship as a function of the sulphur content (Figure 3-12). The coarsening effect of sulphur is noted at small values of R or a fixed growth rate, when the λ value increases as sulphur increases. This is the accepted influence of sulphur on coarsening of graphite eutectic structures. These results show that as R achieves higher values, the value of λ once more decreases, after a peak degree of coarsening, as sulphur increases.

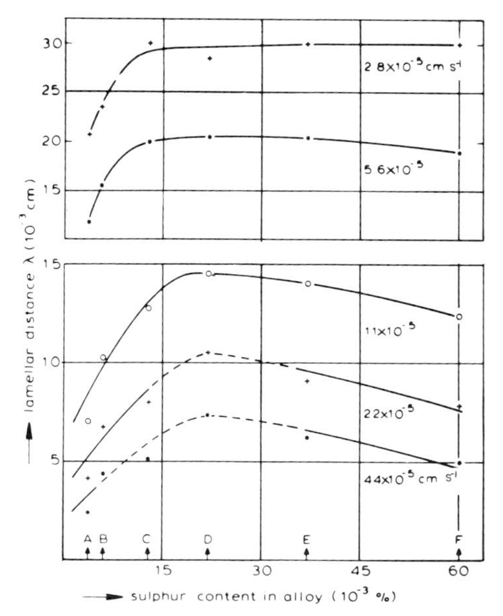

Figure 3-12 Summary by Nieswaag and Zuithoff[21] of λ–R relationships in the gray iron eutectic as a function of sulphur content. At small values of the growth rate, sulphur has a coarsening effect. At high values of the growth rate, there is a maximum in the degree of coarsening as sulphur increases. Growth rate (cm s^{-1}) is indicated on each curve. (Reproduced by permission of Georgi Publishing Co.)

Number of Eutectic Cells

The eutectic cell number is expressed as the number of cells per square centimetre. Theoretically, it should be dependent on the undercooling of the melt, on the type of nucleus, and on the availability of nuclei. Within this framework, a number of empirical factors have been evaluated relating cell number to section size, type of inoculant, sulphur content, silicon content, etc.[7] Nucleation, being dependent on the kinetics of reactions in the melt, is also subject to fading.

Growth of Eutectic Cell and Fineness of the Structure

Morrogh and Oldfield[7] defined two factors which determined the fineness of the graphite eutectic in cast iron. For a given composition, the undercooling for nucleation of a eutectic cell would determine the growth rate and hence the fineness. The greater the undercooling, the finer the graphite. The second factor was the composition of the melt. Morrogh and Oldfield suggested that sulphur increased the coarseness of the eutectic in gray cast iron by reducing the growth rate at a given undercooling.

The use of a directional growth experiments, an example being the results reported in the previous section by Nieswaag and Zuithoff, has given a new insight into growth rate and eutectic spacing in systems like that involving graphite. As observed in Figure 3-11, the spacing λ becomes coarser with increasing sulphur at the same growth rate. Hence the influence of sulphur in a commercial cast iron cannot be to slow down the growth rate of a eutectic cell.

It is more appropriate to consider that the influence of melt composition in cast iron is to change the interface temperature at a given rate of growth. The temperature of growth will vary, as discussed in Section 1-3, according to the conditions which influence nucleation on the $(10\bar{1}0)$ face of graphite. The influence of sulphur at small melt concentrations is to allow graphite growth at a smaller interface undercooling than in a sulphur-free alloy. If graphite branching is dependent on the degree of undercooling, then the effect of sulphur in reducing the undercooling is to coarsen the graphite.

Nieswaag and Zuithoff noted that at large growth rates the influence of increasing sulphur is once more to refine the structure. This is apparently related to a constitutional supercooling effect.

Change from Coarse to Fine Structure Related to Growth Rate

Figure 3-13, from the data of Nieswaag and Zuihoff,[21] shows a feature already described by Hillert and Rao.[18] As the growth rate increases, the λ value changes to a second branch of the λ–R curve, located at lower λ values. This means that there may be an abrupt change from coarse to fine structures at a certain growth rate. Hillert and Rao, and also Nieswaag and Zuithoff, show that both structures, coarse and fine, may coexist.

Hillert and Rao[18] suggested that as the structure became finer and the graphite–γ interface became smoother, the growth would become more similar to that for a purely metallic system and diffusion control would dominate the growth process. The spacing for this type of growth is smaller for any growth rate (curve 5 of Figure 3-11).

Influence of Hydrogen and Nitrogen

In the paper of Morrogh and Oldfield[7] the interesting influence of hydrogen is shown on the undercooling of cast iron. Eventually the undercooling is sufficient to lead to white iron structures.

Both hydrogen and nitrogen gases are surface active, adsorbing at the graphite interface. When present in solution, these gases must compete for sites with sulphur on the graphite interface. In this type of adsorption the atoms are held by van der Waals forces as distinct from the behaviour

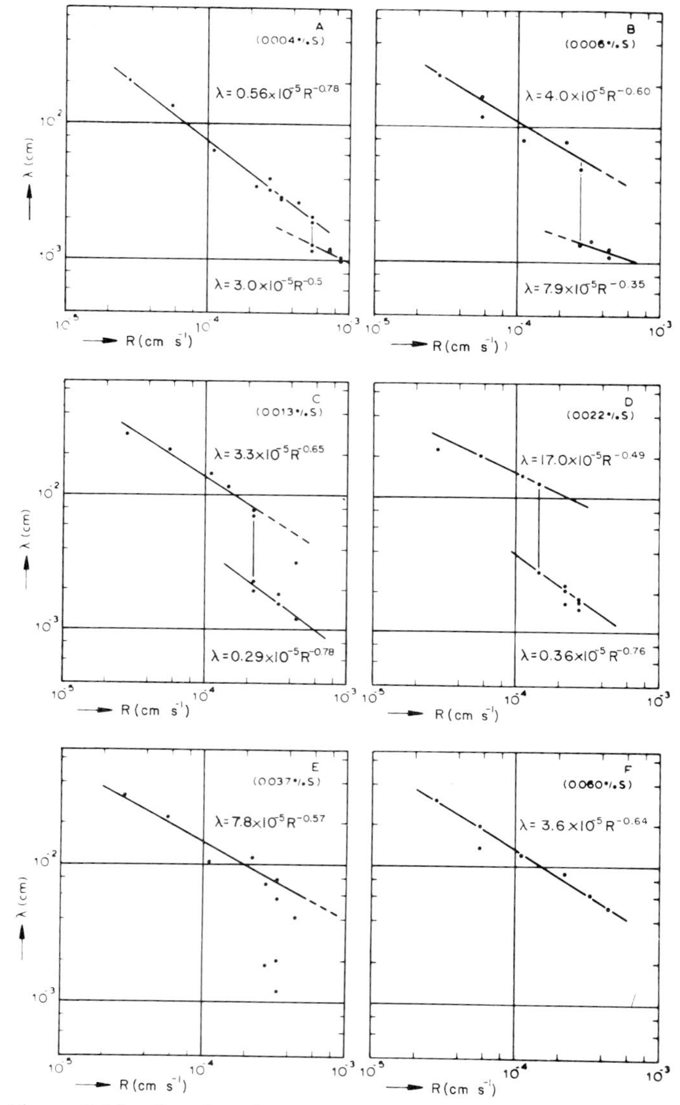

Figure 3-13 Results of Nieswaag and Zuithoff[21] showing two branches of the λ–R curve. At high growth rates the relationship changes to a second curve with smaller λ values. (Reproduced by permission of Georgi Publishing Co.)

described for Mg and Ce which involves electron transfer with graphite. Little data is available for this type of behaviour in cast iron other than curves of undercooling of the type shown by Morrogh and Oldfield. These indicate that hydrogen probably replaces sulphur in the first adsorbed layer. The net effect is to change the interfacial energy. Similar reasoning must apply to the adsorption of the gas nitrogen. In the latter the interface undercooling effect is sufficient to change the growth characteristics so that nitrogen may be introduced into cast iron to produce alloys with intermediate structure of the graphite phase.

The Asymmetric Coupled Zone in Gray Cast Iron

The coupled region in eutectic solidification was described by Kofler[22,23,24] and Scheil.[25] A coupled region indicates the extent of a region of two-phase growth with respect to composition and temperature on a phase diagram. On either side of a coupled region, growth is characterized by that of a single phase.

In metallic systems, the coupled region is uniformly situated between the extensions of the two liquidi (Figure 3-14), while for the graphite–γ system it is asymmetric. This difference can be shown to be due to the forms of temperature dependence of growth for the metallic and non-metallic phases.[26]

To explain an asymmetric coupled zone, Hunt and Jackson[26] used the two growth rate curves of Figure 3-15(a) in which one of the phases α has a marked temperature dependence on the growth rate as for a non-metallic phase. The growth rate of the two-phase structure, as dependent on temperature, was suggested to lie between the curves for the growth rates of the metallic and non-metallic phases. This gives three regions of growth as the undercooling increases (Figure 3-15b). In the first region, primary α grows faster than the eutectic. In the intermediate region, eutectic growth is preferred. In the third region, primary phase growth is again preferred. The coupled zone for this construction is then an asymmetric one, conforming to Figure 3-14(b).

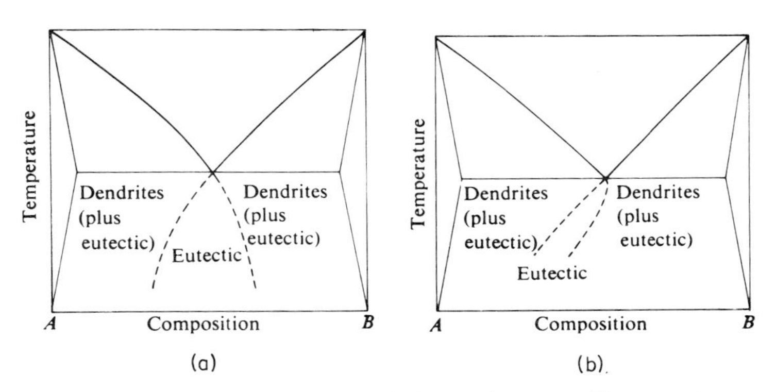

Figure 3-14 (a) Symmetrical coupled zone in a metallic eutectic system. (b) Asymmetrical coupled zone in a eutectic system where one phase is faceted. (Reproduced by permission of The Metallurgical Society of American Institute of Mining Engineers)

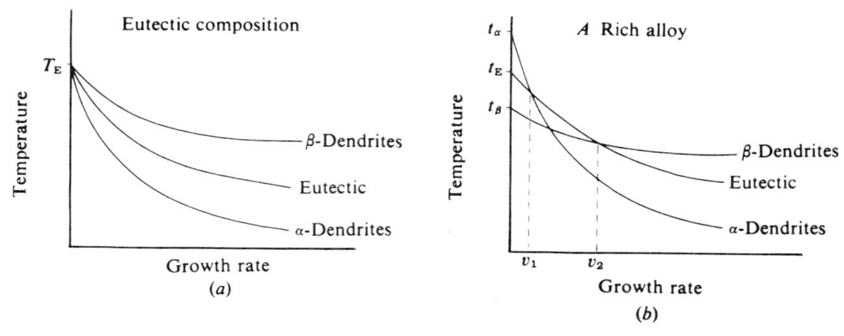

Figure 3-15 (a) Growth rate curves for phases α and β and the α/β eutectic used by Hunt and Jackson[26] to explain an asymmetrical coupled growth region. One phase α must have a more marked temperature dependence on growth than the second phase β. (b) Growth curves for an A-rich hypo-eutectic alloy. The eutectic has a limited growth interval between two regions of primary growth, giving the diagram of Figure 3-14(b). (Reproduced by permission of the Metallurgical Society of American Institute of Mining Engineers)

Asymmetric Coupled Zone; Geometrical Construction; Divorced Eutectic Growth

Jackson and Hunt constructed their asymmetric coupled zone by suggesting that the eutectic growth rate for the particular case being considered was the mean of the growth rate curves for the two phases. Lux, Mollard, and Minkoff[27] constructed a coupled zone for gray iron from the conditions for equal growth rates of the γ and graphite phases. When γ grows faster than graphite, the normal lamellar eutectic growth would cease. Two different growth rate curves were suggested for the two phases, an exponential curve for the graphite and a parabolic curve for γ (Figure 3-16a). The construction is shown in Figure 3-16(b).

In Figure 3-16(a), the growth curve for graphite and the growth curve for γ meet at a point with an undercooling $\Delta T_{\gamma/G}$. If both phases grow at this undercooling, they would have equal growth rates. Two further constructed points for equal growth rates of the two phases are shown marked in circles 1 and 2. These three points give P, Q, R the resulting coupled zone shown in Figure 3-16(b). With this construction, liquid alloys crystallizing to the left of the

---

Figure 3-16 (a) Growth rate curves suggested by Lux, Mollard, and Minkoff[27] to construct an asymmetrical coupled zone for gray cast iron. (b) Construction of an asymmetrical coupled zone for gray cast iron using the growth rate curves and criterion of (a).[27] (Reproduced by permission of Georgi Publishing Co.) (c) Growth rate curves for graphite $(10\bar{1}0)$ and γ showing a suggested curve for (0001) graphite surfaces growing by a screw dislocation mechanism. If the γ growth curve and the (0001) graphite curve do not meet, γ growth is always faster than graphite. Then no coupled region can exist. This is suggested as the case for spherulitic graphite growth

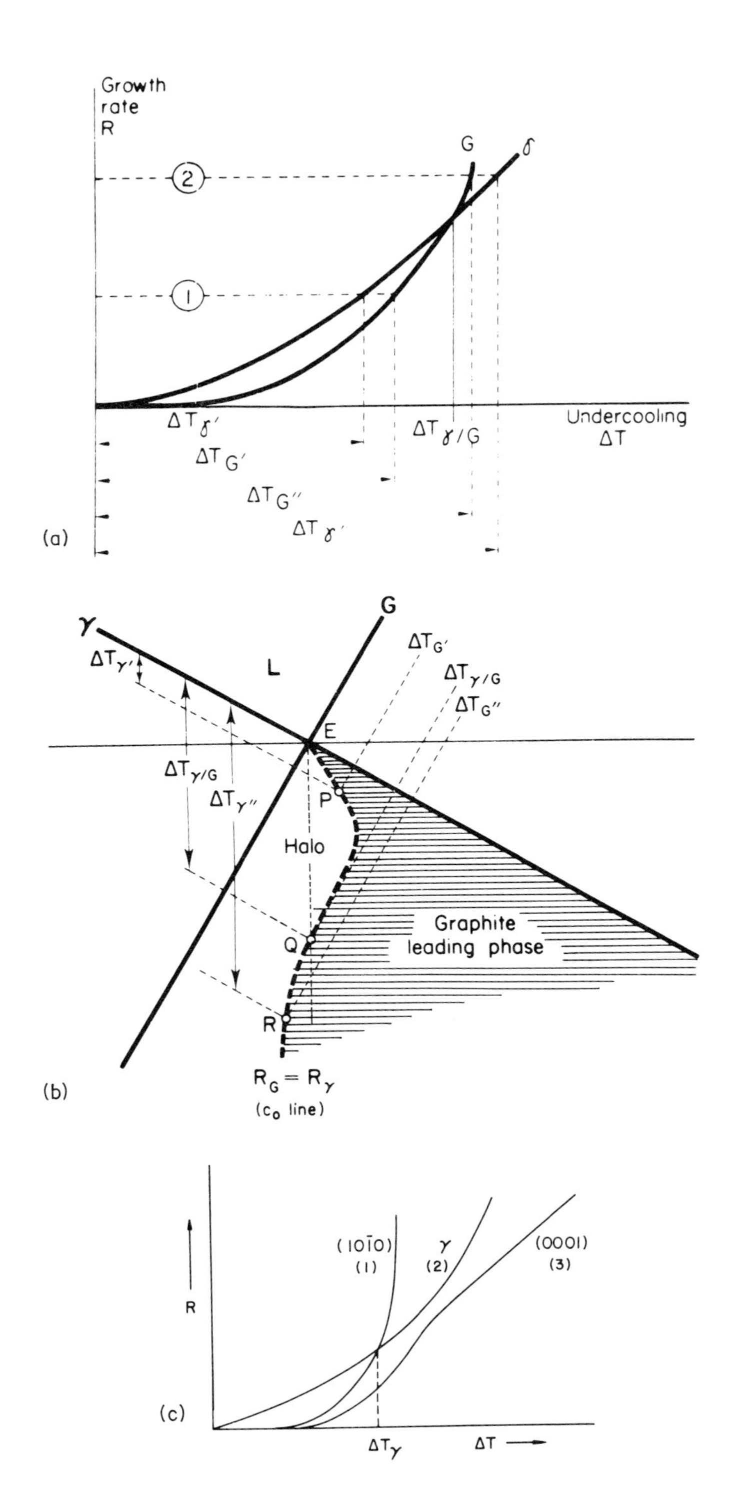
Growth rate R
G
γ
2
1
ΔTγ′
ΔTG′
ΔTG″
ΔTγ′
ΔTγ/G
Undercooling ΔT
(a)
γ
G
ΔTγ′
L
ΔTG′
ΔTγ/G
ΔTG″
E
ΔTγ/G
ΔTγ″
P
Halo
Graphite leading phase
Q
R
RG = Rγ
(c0 line)
(b)
R
(10Ī0) (1)
γ (2)
(0001) (3)
(c)
ΔTγ
ΔT

bounding curve have γ growing faster than graphite and no probability of achieving a two-phase eutectic type of structure. Within the coupled zone construction, the graphite phase grows faster than γ, typifying the γ-graphite eutectic.

Figure 3-16(c) shows an additional growth rate curve for a graphite crystal surface growing by a screw dislocation mechanism. The characteristics of this type of curve are that, for small undercoolings, the growth rate is small and parabolic and for increased undercoolings the growth rate becomes linear. If the γ growth rate curve controlled by diffusion and the growth rate curve for graphite growing by a screw dislocation do not meet, the γ growth rate is always faster than graphite. Then no coupled zone exists. Where γ nucleates and grows without graphite, the liquid changes in composition. Graphite can then nucleate. This is followed once more by γ, but not in a correlated (cooperative) growth manner. This is a form of divorced eutectic growth in which the two phases grow by their individual mechanisms and independently.

Competitive versus Cooperative Growth

The model for competitive versus cooperative growth in eutectic systems (see section 1-5) suggests that the structure formed depends on the competition between (a) the dendrites of the two phases growing separately and (b) the growth of the two phases cooperatively. The structure observed is the one which grows the fastest. This model was given by Botschvar,[28] Tammann,[29] and others. It gives an uncomplicated picture of the types of behaviour possible.

Analysis of Diffusion Controlled Growth of Gray and White Eutectics in Cast Iron

Calculations to test the cooperative growth model of the gray and white eutectic structures in cast iron were made by Hillert and Rao.[18] The approach to understanding the mechanism of growth for the two eutectics was to calculate the rates for cooperative growth based on a diffusion model. Calculation was also made of the rate of dendritic growth of the individual phases. To test the diffusion model, comparison was made between the calculated growth rates of the eutectics and the experimentally observed growth rates.

The results of Hillert and Rao[18] will be given to demonstrate some characteristics of the nature of solidification in this system. Hillert used the Zener theory of two-phase cooperative growth[30] to predict the critical lamellar spacing λ^*, and from this the spacing for the maximum cooperative growth velocity at $\lambda = 2\lambda^*$.

From the Zener theory:

$$\lambda^* = \frac{2\sigma V_m^L}{-\Delta G_m} = \frac{2\sigma V_m^L T_0}{\Delta H_m \Delta T}$$

where σ = interfacial energy between the phases
V_m^L = molar volume of the liquid
T_0 = equilibrium growth temperature
ΔH_m = enthalpy of melting
ΔT = undercooling

V_{max} is given by:

$$V_{max} = \frac{D}{4a\lambda^*} \frac{1}{f^{Gr}f^{\gamma}} \frac{X_c^{L/\gamma} - X_c^{L/Gr}}{X_c^{Gr} - X_c^{\gamma}}$$

where f^{Gr} and f^{γ} are mole fractions of phases in the eutectic, X_c^{Gr} and X_c^{γ} are compositions of the eutectic phases $X_c^{L/Gr}$ and $X_c^{L/\gamma}$ are compositions of the liquid in equilibrium with graphite and γ at ΔT, D is the diffusion coefficient, and a is approximately 0.5.

Figure 3-17 is the graphical summary for the gray iron system. The equa-

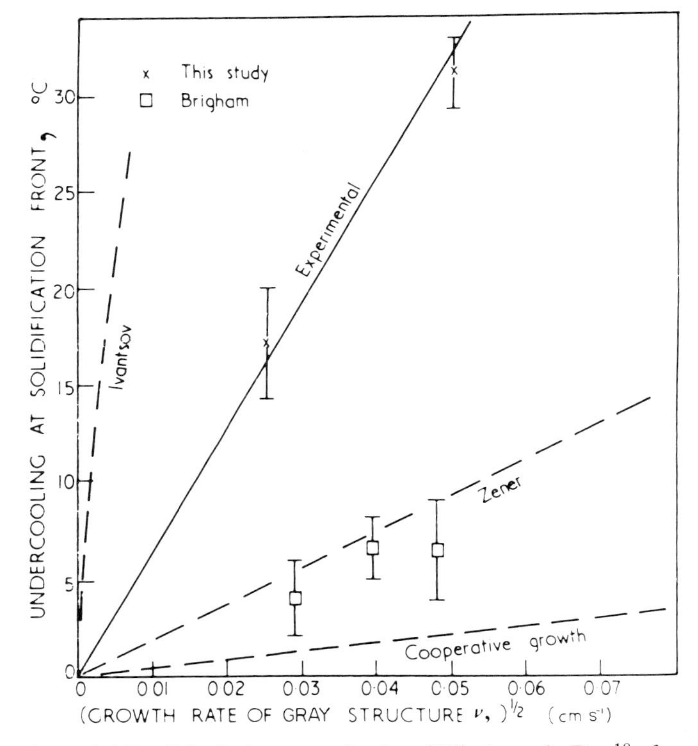

Figure 3-17 Calculations made by Hillert and Rao[18] for cooperative growth of the graphite eutectic compared with experimentally determined values. The correlation was not good. Also included are theoretical curves for the growth rate of the graphite phase based on diffusion control using analyses by Ivantsov[31] and Zener.[30] (Reproduced by permission of The Metals Society)

tions of growth for graphite were calculated from the theory of Ivantsov[31] and from the theory of Zener.[30] The calculated growth rate curve for graphite on the model of Ivantsov lies above the experimentally obtained growth curve for the gray iron eutectic, i.e. the experimental value of the gray iron eutectic growth is greater than the Ivantsov theoretical rate for graphite. However, the graphite growth rate based on Zener's model is faster than the experimentally observed eutectic growth.

The theoretical calculation for cooperative growth of the eutectic shows faster rates than that experimentally observed. From the calculations, Hillert suggested that the cooperative model for the growth of the graphite eutectic was in error.

Figure 3-18 shows Hillert's data for the white iron eutectic. Here, again, the calculations for cooperative growth show faster rates than those experimentally determined for the ledeburite eutectic. Hillert suggested that here, too, the cooperative model did not apply.

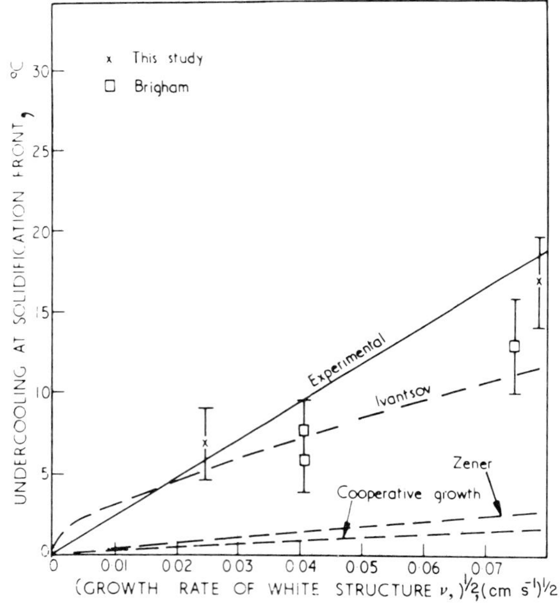

Figure 3-18 Hillert's calculations[18] for growth of the white iron eutectic compared with experimentally obtained values for ledeburite. The agreement for the eutectic growth is poor but Ivantsov's calculations, applied to Fe_3C, appear to correlate with the experimental values of eutectic growth. Therefore Hillert suggested that the white iron eutectic growth might be controlled by the growth of the Fe_3C phase. (Reproduced by permission of The Metals Society)

The experimental data for growth of the white iron eutectic seemed to be approximated by the curve for growth of Fe_3C according to Ivanstsov. Therefore, Hillert suggested that the white iron eutectic might be controlled by the growth of Fe_3C.

Model of Eutectic Growth Involving Interface Attachment Kinetics

Fredriksson and Weterfall[32] used similar equations to those applied by Hillert in analysing the graphite eutectic but employed an interface attachment coefficient, μ, for graphite growth. In this way, they could span the gap between the theoretical diffusion controlled calculations and the rates experimentally observed. The kinetic attachment coefficient, μ, takes into account the interface controlled growth mechanism for graphite. Their analysis was applied to data of Brigham, Purdy, and Kirkaldy[33] and Lux and Kurz,[34] as well as to data of their own. Figure 3-19 shows the theoretical curve, marked $\mu = \infty$. The other experimental data fell in a region of the diagram far removed from the theoretical curve.

The equation suggested for graphite growth simply related velocity to the carbon supersaturation by the attachment coefficient:

$$v = \mu(X_i^{L/Gr} - X_0^{L/Gr})$$

where μ = rate constant for the reaction at the interface

$X_0^{L/Gr}$ = equilibrium value of the carbon content in the liquid at the graphite–liquid interface

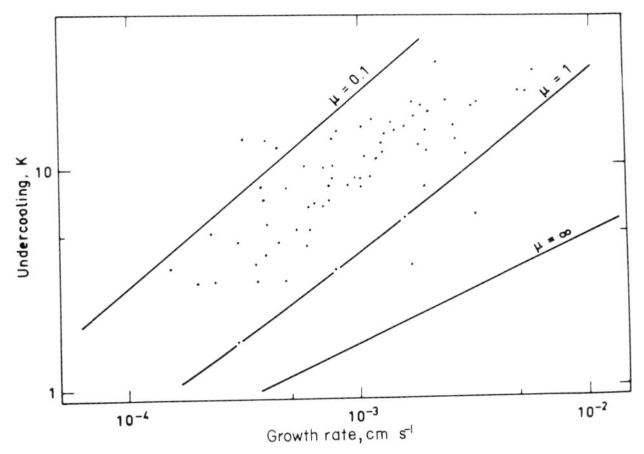

Figure 3-19 Graph of Fredriksson and Weterfall[32] for graphite eutectic growth, employing calculations with an interface attachment coefficient μ. All the data analysed fell within the range $\mu = 0.1$ and $\mu = 1$. The curve marked $\mu = \infty$ is the theoretical curve for diffusion controlled growth. (Reproduced by permission of Georgi Publishing Co.)

$X_i^{L/Gr}$ = actual value of the carbon content at the interface

The following equation was introduced:

$$X^{L/\gamma} - X^{L/Gr} = (X_0^{L/\gamma} - X_0^{L/Gr})(1 - S^*/S) - (X_i^{L/Gr} - X_0^{L/Gr})$$

where S is the lamellar spacing and S^* is the critical spacing where the growth rate is zero. By eliminating $X_i^{L/Gr}$, they obtain:

$$v = \frac{(X_0^{L/\gamma} - X_0^{L/Gr})(1 - S^*/S)}{2f^{Gr}f^{\gamma}(X^{Gr} - X^{\gamma})S/D + 1/\mu}$$

The maximum growth rate is given for

$$\frac{S}{S^*} = 1 + \sqrt{[1 + D/2f^{Gr}f^{\gamma}(X^{Gr} - X^{\gamma})S^*\mu]}$$

Using different values of μ, Figure 3-20 shows that the experimental data lie between $\mu = 0.1$ and $\mu = 1.0$. They suggested that the considerable scatter in the experimental points was due to different values of μ in the experiments. The data showed that only the fine graphite eutectic could be analysed using an interface coefficient.

On the basis of the Fredriksson–Weterfall analysis, the effect of sulphur is to decrease the interface attachment coefficient. For low sulphur alloys, μ was suggested to be equal to ∞. For high sulphur alloys, $\mu = 0.1$.

Kinetic Effect in Graphite Eutectic Growth

Lesoult and Turpin[35] analysed undercooled type structures in gray iron with a plane interface and introduced a kinetic effect. While Fredriksson and Weterfall used the Zener–Hillert equation modified by a kinetic factor to calculate growth velocities, Lesoult and Turpin commenced from the Jackson–Hunt analysis of eutectic growth. The interface temperature was, however, calculated by kinetic equations instead of employing equilibrium.

A λ–v relationship was calculated by maximizing the temperature of the interface with respect to λ, and a relationship was established between the mean composition of the interface and the velocity v. The analysis showed that the average composition of the interface must be strongly shifted to hyper-eutectic values if cooperative growth with a plane front is to be maintained. They explain in this way the displacement and asymmetry of the cooperative growth region.

Kurz and Fisher Analysis of Eutectic Growth

Kurz and Fisher published papers analysing growth in faceted non-faceted systems. These included the flake graphite eutectic in cast iron. A particular aspect dealt with was the location of the coupled zone[36] which was calculated for the lederburite eutectic as well as for the flake graphite eutectic In a second paper, the spacing between eutectic phases was calculated.[37]

In their calculation of the coupled zone[36] the approach was that of competitive growth between primary phases and the eutectic, discussed in Section 1-5 and in this section for the paper by Hillert and Rao.[18] For the γ phase of the eutectic, Kurz and Fisher used equations based on needle growth. For the graphite phase, the equations employed were for the growth of a plate. The solutions used were for diffusion controlled growth in a positive temperature gradient as developed by Burden and Hunt.[38] These temperature conditions are applicable to directional solidification experiments and are those for which most of the data on eutectic growth have been obtained. For comparison with the growth of the eutectic, the Jackson–Hunt equation with a modified constant was employed, this also being applicable to diffusion control.

The paper of Burden and Hunt[38] gave the calculation of the growth temperature of dendrite tips and the eutectic interface as a function of the growth velocity v and temperature gradient G. Figure 3-20 shows the curves of

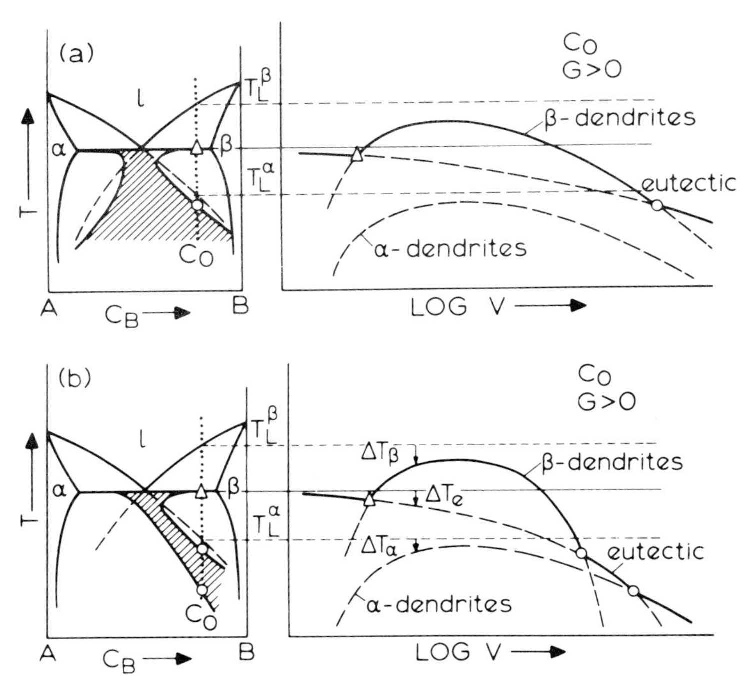

Figure 3-20 Application by Kurz and Fisher[36] of theory by Burden and Hunt[38] for growth of primary and eutectic phases in a positive temperature gradient to the graphite eutectic. The form of the growth rate dependence on temperature goes through a maximum as the growth velocity increases. Phases growing as needles would have the growth curves shown in (a). This gives a symmetrical couples region. If one phase grows as a disc and the eutectic is undercooled, (b) results, which is the case for asymmetrical coupled growth. (Reproduced by permission of the Metals Society)

temperature T and log v, together with phase diagrams for two cases of eutectic growth. In Figure 3-20(a), the eutectic consists of two phases, both of which obey the growth laws of a needle. In Figure 3-20(b), the eutectic consists of one phase α growing as a needle and the second phase β growing as a plate. The two types of coupled zone conforming to this situation are shown 'hatched' in the phase diagrams.

For growth in a positive temperature gradient, the curves for dendrite growth show a maximum in the T–log v relationship. For a composition C_0 in Figure 3-20(a), the β growth curve cuts the eutectic curve twice. Since the form of the α curve is similar to that for β, but is displaced only along the temperature axis, a symmetrical form of coupled zone results. In Figure 3-20(b), for the case of β growing as a plate with a different and much steeper temperature dependence on the growth rate, the β curve cuts both the eutectic and the α curves. A coupled zone of the asymmetrical type results. Kurz and Fisher relate the plate-like growth of the β phase to faceted crystals, arising from growth anisotropy. Coupled zones are calculated for several systems and these diagrams have been shown to correlate with experimentally determined curves.

In the second paper,[37] Kurz and Fisher develop a method for calculating the spacing between phases in faceted non-faceted systems. The problem tackled is one observed in a number of eutectic systems classified as 'irregular' or class II, in which the spacing between phases, λ, does not obey that theoretically obtained from the Jackson–Hunt analysis. Kurz and Fisher treat the spacing problem in its dependence on the difficulty of branching of the faceted phase. They present an analysis of these class II eutectics using a stability criterion. These eutectics have to grow at the limit of stability of the faceted phase so as to permit this phase to branch. This leads to typically coarse and irregular structures, associated with much higher undercooling than those of class I.

The Kurz and Fisher papers pose a number of interesting questions. The authors point out the large uncertainties which exist in their analysis, in particular the attempt to relate theory to mechanisms of growth of irregular eutectics which are not clearly characterized.

For cast iron metallurgy, these problems are of particular importance since the γ-graphite eutectic, the mode of graphite growth, and the mode of cooperation of graphite in growth with the γ phase lie at the basis of understanding the microstructure. The analysis of Kurz and Fisher pays only minor attention to the mechanism of growth of the graphite phase. The diffusion controlled growth of graphite is open to question. The competitive growth analysis of both gray and white iron euectics has been published,[18] and the calculations made, when compared with experiment, show a lack of agreement between the cooperative growth model and experimental measurements. The diffusion controlled growth of graphite is then to be queried.

Class II eutectic structures are much more complex than can be dealt with on the basis of the spacing between phases, or on the form of the coupled zone. One particular example of this in class II is the spiral eutectic (see

Chapter 1). The geometry is a more interesting characteristic of this type of eutectic growth than the spacing. This eutectic in the Al–Mg–Si system was shown to grow by the independent development of the Mg_2Si phase.

For the graphite eutectic of cast iron, different forms exist and all the structures can be related to independent growth of the graphite phase. Thus it is possible to understand the rod-like growth of coral graphite (see Chapter 7) which does not seem possible to grow as a eutectic by a cooperative mechanism with the γ phase. Similarly, the chunky or vermicular graphite structure, which is interconnected and thus related to eutectic growth, must grow with independent development of the graphite phase. The case of spherulitic growth has been characterized as a case of divorced eutectic growth[27] in which the graphite phase growing independently in spherulitic form is encapsulated by the later independent growth of the γ phase.

3-3 The Ledeburite Eutectic

The growth of the γ–Fe_3C eutectic was first investigated by Benedicks.[39] It was reexamined by Hillert and Steinhauser[40] as part of an investigation aiming to compare the kinetics of gray and white solidification. Hultgren, Lindblom, and Rudberg[13] had observed that the ledeburite eutectic grows with a much higher speed than the graphite eutectic.

The structure of Fe_3C is orthorhombic. Hillert noted that solidification of the eutectic commenced by flat plates of Fe_3C growing into the melt, branching in the process. The γ phase then grows in dendritic manner over the Fe_3C plate. This unstabilizes the Fe_3C which grows through the γ. Two types of eutectic structure then develop (Figure 3-21): a lamellar eutectic structure in the direction of growth of the flat Fe_3C plates and rod-like structure growing normal to the plates. The latter originates by growth of the Fe_3C through the γ phase to form a cooperative growth front.

The rate of cooperative growth of γ and Fe_3C normal to the plate is slower than the rate of growth of the eutectic in the direction of the plate. In the plate direction, the following relationships occur between the two phases:

$$(104)Fe_3C \parallel (101)_\gamma$$
$$\langle 010\rangle Fe_3C \parallel \langle \bar{3}\bar{1}0\rangle_\gamma$$

3-4 Gray or White Solidification of Cast Iron

The gray or white solidification mode of cast iron is dependent on the relative nucleation possibility and growth rates of the graphite and Fe_3C phases. This will depend on alloy chemistry of the melt and on the phase growth in the conditions established. Figure 3-22 shows the range of existence of the gray and white structures plotted graphically as temperature versus growth rate.[18]

The equilibrium temperature for the graphite eutectic is 1153 °C and that for the Fe_3C eutectic is 1148 °C. Between these temperatures, only the

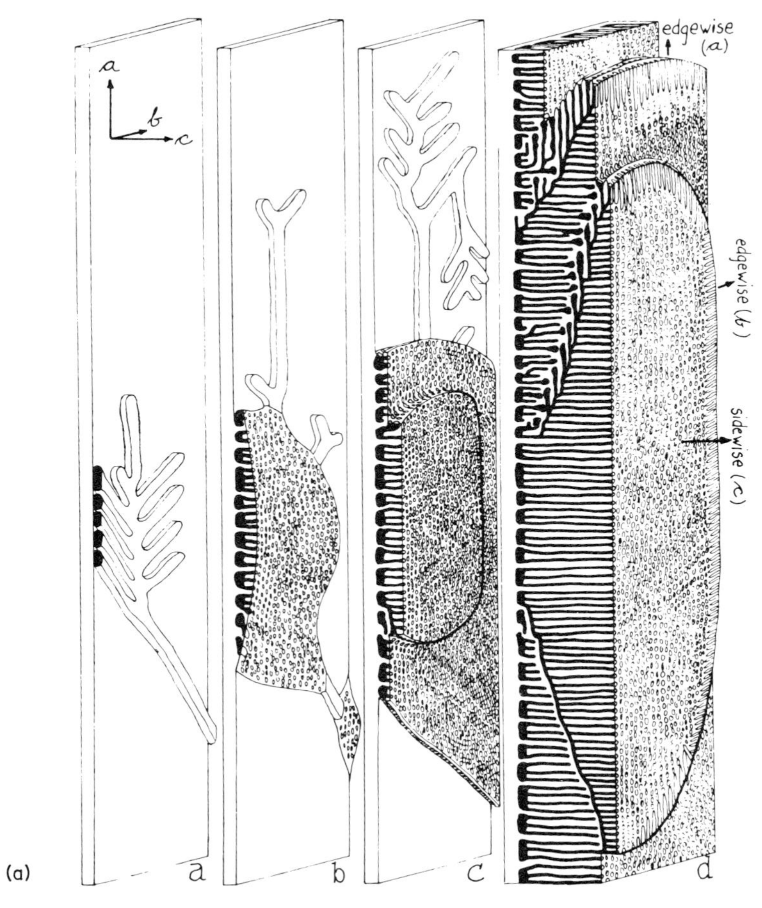
a
b
c
edgewise
(a)
edgewise (b)
sidewise (c)
(a)
a
b
c
d

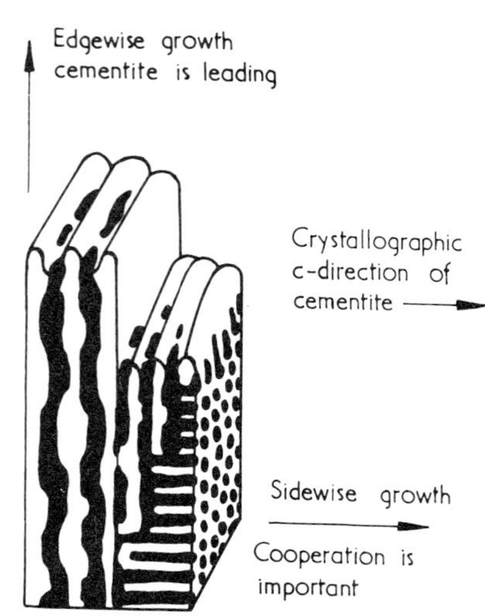
Edgewise growth
cementite is leading
Crystallographic
c-direction of
cementite
Sidewise growth
Cooperation is
important
(b)

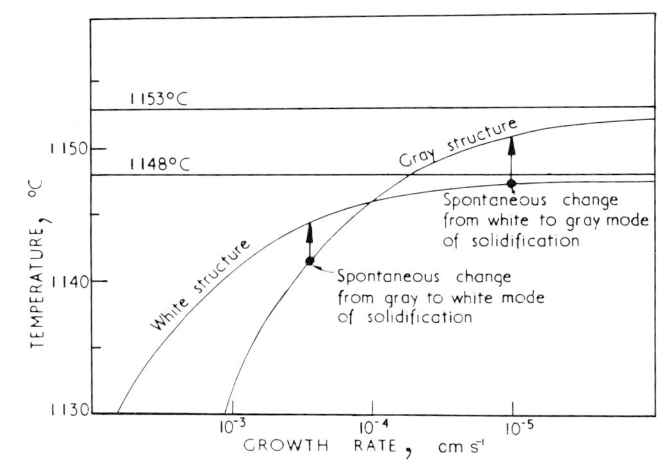

Figure 3-22 Range of existence of gray and white cast iron structures plotted graphically as temperature versus growth rate.[18] (Reproduced by permission of The Metals Society)

graphite eutectic can nucleate and grow. Below 1148 °C, the Fe_3C eutectic can nucleate and grow. The growth rate of this eutectic rapidly exceeds that of the gray iron eutectic and at a temperature of approximately 1140 °C there should be a change from gray to white solidification.

Influence of Melt Chemistry on White Solidification

Alloying elements in cast iron which enter the carbide phase affect the tendency to white solidification.[3] They stabilize the carbide and promote the γ–Fe_3C eutectic.

This can be predicted from the distribution coefficient of the element between Fe_3C and γ. If $k = M_{Fe_3C}/M_\gamma > 1$, where:

$$M_{Fe_3C} = \text{solubility of element in } Fe_3C$$
$$M_\gamma = \text{solubility of element in } \gamma$$

then the element will enter into and stabilize the Fe_3C phase.

Figure 3-21 Growth of the ledeburite eutectic from observations made by Hillert and Steinhauser.[40] In (a), four stages are shown of the eutectic development. Initially a plate of Fe_3C must grow and a γ dendrite grows over the Fe_3C plate. This unstabilizes the Fe_3C which grows through the γ. Two types of eutectic structure then develop, a lamellar eutectic in the Fe_3C growth direction and a rod-like eutectic normal to the plates. (Reproduced by permission of Jernkontoret.) This is shown in more detail in (b).[18] (Reproduced by permission of The Metals Society)

Inverse Chill

The explanation of inverse chill in cast iron was given by Hillert and Rao[18] on the basis of kinetics. In inverse chill, the gray iron structure appears at the outer surface and the white iron structure in the interior. This is the reverse of the common chilled iron occurrence. It is expected that the most rapidly freezing iron will enter the temperature range of rapid Fe_3C growth and therefore the outside of a cast part will be white. At the slower rates of the interior, gray iron freezing will proceed.

Hillert assumed that, for inverse chill, graphite nucleated in the mould wall since it can do this more easily than the Fe_3C phase. However, when Fe_3C nucleates, it grows faster. The location of the two structures in this case is then reversed.

References

1. Sadocha, J. P., and J. E. Gruzleski: *The Metallurgy of Cast Iron* (Eds. B. Lux, F. Mollard, and I. Minkoff), Georgi Publ. Co., Switzerland, 1975.
2. Hirth, J. P., and G. M. Pound: *Condensation and Evaporation*, Progr. Mater. Sc. Vol. 11, Pergamon, 1963.
3. Hillert, M.: *Recent Research on Cast Iron* (Ed. H. Merchant), Gordon and Breach, New York, 1968.
4. Van der Merwe, J. H.: *Proc. Phys. Soc. (London)*, **A63**, 616, 1950.
5. Christian, J. W.: *The Theory of Transformations in Metals and Alloys*, Pergamon, Oxford, 1965.
6. Smalley, O.: *Foundry Trade J.*, **136**, 227, February 1974.
7. Morrogh, H., and W. Oldfield: *Iron Steel*, **32(11)**, 479, 1959.
8. Lux, B. *Mod. Cast*, **45**, 222, 1964.
9. Rosenstiel, A. P., and H. Bakkerus: *Giesserie Tech. Wiss. Beih.*, **16(3)**, 1, 1964.
10. Zeedijk, H. B.: *J. Iron Steel Inst. London*, **203**, 737, 1965.
11. Deuchler, W.: *Giess. Tech. Wiss. Beih.*, **14**, 745, 1964.
12. Jacobs, M. H., T. J. Law, D. A. Melford, and M. J. Stowell: *Met. Technol.*, **1**, 490 1974.
13. Hultgren, A., Y. Lindblom, and E. Rudberg: *J. Iron Steel Inst. London*, **176**, 365, 1954.
14. Morrogh, H., and W. Oldfield: *Iron Steel*, **32(11)**, 431, 1959.
15. Bunin, K. P., Y. Malinochka, and S. Fedorova: *Litenoe Proizvod*, **4**, 21, 1953.
16. Kusakawa, T., and E. Nakata: *Imono*, **35**, 470, 1963.
17. Minkoff, I., and B. Lux: *Cast Met. Res. J.*, **6**, 181, 1970.
18. Hillert, M., and V. V. Subba Rao: *The Solidification of Metals*, Iron Steel Inst. Publ. 110, London 1968.
19. Jackson, K. A., and J. D. Hunt: *Trans. Metall. Soc. AIME*, **236**, 1129, 1966.
20. Lakeland, K. D.: *J. Australian Inst. Metals* **10**, 55, 1965.
21. Nieswaag, H., and A. J. Zuithoff: *The Metallurgy of Cast Iron*, (Eds. B. Lux, F. Mollard, and I. Minkoff), Georgi Publ. Co., Switzerland, 1975.
22. Kofler, A.: *Z. Metallkd.*, **41**, 221, 1950.
23. Kofler, A.: *Microchem.*, **40**, 311, 1953.
24. Kofler, A.: *Microchem.*, **40**, 405, 1953.
25. Scheil, E.: *Z. Metallkd.*, **45**, 298, 1954.
26. Hunt, J. D., and K. A. Jackson: *Trans. Metall. Soc. AIME*, **239**, 864, 1967.
27. Lux, B., F. Mollard, and I. Minkoff (Eds.): *The Metallurgy of Cast Iron*, Georgi Publ. Co., Switzerland, 1975.

28. Botschvar, A. A.: *Z. Anorg. Allg. Chem.*, **220**, 334, 1934.
29. Tammann, G.: *Z. Metallk.*, **25**, 236, 1933.
30. Zener, C.: *Trans. AIME*, **167**, 550, 1946.
31. Ivantsov, G. P.: *Growth of Crystals*, Consultants Bureau, New York, 1960.
32. Fredriksson, H., and S. E. Weterfall: *The Metallurgy of Cast Iron* (Eds. B. Lux, F. Mollard, and I. Minkoff), Georgi Publ. Co., Switzerland, 1975.
33. Brigham, R. J., G. R. Purdy and J. S. Kirkaldy: *Proc. Int. Conf. Crystal Growth*, Boston, 1966.
34. Lux, B., and W. Kurz: *The Solidification of Metals*, Iron Steel Instl Publ. 110, London, 1968.
35. Lesoult, G., and M. Turpin: *The Metallurgy of Cast Iron* (Eds, B. Lux, F. Mollard, and I. Minkoff), Georgi Publ. Co., Switzerland, 1975.
36. Kurz, W., and D. J. Fisher: *Int. Metall. Rev.*, **24**, 177, 1979.
37. Kurz, W., and D. J. Fisher: *Acta Metall.*, **28**, 777, 1980.
38. Burden, M. H., and J. D. Hunt: *J. Cryst. Growth*, **22**, 328, 1974.
39. Benedicks, C.: *Intern. Z. Metallographie*, **1**, 184, 1911.
40. Hillert, M., and H. Steinhauser: *Jernkontorets Ann.*, **144**, 520, 1960.

Bibliography

Boyles, A.: *The Structure of Cast Iron*, Am. Soc. Met., 1949.
Kurz, W., and P. R. Sahm: *Gerichte Erstarrte Eutektische Werkstoffe*, Springer-Verlag, 1975.
Fine, M. E.: *Introduction to Phase Transformations in Condensed Systems*, Macmillan, 1965.

Chapter 4
Thermodynamics of the Iron–Carbon System

The thermodynamics of the Fe–C system have been intensively studied and used to understand different aspects of steel and cast iron metallurgy. Cast iron is a complex material with stable and metastable phases and having elements in solution which influence the extent and stability of the phases. In this chapter, some of the thermodynamics of binary systems are presented and extended to interpreting solidification and solid state transformations in ternary systems.

Fundamental to understanding the overall systems is the provision of reliable phase diagrams which for cast iron might give the following:

1. The regions of stability of graphite and carbide phases.
2. The influence of alloying elements on the extent of the phase regions and the relative stability of the graphite and carbide phases.

There are in addition structure diagrams which attempt to define final cast structure related to composition and cooling rate. The provision of these diagrams has depended on experiment as well as on thermodynamic analysis.

Initially in this chapter, some fundamental thermodynamic relationships are defined. These concern the activity of an element in a binary solution, Raoult's and Henry's laws, deviations from ideality, and activity coefficients. The Gibbs–Duhem equation is given for determining the activity coefficient of a component in a binary solution from knowledge of the activity coefficient of the second component.

Thermodynamics of ternary systems are next described for the system Fe–C–M. Interpolation formulae of Wagner for determining activity coefficients are presented and an equation due to Hillert for activity changes in ternary systems is described.

Different ways are given for expressing the influence of third elements with respect to the solubility of graphite in liquid iron at the eutectic temperature and calculations of Neumann for the saturation ratio are presented.

The final part of the chapter reviews theory for Fe–C–Si structural diagrams and possible interpretations of the liquid and solid state transformation based on thermodynamics. The role of silicon on graphitization in the liquid state is dealt with. Current evidence points to this being related to the nucleation problem of graphite. Recent work on the solid state transformation is concerned with the influence of alloy elements on the pearlite transformation. Theory of the partitioning of solute in this transformation and its influence on kinetics is described. The influence of alloy elements on austenite transformation to pearlite and the thermodynamics of pro-eutectoid transformation affecting the matrix structure of cast iron are described.

4-1 The Iron–Carbon System; Fundamental Thermodynamic Relationships

Raoult's law is illustrated by curve I of Figure 4-1(a). In an ideal solution of two components, the activity of an individual component is equal to its mole fraction. Departures from ideality occur, shown by curves II and III of Figure 4-1(a). These departures may be positive or negative. The relationship between the activity and the mole fraction gives the activity coefficient:

$$\frac{a_i}{x_i} = f_i$$

where x_i is the mole fraction of component i. For non-ideal solutions, as the mole fraction of the solute element approaches 1, the activity of the solute a_i approaches 1 and Raoult's law is obeyed. This is seen in curves II and III of Figure 4-1(a).

For non-ideal solutions, Henry's law is obeyed. This is shown in the diagram of the activity of Si in liquid Fe at 1873 K (Figure 4-1b).[1] At small values of x_{Si}, a_{Si} is proportional to x_{Si} and the activity coefficient f_{Si} is independent of x_{Si}.

Positive deviations from Raoult's law are related to systems in which heat is absorbed on mixing the constituents, while negative deviations are found when heat is liberated on mixing.

The Iron–Carbon and Iron–Silicon Activity Diagrams

The activity of carbon in liquid iron has been studied by Neumann and Schenk,[2] Rist and Chipman,[3] and Turkdogan, Lecke, and Masson.[4] Figure 4-2 shows a curve for the carbon activity in liquid iron at 1823 K due to Rist and Chipman.[3] A positive deviation from Raoult's law is shown. Figure 4-1(b) shows the activity of silicon in liquid iron at 1873 K[1] and the negative deviation from Raoult's law.

Gibbs–Duhem Relationships

The Gibbs–Duhem equation enables the activity coefficient f_1 of one component of a binary solution to be obtained if the activity coefficient f_2 of the

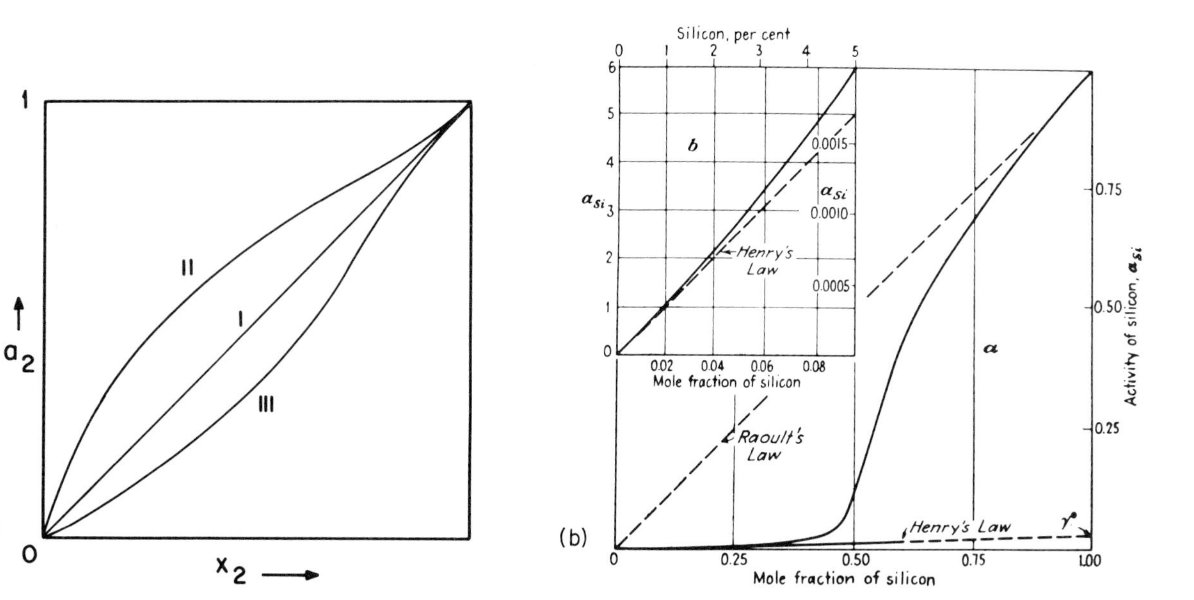

Figure 4-1 (a) Curve I, Raoult's law. Activity of component 2 is equal to its mole fraction x_2. Curve II, positive departure from Raoult's law. Curve III, negative departure.[5] (Reproduced from *Thermodynamics of alloys*, 1952, written by Carl Wagner, with permission of the publishers, Addison-Wesley, Advanced Book Program, Reading, Massachusetts, U.S.A.) (b) Activity of silicon in liquid iron at 1873 K.[1] (Reproduced by permission of The Metallurgical Society of American Institute of Mining Engineers)

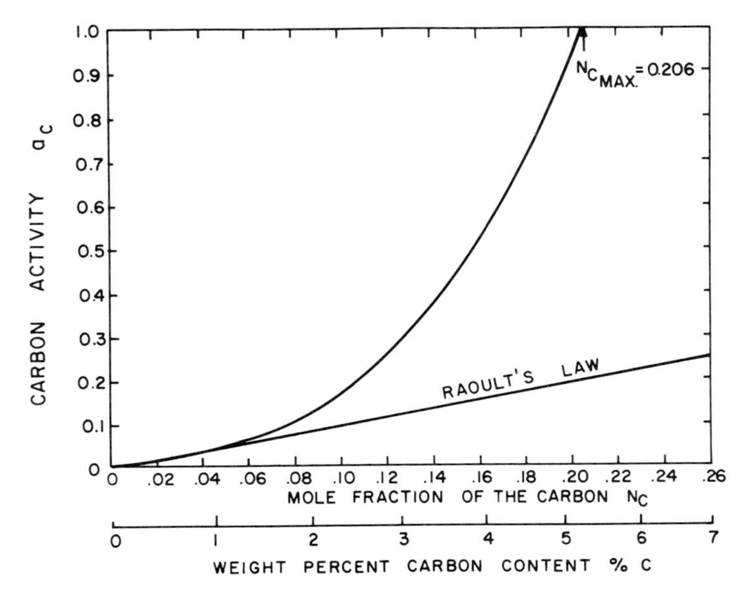

Figure 4-2 Activity of carbon in liquid iron at 1823 K due to Rist and Chipman,[3] taken from the publication by Neumann.[7] (Reproduced by permission of Gordon and Breach Science Publishers Inc.)

second component is known over a range of compositions. This is stated as:

$$x_1 \, \mathrm{d} \log f_1 + x_2 \, \mathrm{d} \log f_2 = 0$$

or

$$\mathrm{d} \log f_2 = -\frac{x_1}{x_2} \, \mathrm{d} \log f_1$$

where x_1 and x_2 are mole fractions of components 1 and 2 respectively. Integration can be performed to give f_2 from a knowledge of f_1.

4-2 Multicomponent Systems

The solubility of carbon in Fe–C and the decrease or increase of this solubility is related to the addition of another element to the system. The influence of the addition element is closely related to the activity of the carbon. The problem is to derive this behaviour by direct mathematical relationships. Some important effects can then be calculated, e.g. the influence of different elements in a liquid iron melt on the solubility of graphite in the liquid.

Interpolation Formulae of Wagner

Wagner[5] calculated the activity coefficient of a component 2 in a multicomponent system (1, 2, 3, 4, etc.) from its value in the system 1–2 and its limiting value in the systems 1–3, 1–4, etc., as these systems dissolve component 2 with the given mole fractions x_3, x_4 respectively. A Taylor series expansion was used for deriving an expression for the logarithm of the activity coefficient for component 2 in a solution with mole fractions x_2, x_3, x_4 of the various solutes. The following expression was obtained:

$$\ln f_2(x_2 x_3 x_4) = \ln f_2^0 + \left(x_2 \frac{\partial \ln f_2}{\partial x_2} + x_3 \frac{\partial \ln f_2}{\partial x_3} + x_4 \frac{\partial \ln f_2}{\partial x_4} + \cdots\right) + \left(\frac{1}{2} x_2^2 \frac{\partial^2 \ln f_2}{\partial x_2^2} + x_2 x_3 \frac{\partial^2 \ln f_2}{\partial x_2 \partial x_3} + \cdots\right) + \cdots$$

where f_2^0 is the activity coefficient of component 2 in the pure solvent. The derivatives are taken for the limiting case of zero concentration of all solutes. Disregarding terms involving second and higher derivatives, the logarithm of the activity coefficient becomes a linear function of the mole fractions of the various solutes.

The following equation is obtained:

$$f_2(x_2 x_3 x_4) = f_2(x_2 x_3 = 0, \quad x_4 = 0) f_2^{(3)} f_2^{(4)}$$

The first factor on the right-hand side is the activity coefficient of component 2 in the binary system 1–2 with the mole fraction x_2. The multipliers which follow are correction factors for components 3, 4, etc. These correction factors $f_2^{(3)}, f_2^{(4)}, \ldots$ are given by

$$f_2^{(3)} = \frac{f_2(x_2 x_3 x_4 = 0, \ldots)}{f_2(x_2 x_3 = 0, x_4 = 0, \ldots)}$$

$$f_2^{(4)} = \frac{f_2(x_2 x_3 = 0, x_4, \ldots)}{f_2(x_2 x_3 = 0, x_4 = 0)}$$

The values of x_2 in the numerator and denominator must be the same, but can be chosen arbitrarily. Analogous equations may be noted for the activity coefficient of any component other than 2.

Calculations of Hillert

Hillert[6] derived an equation for the change of activity of one component in a two-phase region of a ternary system as a result of a change of concentration of the third element M. It is necessary to know the partition coefficient for the alloying element between the two phases. For the Fe–C system, important predictions of alloying behaviour can be made from a knowledge of how the carbon activity changes along a liquidus surface as the composition changes.

The expression obtained by Hillert could be employed to follow activity changes in the liquid iron in contact with Fe_3C. For this, the expression becomes:

$$\frac{d \ln a_c}{dx_M^L} = \frac{K_M^{Fe_3C/L} x_{Fe}^L - x_{Fe}^{Fe_3C}}{x_C^L x_{Fe}^{Fe_3C} - x_C^{Fe_3C} x_{Fe}^L}$$

where $K_M^{Fe_3C/L}$ is the partition coefficient for the alloying element between Fe_3C and L.

For low alloy content, the equation can be simplified. After integration, the following is obtained:

$$\Delta \ln a_c = \frac{K_M^{Fe_3C/L} x_{Fe}^L - x_{Fe}^{Fe_3C} \Delta x_M^L}{x_{Fe}^L - x_{Fe}^{Fe_3C}}$$

Hillert showed two applications of this relationship to calculate alloying influences in cast iron. In the first example, calculation is made of the alloying addition necessary to stabilize Fe_3C instead of graphite. From the activity line of the binary equilibrium diagram of Figure 2-2, the activity of carbon in liquid iron in equilibrium with Fe_3C would be 1.2 at 1445 K. The activity of carbon in liquid iron in equilibrium with graphite at that temperature is 1.0. For an element entering preferentially into the Fe_3C lattice so that $K_M^{Fe_3C/L} = 5$, the value of X_M^L to make the activity change from stability with graphite to stability with Fe_3C is 0.4 atomic per cent. This suggested that only small alloying additions were necessary to have marked influences on gray or white solidification. In the second example quoted, the equation was used to calculate the amount of phosphorus required to lower the γ–Fe_3C eutectic temperature to 1223 K where the ternary γ–Fe_3C–Fe_3P eutectic crystallizes.

Carbon Solubility in Multicomponent Systems

Neumann[7] reviewed the data for the influence of elements on the solubility of carbon in liquid iron. This is shown graphically in Figure 4-3. The slopes of the lines are called the solubility factors. For carbide forming elements, the slopes are positive. For graphite promoting elements, the slopes are negative. The change of solubility of carbon in iron for the presence of any one element at low concentration can be represented by temperature independent linear equations of the form:

$$\Delta N_C^{(x)} = m N_x$$

where x is the added element, N_C is the mole fraction of carbon in Neumann's notation, and m is the slope of the solubility curve.

Expressing the change of solubility in weight per cent.:

$$\Delta\% \ C(x) = m' \cdot \% \ x$$

m' is the solubility factor expressed in weight per cent and $\%x$ is the concentration of x. Elements which increase the solubility of carbon are Mn, Cr, Mo, W, Ta, V, Nb, and Ti. Elements which reduce the solubility of carbon are Si, Al, Cu, Ni, Co, Zr, P, and S.

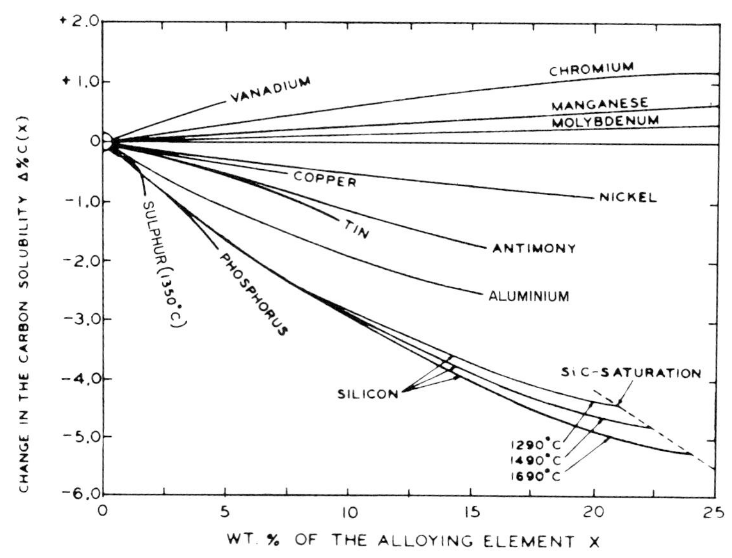

Figure 4-3 The influence of elements on the solubility of carbon in molten iron.[7] (Reproduced by permission of Gordon and Breach Science Publishers Inc.)

In the Fe–C system, Neumann described the solubility of carbon in molten iron from the equilibrium diagram as follows:

$$\%C_{max} = 1.3 + 2.57 \times 10^{-3} \cdot T\text{K (between 1425–2273 K)}$$

Using mole fraction and the absolute temperature, this equation becomes

$$\log N_{C_{max}} = -\frac{12.7276}{T} + 0.7266 \log T - 3.0486$$

For the eutectic temperature 1152°C (1425 K), the maximum solubility of carbon in iron is 4.26 per cent or $N_{C_{max}} = 0.1714$.

For a multicomponent system, the solubility of carbon can be calculated from the solubility factors. For a Fe–C–Mn–Si–P–S alloy this is:

$$\%C_{max} = 1.3 + 2.57 \times 10^{-3} T\text{°C} + 0.027\ \%\text{Mn} - 0.31\ \%\text{Si} - 0.33\ \%\text{P} - 0.4\ \%\text{S}$$

To calculated the carbon content of the eutectic, the eutectic temperature must be inserted. Different investigations have been directed to confirming the additivity of the coefficients in this equation. This has been verified for different systems at low element concentration, e.g. Fe–C–Al–Si–P–Mn–S.[8]

Saturation Ratio and Degree of Saturation

Knowledge of the carbon solubility in the liquid at the graphite eutectic temperature and the carbon solubility in austenite at the eutectic temperature

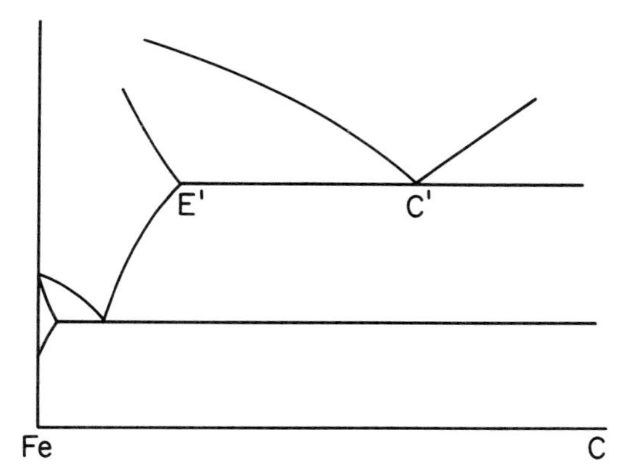

Figure 4-4 Fe–C phase diagram marked to show the points C' and E'

enable an estimate to be made of the amount of eutectic. Figure 4-4 shows the phase diagram marked with the two compositions of interest C' and E'. C' is the carbon solubility in liquid iron saturated with graphite at the eutectic temperature and E' is the carbon solubility in austenite at that temperature.

The saturation ratio, S_r, is defined as:

$$S_r = \frac{C - E'}{C' - E'}$$

where C is the analysed carbon content of the alloy. A value of $S_r = 1$ implies a completely eutectic structure. A value of $S_r = 0.9$ is taken to mean that the structure will contain 90 per cent of eutectic. The value of C' is estimated as in the equation for $\%C_{max}$ above and becomes:

$$C' = 4.26 - 0.31\ \%\text{Si} - 0.33\ \%\text{P} - 0.40\ \%\text{S} + 0.027\ \%\text{Mn}$$

For a change of eutectic temperature, the value of 4.26 should be changed correspondingly. The problem is to calculate E' since limited data are available for the influence of alloying elements on the solubility of carbon in austenite.

The difficulty is overcome by simplifying the calculation of the saturation ratio S_r by using the degree of saturation S_c:

$$S_c = \frac{C}{C'}$$

Neumann indicates that, in practice, it is sufficient to use the simplified form

$$S_c = \frac{C}{4.26 - (\text{Si} + \text{P})/3.2}$$

Carbon Equivalent

The value termed carbon equivalent is also employed to relate a cast iron alloy to the percentage of eutectic in the solidified structure. It is calculated as follows:

$$\mathrm{CE} = \%\mathrm{C} + 0.31\ \%\mathrm{Si} + 0.33\ \%\mathrm{P} - 0.027\ \%\mathrm{Mn} + 0.40\ \%\mathrm{S}$$

It is equal to the carbon content of the alloy plus the amount of carbon equivalent to the added elements. In this equation, the signs of the solubility factors are reversed. The difference between CE and 4.26 shows the position of the alloy on the Fe–C diagram in relation to the eutectic.

4-3 Structural Diagrams for Cast Iron

Different approaches have been made to indicate by a diagram the final structure of cast iron after solidification. Laplanche[9] corrected the diagram of Maurer,[10] presenting the location of white, gray, and mottled structures in relation to the carbon and silicon contents of cast iron. He also indicated the influence of section size of casting on structure, and in a later series of diagrams[11] showed the influence of alloy elements, e.g. Cr, Mo, and Mn, on the matrix structure.

Figure 4-5(a) shows a Maurer type diagram. In the structure of cast iron, two possibilities exist for the high carbon phase solidifying from the melt—either carbide or graphite. An overlapping region exists where both phases may appear in the structure. The matrix in the normal analysis is the result of a solid state transformation of austenite and has either ferrite or pearlite as two possibilities, with an overlapping region of both phases. The diagram published by Laplanche is shown in Figure 4-5(b).

Laplanche criticized the Maurer diagram, arguing that all irons of carbon content greater than 4.30 per cent would have a ferritic matrix. The structure of all irons in the vicinity of 4.30 per cent carbon would be indeterminate. In Laplanche's diagram, the boundaries are curved and relate to what was termed a graphitization coefficient K. For cylindrical test bars of 30 mm diameter, K is given by:

$$K = \frac{4 \times \mathrm{Si}}{3}\left(1 - \frac{5}{3\mathrm{C} + \mathrm{Si}}\right)$$

All irons having the same value for the coefficient K will have the same structure after cooling in identical conditions. The diagram shows the lines with $K = 0.2, 0.5, 1.0, 1.5, 2.0, 3.0, 4.0, 5.0, 6.0$. These were called lines of equal graphitizing tendency. The dashed lines on the diagram correspond to different values of S_c and are given by:

$$S_c = \frac{\text{total carbon}}{4.30 - 0.3 \times \mathrm{Si}} = \frac{C}{C'} = \frac{TC}{E}$$

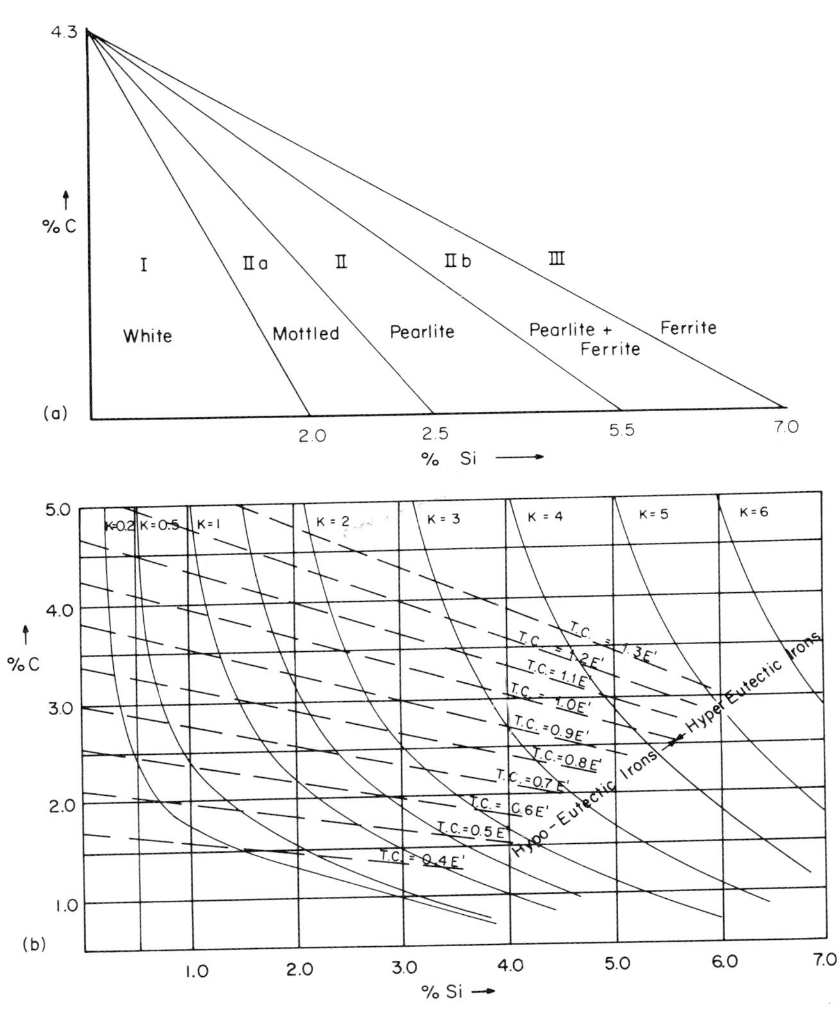

Figure 4-5 (a) Schematic Maurer diagram. (b) Laplanche diagram. Curved lines are for K. Dashed lines are for S_c with different values of total carbon[9] (Reproduced by permission of American Society for Metals)

Thus all cast iron structures are determined by the two coefficients S_c and K. The degree of saturation S_c is selected by the desired position of the alloy in relation to C'. The coefficient K is determined by the desired structure. The intersection of K and S_c gives the desired C and Si contents.

Contribution of Patterson and Döpp

In a broad review intended to evaluate the properties of malleable iron, Patterson and Döpp[12] discussed the different types of structural diagram available and also the methods of locating composition in relation to the eutectic. With respect to the graphitization factor K of Laplanche, it was suggested that the theoretical basis was debatable. However, the empirical relationships are adequately described by the calculations.

Patterson and Döpp showed Figure 4-6 as a relationship between compara-

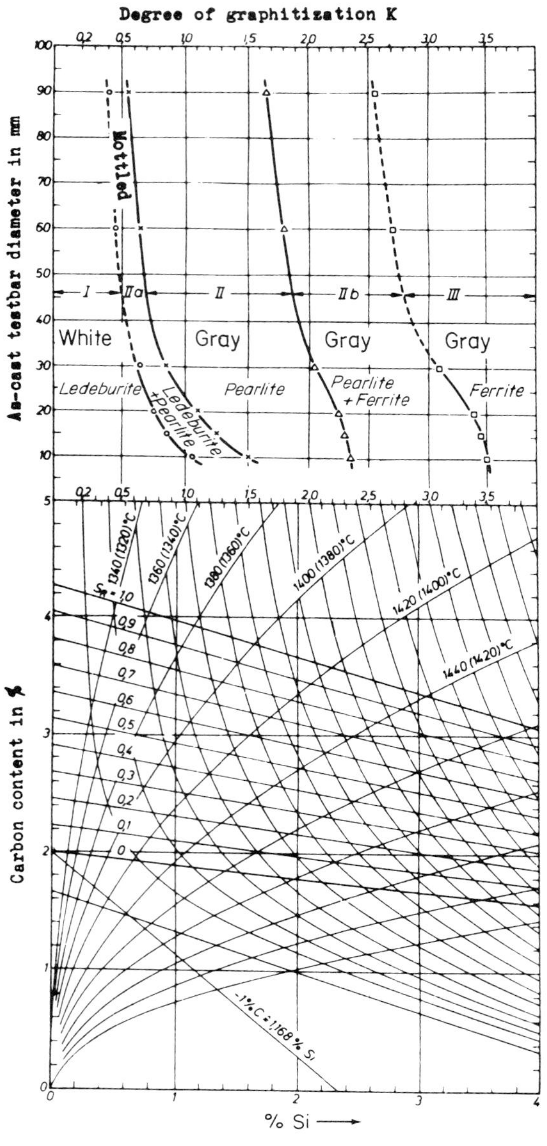

Figure 4-6 Structural diagrams due to Patterson and Döpp.[12] (Reproduced by permission of Giesserei-Verlag)

tive cooling conditions and structure formed. In this diagram, the upper figure relates test bar diameter and structure. The lower figure shows the lines for K values intersecting lines for S_r where S_r is the saturation ratio, given in this instance as the standard formula:

$$S_r = \frac{\%C - 2.0 + 0.10\ \%Si}{2.3 - 0.2\ \%Si}$$

4-4 Thermodynamics and Structural Diagrams

Structure diagrams offer guidelines in obtaining the required structures in gray and malleable cast iron. Looking at the generalized scheme of Figure 4-7, the diagrams are concerned with two aspects of phase transformations in the Fe–C–Si system. For simplification the dividing lines of structure in this diagram are held vertical and parallel with the Fe–C boundary. The vertical lines indicated represent structural changes related to the element silicon. Close to the Fe–C boundary, the structural change of importance is the liquid–solid transformation. The feature of interest is the relative nucleation and growth tendency of the graphite phase as compared with the carbide phase.

Further removed from the Fe–C boundary, the influence of silicon is on the solid state transformation of γ. Silicon in the ternary Fe–C–Si system determines the relative pearlitic or ferritic nature of the matrix, which effect might be considered in relation to the nucleation and growth rates of pearlite and the α phase. The range of existence of the three-phase area, γ, α, and carbide, is extended by silicon and the final resulting structure is related to these features.

Both the liquid and solid transformations are influenced by cooling rate. As seen from Figure 4-6, the influence of increasing cooling rate on the

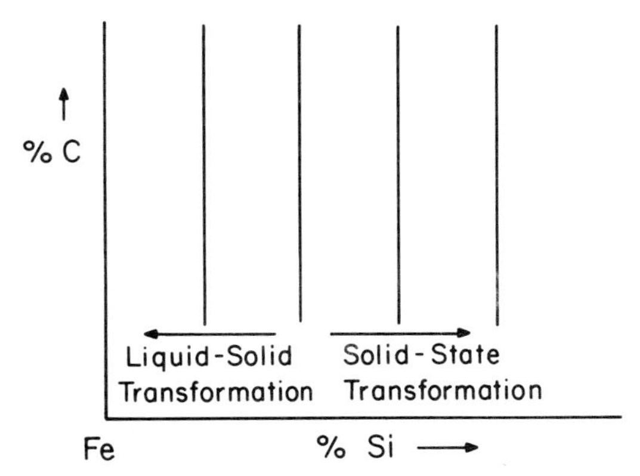

Figure 4-7 Representation of a structural diagram for the Fe–C–Si system

liquid–solid transformations is to favour the cementite transformation and extend the white structure to a higher silicon content. The effect of increasing cooling rate on the solid state transformation favours the pearlite reaction and extends this similarly into a region of higher silicon content.

The influence of other elements in solution can be rationalized by their effects on these two (liquid and solid) transformations. Some of the diagrams due to Laplanche[11] show contributions of individual elements, and enable calculations of composition where interest lies in retaining a pearlite matrix.

4-5 The Liquid–Cementite or Liquid–Graphite Transformation

In the absence of silicon, the phase which appears in an Fe–C melt at normal freezing rates is cementite, and not graphite. This appears to be related to nucleation, and not directly to the thermodynamics of Fe–C–Si.[13]

Hillert[13] made observations on eutectic cell formation in cast iron, adding small amounts of silicon. The additions would change solidification from completely white to completely gray. For a small number of cells formed, regions between the cells solidify white. As discussed in Chapter 3, the white solidification mode is more rapid than gray and will dominate the structure once it grows. Further silicon additions increased the number of graphite nuclei formed. It was reasoned that from the thermodynamic point of view, although silicon decreases the stability of cementite, this has no bearing on the graphite growth problem. Rather, the appearance of the gray structure is dependent on graphite being nucleated as a function of the addition of silicon.

4-6 The Transformation of Austenite to Pearlite or Ferrite in Cast Iron

The transformation of austenite to ferrite or pearlite in a binary Fe–C system is relatively well understood, but a number of complexities exist for the ternary Fe–C–M. These stem from the requirements of diffusion of the alloying element during the transformation and from the range of temperature and composition within which two-phase equilibria can exist in a ternary system. Gray cast iron is essentially an Fe–C–Si alloy. The matrix structure can be ferrite, pearlite, or a mixture of these two structures, depending on composition and cooling rate. Some of theoretical background for understanding the possible structures which are found is given in the following.

Partitioning of Solute; No-partition Temperature

Hillert[14] reviewd the research of Hultgren[15] in which the observation was made of two transformation curves for ferrite forming from austenite. Hultgren suggested two different modes of phase formation as a function of temperature. At high temperature, the alloying elements in austenite can partition between the matrix and the growing phase or phases. At low temperature, insufficient time may be present for diffusion of the alloying elements in austenite and the growth rate then depends on the rate of carbon diffusion. In this case, the growing phase in the case of ferrite, or phases in the growth of pearlite, may inherit the alloy content of the matrix.

Hultgren called this case paraequilibrium and the transformation products were called paraferrite, paracementite, and parapearlite. Paraequilibrium signified that the alloying elements were distributed in a non-equilibrium or no-partition manner. This could explain the deep bay in austenite transformation kinetics, particularly for the effect of Mo on the TTT diagram which shows two C curves for the formation of α. With respect to partitioning of solute other than carbon this would slow down a transformation and shift the knee of the pearlite transformation to longer times. For the case of Si, Bain and Paxton[16] ascribed a moderate effect of this element on hardenability as it slowed down the rate of the γ–pearlite transformation.

Cahn and Hagel on Partitioning

Cahn and Hagel[17] used the pseudo-binary section of a Fe–C–M ternary diagram shown in Figure 4-8. Two separate diagrams are superimposed, the dash-p diagram being related to transformation with no partitioning and the full-line diagram to transformations with alloy partitioning. The range of temperature and composition for the existence of $\alpha + \gamma +$ carbide is shown in the full-line diagram. The no-partitioning eutectoid T is T_e^p, shown by broken lines. They examined the ternary system for the influence of the ternary

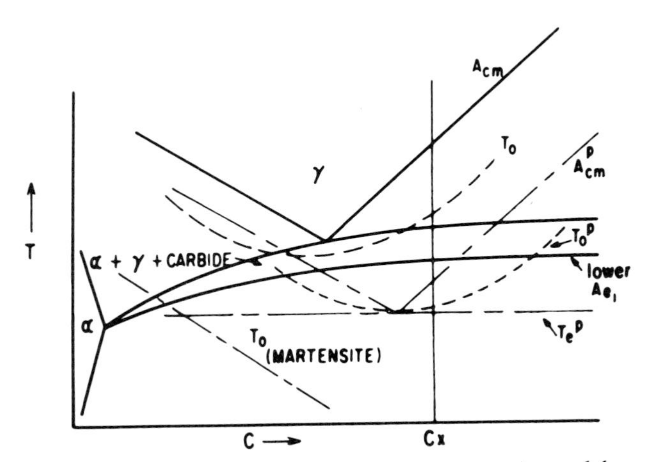

Figure 4-8 Binary section of a ternary system used by Cahn and Hagel.[17] The continuous curves are for equilibrium transformation, by alloy partitioning. The dash-p curve is for transformation by no partitioning of solute. T_e^p is a no-partition eutectoid temperature. (Reproduced by permission of The Metallurgical Society of American Institute of Mining Engineers)

element on the temperature interval down to T_e^p and the precipitation of phases in this interval.

The ternary element changes the temperature of the eutectoid transformation, which is T_e for a binary system of composition C_e. T_o is the temperature at which $\Delta G_o = 0$, being the difference in free energy between austenite and a phase mixture of $\alpha + Fe_3C$. For any composition C, the pearlite equilibrium temperature becomes:

$$T_o = T_e \frac{RT_e}{2C_e \Delta S_o}(C - C_e)^2$$

where ΔS_o is the entropy change.

There is an interval between T_o and T_e which Cahn and Hagel ascribed (for non-alloy austenite) to the instability of the pearlite phase with respect to a two-phase mixture of γ and α, or γ and carbide, or γ and graphite. This is the case for the binary system. For a ternary system the alloy element does or does not partition. This introduces changes in the three phase region in stability and transformation sequence.

Estimate of Temperature Interval between T_o and T_o^p

The approach to calculating the interval between T_o and T_o^p was based on thermodynamics. In Fe–C, carbon partitions and α plus Fe_3C form with compositions close to equilibrium. The pearlite growth rate is related to carbon diffusion in austenite. With alloying elements and partitioning at high temperature, α plus carbide will have these elements according to equilibrium segregation. At lower temperatures, there is no partitioning.

At high temperature, the value of ΔG_o is small. If segregation of the elements does not occur, pearlite formed with non-equilibrium composition would have a high free energy. However, at a lower temperature, pearlite may form without partitioning on thermodynamic grounds. The excess free energy of pearlite with non-equilibrium segregation can be calculated, using the following definitions:

$$K = \text{segregation, i.e. } \frac{\text{Concentration of alloy in carbide}}{\text{Concentration of alloy in ferrite}}$$

K^0 = equilibrium segregation

$\Delta G_p(K)$ = difference in free energy between partially segregated and equilibrium pearlite

For the case where $K = 1$, i.e. no partition:

$$\Delta G_p(1) = C_a RT\{\ln[1 + (K^0 - 1)V_c] - V_c \ln K^0\}$$

where C_a = concentration of alloy element
V_c = mole fraction of metal in austenite to become carbide

This gives an upper limit T_o^p to the temperature at which pearlite can form with no partition. The dash-p diagram of Figure 4-8 was calculated for no segregation. For the double diagram, pearlite can form below T_o^p without alloy partitioning. Above T_o^p, it is possible to have pro-eutectoid reaction which could proceed without partitioning and with a greater decrease of free energy than for the pearlite transformation.

References

1. *Basic Open Hearth Steelmaking*, Am. Inst., Min. Metall. Eng., 1951.
2. Neumann, F., and H. Schenk: *Arch. Eisenheuttenw.*, **30**, 477, 1959.
3. Rist, A., and J. Chipman: *Rev. Metallurg. Mem.*, **53**, 796, 1956.
4. Turkdogan, E. T., J. E. Lecke, and G. R. Masson: *Acta Metall.*, **4**, 396, 1956.
5. Wagner, C.: *Thermodynamics of Alloys*, Addison-Wesley, 1952.
6. Hillert, M.: *Acta Metall.*, **3**, 34, 1955.
7. Neumann, F.: *Recent Research on Cast Iron* (Ed. H. Merchant), Gordon and Breach, New York, 1968.
8. Turkdogan, E. T.: *J. Iron Steel Inst. London*, **182**, 66, 1956.
9. Laplanche, H.: *Metal Progr.*, **52(6)**. 991, 1947.
10. Maurer, E.: *Stahl Eisen*, **44**, 1522, 1924.
11. Laplanche, H.: *Metal Progr.*, **55**, 839, 1949.
12. Patterson, W., and R. Döpp: *Giessereiforschung* (in Engl.), **21(2)**, 91, 1969.
13. Hillert, M.: *Trans. Am. Soc. Met.*, **53**, 555, 1961.
14. Hillert, M.: *Phase Transformations*, p. 181, Am. Soc. Met., 1970.
15. A. Hultgren: *Jernkontorets Ann.,* **135**, 403, 1951.
16. Bain, E. C., and H. W. Paxton: *Alloying Elements in Steel*, 3rd printing, Am. Soc. Met., 1966.
17. Cahn, J. W., and W. C. Hagel: *Decomposition of Austenite by Diffusional Processes* (Eds. V. F. Zakay and H. I. Aaronson), p. 131, Wiley–Interscience, New York, 1962.

Bibliography

Darken, L. S., and R. W. Gurry: *Physical Chemistry of Metals*, McGraw-Hill, 1953.

Chapter 5
Solidification of Spheroidal Graphite Cast Iron

The instability in growth of a graphite flake is described and related to undercooling. The different ways in which the undercooling of graphite can be achieved by the influence of the composition of liquid iron are presented. Three types of interaction are treated:

1. Strong interaction by reactive elements of the type Mg, Ce, and rare earths. This leads to kinetic undercooling. Different theories of this type of interaction are described and calculation presented of the undercooling achieved.
2. Neutral interaction by elements which form a boundary layer and contribute to the constitutional supercooling at the growing crystal surface. A typical element of this type is Si. The constitutional supercooling is calculated.
3. Weak interaction by elements which form an adsorbed boundary layer. The element sulphur is noted to lower the graphite melt interfacial energy when present in solution. It therefore allows graphite crystallization at temperatures closer to the equilibrium one and thus acts in a manner opposite to those elements promoting kinetic and constitutional supercooling.

The overall effect of the melt chemistry is presented in the form of calculated contributions to the undercooling. This is treated in the present chapter when defining a narrow range of undercooling required for spherulitic growth.

Experimental evidence of the solidification sequence of spheroidal graphite cast iron is discussed and a summary is given of the employment of temperature measurements during solidification as a means whereby the structure of such a cast iron can be monitored.

5-1 The Original Papers by Morrogh and Williams[1,2]

In their 1947 paper,[1] Morrogh and Williams described their research on graphite formation in Ni–Fe–C, Ni–C, and Co–C alloys. In this, it was noted that spherulitic graphite formed in cast iron in the absence of sulphur when Ca and/or Mg was added. They suggested, in addition, the presence of Ba and Sr.

Morrogh and Williams also observed that rapid cooling of Ni–C alloys gave spherulitic graphite structures. Taking into account the rapid cooling effects, they suggested that the influence of spheroidizing elements was to cause the formation of graphite at a lower temperature than that required to give normal undercooled graphite.

The effects in Ni–C alloys were suggested as being due to stabilization of the Ni_3C phase. This phase would subsequently decompose at a lower temperature than that at which undercooled graphite normally grows. The decomposition process of a solid carbide served to correlate the spherulitic graphite growth process in a liquid alloy treatment, with the solid state decomposition and graphite forming process of malleable iron. The study was extended to cast irons where the idea was held that completely spherulitic structures in cast iron required the initial solidification of acicular white iron structures. Alloying elements were necessary which would allow decomposition at low temperatures, suggesting something in the neighbourhood of 800 °C.

These early ideas on spherulitic growth give an indication of the problems of that time in understanding the complicated processes of graphite growth at large undercoolings. The first theories were not related to crystal growth mechanisms.

Having little fundamental knowledge of these mechanisms, the task of controlling the production of these complex cast alloys with unusual graphite forms was performed empirically. This slowed the commercial development until solutions could be obtained by experiment to allow production with reliability of structure and mechanical properties. A second attack on a new series of problems can now be undertaken with new theoretical knowledge. This should allow solution of such problems as segregation effects, particularly in thick castings. It is of importance that prior knowledge of likely composition effects should be available at melt-down to determine the structures which may result from composition and to learn how to control analyses within required limits.

The 1948 paper[2] reported the cerium treatment of gray cast iron to produce spherulitic graphite structures. In this paper, some doubt was expressed as to whether hyper-eutectic spherulites formed directly from the melt or via a carbide phase. The problem of accounting for eutectic crystallization in such a system was solved by suggesting that eutectic graphite crystallized in spherulitic form on an already existing hyper-eutectic spherulite.

5-2 Spherulitic Morphology and Surface Energy

Energy arguments were presented as an explanation of forms which are spherulitic, although the energy theorems are strictly applicable to equilibrium growth. This is far from being the case for growth of a spherulite in cast iron.

An equilibrium form is one which should be assumed by a crystal under conditions which allow the total surface energy to have the smallest value in relation to the volume. This is expressed as follows:

$$\phi = \int \sigma \, ds \qquad (5\text{-}1)$$

where ϕ = total surface free energy
σ = specific surface energy of any surface element of area ds

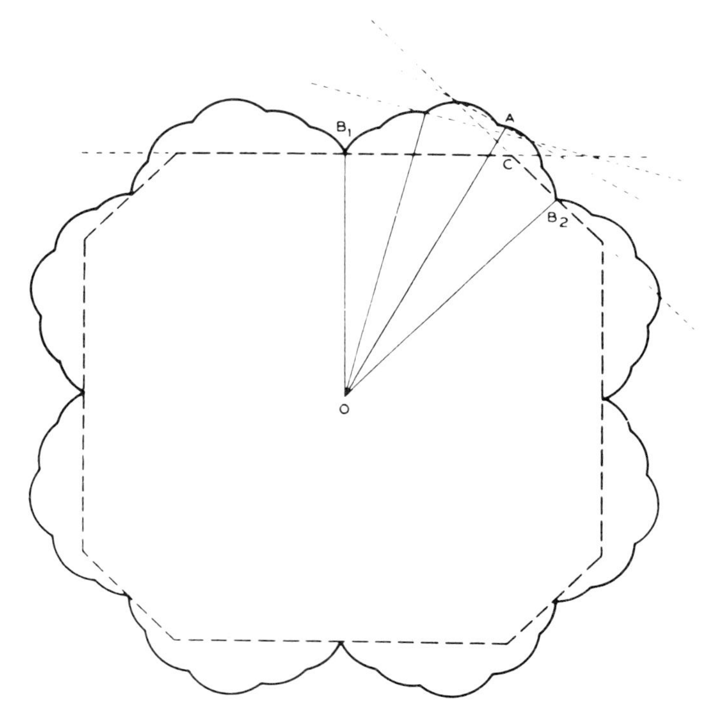

Figure 5-1 Construction of equilibrium form for a crystal using the Gibbs–Wulff theorem. Full line is the polar plot of surface free energy. Light dashed lines are planes normal to radius vectors of this plot. Heavy dashed lines form the equilibrium polyhedron which is formed by lines which are not crossed by any other normal to a radius vector. From C. Herring.[3] (Reprinted from Structure and Properties of Solid Surfaces edited by R. Gomer and C. S. Smith by permission of The University of Chicago Press. Copyright © 1953 The University of Chicago Press)

The integration of Equation (5-1) is performed over all orientations. The construction of an equilibrium form for a crystal can be made using the Gibbs–Wulff theorem[3] (Figure 5-1). This gives the equilibrium form as being geometrically similar to the envelope of perpendiculars to radius vectors in a polar plot of specific surface free energy for the crystal being examined. From this theorem and construction, it is also possible to show that the perpendiculars from the centres of crystals to the bounding faces are proportional to the surface energies of the faces:

$$\frac{R_1}{R_2} = \frac{\sigma_1}{\sigma_2}$$

The theory and construction are not ideally applicable to growth forms since these must result from kinetic factors. After growth, a faceted crystal will be bounded by faces of low index and hence of lowest surface energy (see the next section) but the radii are not necessarily proportional to the energies. There is no theorem for growth similar to the Gibbs–Wulff theorem for equilibrium forms.

For spherulitic or other forms which are polycrystalline, it is however possible to show under what conditions the polycrystalline equilibrium forms may have lower total energies than a single crystal equilibrium form. This depends on the relationship between the energies of bounding faces and the energies of internal boundaries. The polycrystalline form can be totally bounded by faces having the lowest surface energies. More exact formulation of this problem has been worked out in the following analysis by Fullman.[4]

Fullman's calculations were for crystalline materials with anisotropic surface free energy. A comparison was made of the total boundary free energy for single crystal and polycrystalline forms using two-dimensional and three-dimensional models having radial or tangential intercrystalline boundaries. The total boundary free energy Γ is made up as follows:

$$\Gamma = \phi + \psi$$

where ϕ = total surface free energy
ψ = total intercrystalline boundary free energy

The equilibrium form now is that for which Γ has the smallest value. Figure 5-2 shows the quantities of interest in the calculations as related to a single two-dimensional crystal.

The total free energy ϕ is equal to the free energy Γ_1:

$$\Gamma_1 = \phi = 2R_1\sigma_2 + 2R_2\sigma_1 \tag{5-2}$$

and

$$\text{Area } A = 4R_1R_2 \tag{5-3}$$

From Wulff theory:

$$\frac{R_1}{R_2} = \frac{\sigma_1}{\sigma_2} \tag{5-4}$$

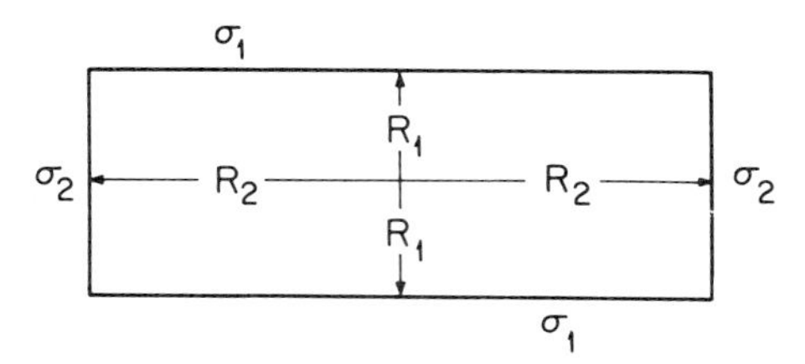

Figure 5-2 Single two-dimensional crystal used in calculation of Fullman.[4] (Reproduced by permission of Pergamon Press Ltd.)

Combining Equations (5-2) and (5-4), the dimensionless quantity $\Gamma_1/A^{1/2}\sigma_1$ is found. This is independent of crystal size and was used by Fullman as a means for comparison of the stability of various geometrical arrangements. Small values correspond to high stability for the single two-dimensional crystal concerned:

$$\Gamma_1/A^{1/2}\sigma_1 = 4(\sigma_2/\sigma_1)^{1/2} \tag{5-5}$$

For three-dimensional crystals, Fullman examined the quantity $\Gamma_1/V^{2/3}\sigma_1$, while for the polycrystalline arrangement of the two-dimensional crystal, the model shown in Figure 5-3 was employed (termed a cylindrulite). This has n_2 identical triangles oriented so that the aggregate is bounded completely by surface σ_1. The crystals are joined by n_2 intercrystalline boundaries with length R and specific interfacial free energy σ_b. The total surface and intercrystalline boundary free energy is

$$\Gamma_2 = 2n_2 \sin(\pi/n_2)R\sigma_1 + n_2R\sigma_b \tag{5-6}$$

The area A_2 is:

$$A_2 = n_2 \sin(\pi/n_2)\cos(\pi/n_2)R^2 \tag{5-7}$$

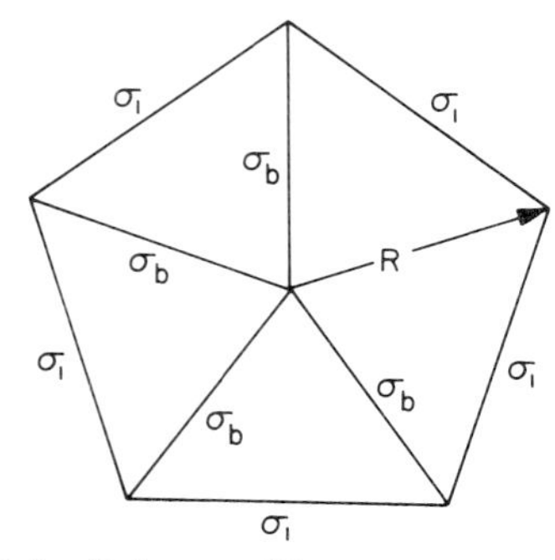

Figure 5-3 Polycrystalline arrangement of two-dimensional crystal used in calculation of Fullman,[4] and termed a cylindrulite. (Reproduced by permission of Pergamon Press Ltd.)

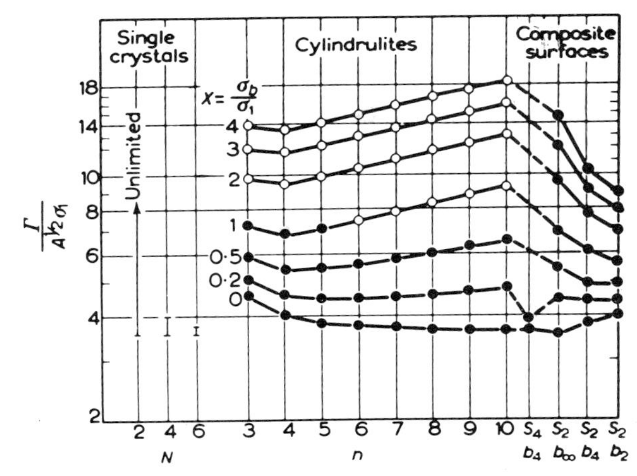

Figure 5-4 Plot showing relative stability of single crystals and polycrystalline configurations for two-dimensional crystals. For the latter, small values of χ lead to more stable forms than single crystals.[4] (Reproduced by permission of Pergamon Press Ltd.)

The dimensionless quantity $\Gamma_2/A_2^{1/2}\sigma_1$ is given by

$$\Gamma_2/A_2^{1/2}\sigma_1 = [n_2/\sin(\pi/n_2)\cos(\pi/n_2)]^{1/2}[2\sin(\pi/n_2) + \chi] \tag{5-8}$$

where $\chi = \sigma_b/\sigma_1$

The solutions of Equation (5-8) are plotted in the lower central part of Figure 5-4 for various values of n and χ (filled in circles).

For sufficiently small values of σ_b relative to σ_1, the polycrystal may have a lower free energy than the single crystal. Fullman's calculations applied to cylindrulites showed that the total free energy is less than that of a single crystal when the interfacial free energy is less than 0.2, the minimum surface free energy, i.e. $\sigma_b/\sigma_1 < 0.2$.

5-3 Surface Energy Models of Spherulitic Graphite Growth in Cast Iron

In papers published by Buttner, Taylor, and Wulff[5] and Keverian, Taylor, and Wulff[6] the equilibrium arguments for the spherulitic form of a graphite crystal, based on the anisotropic surface energy, were used to define the growth form. The energy between graphite crystal faces and the melt depends on the presence of sulphur. This element is surface active. When it is removed from the melt by the presence of reactive additions like Mg, the melt–graphite interfacial energy is increased. These researches suggested that the graphite then grows in a spherulitic form which is energetically more favourable. The crystal becomes bounded only by (0001) surfaces which have the lowest energy. This is an application of the equilibrium theory.

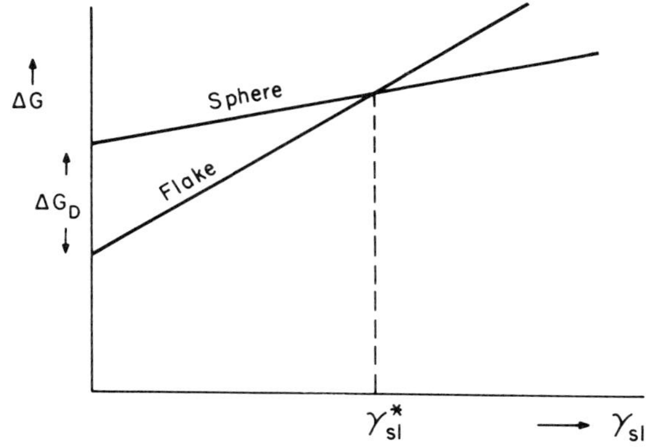

Figure 5-5 Change of free energy ΔG for crystallization of a flake or spherulite as a function of the interfacial free energy between melt and solid γ_{se}. ΔG_D represents the energy stored in the interior of the sphere by the low angle boundaries of the graphite spherulite. The lamella is more stable below a critical value of the interfacial energy γ_{sl}^*. (From Buttner, Taylor, and Wulff.[5] Reproduced by permission of American Foundrymen's Society)

An alternative growth theory was also proposed. The increase of surface energy in the absence of sulphur required greater undercooling for growth. A spherulitic crystal resulted from the ensuing changes in the growth rates.

Theories expressing the equilibrium form have been given in other papers, e.g. Geilenberg.[7] A representation of the views of Buttner, Taylor and Wulff is given in Figure 5-5.

5-4 Growth Models of Spherulites

Some of the models concerned with growth processes of spherulites in cast iron are described in this section.

Model of Hillert and Lindblom

Hillert and Lindblom[8] suggested the growth of graphite spherulites from screw dislocations (Figure 5-6). According to their scheme, screw dislocations would be generated by the inclusion of foreign atoms in the graphite lattice. They suggested that Ce and Mg atoms attached to the carbon atoms at the growing edge of a close-packed plane. This would provide disturbances which led to the development of new screw dislocations. In the model illustrated in Figure 5-6, after a certain growth interval, the spirals would no longer fit together and there would be a continuous tendency to divide into new branches. The growth would then become spherulitic.

Figure 5-6 Model of Hillert and Lindblom[8] for growth of graphite spherulites from screw dislocations. (Reproduced by permission of The Metals Society)

This model was an early one. Screw dislocations do play a role in graphite growth,[9] but it is not necessary to seek their creation by reactive elements to account for spherulitic growth.

Helical Growth Mechanism of Double and Hellawell

Figure 5-7(a) shows a model for spherulitic graphite growth suggested by Double and Hallawell.[10] This followed from their observations on vapour-grown fibres of graphite (Figure 5-7b) which have conical apex angles with fixed values. Basing their model on their own investigations of rotation boundaries in graphite[11] they suggested that the fixed values of the apex angle were related to coincidence boundary theory. The angle of overlap θ of the graphite layers in a fibre (Figure 5-7c) was such as to give coincidence of the two overlapping graphite planes. An observed apex angle ϕ of $140°$[12] corresponded with an overlap angle $\theta = 21.8°$ where the relation between ϕ and $\theta = 2\pi \cos \phi/2$.

Their model for a graphite spherulite might account for a herring-bone structure of spherulites shown by Tsuchikura, Kusakawa, and Okumoto[13] in sections examined by electron microscopy after ion etching.

Model of Hunter and Chadwick

Hunter and Chadwick[14,15] examined ion etched sections of spherulitic graphite crystals in an electron microscope. They observed the patterns of chevron markings noted by Tsuchikura, Kusakawa, and Okumoto,[13] discussed in the previous section on the research of Double and Hellawell. In the observations of Hunter and Chadwick, the chevron angle changed towards the outer circumference of the spherulite. Towards the centre of the spherulite, markings reminiscent of planar graphite crystals were noted. They sug-

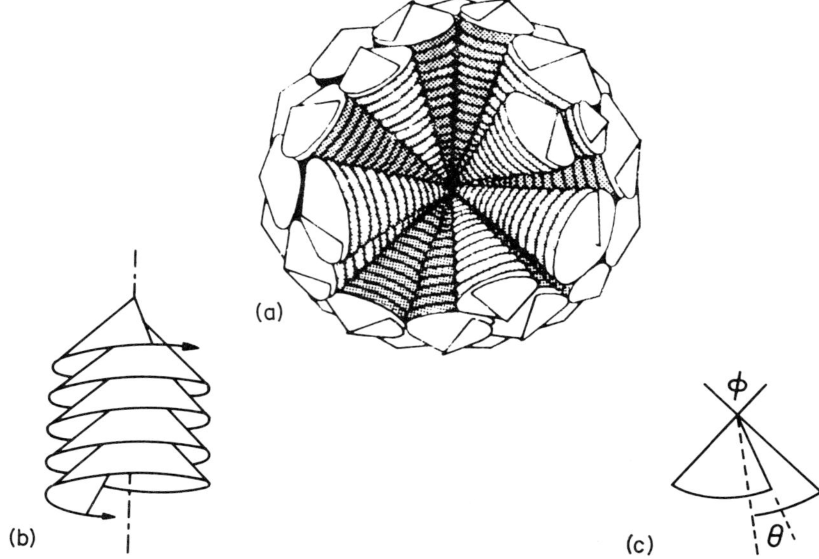

Figure 5-7 (a) Model for spherulitic graphite growth of Double and Hellawell.[10] (Reproduced by permission of Georgi Publishing Co. (b) Fibrous growth of graphite with conical apex angle of fixed value.[12] (Reproduced by permission of Pergamon Press Ltd. (c) Fixed value of the apex angle is related to coincidence boundary theory. If θ is the overlap angle, θ and ϕ are related by $\theta = 2\pi \cos \phi/2$.[12] (Reproduced by permission of Pergamon Press Ltd.)

gested that the spherulite structure commenced as a flake and underwent dendritic branching (see Section 2-5).

Investigations of this nature, examining the growth structure of a spherulitic crystal in a systematic manner, might be important in understanding the way in which instability of surfaces develops in a spherulitic crystal. Hunter and Chadwick suggested that the first growth is in flake form, and this is followed by dendritic branching. The chevron markings were not accounted for, however. They might represent a pyramidal interface which disappears towards the end of growth. Such an interface is in agreement with growth pyramids as observed on graphite surfaces.[16]

Shubnikov's Spherulite

In Shubnikov's model[17] (Figure 5-8a) a spherulite commences as an acicular crystal and branches during growth. A crystal which grows by branching, and in which growth continues from the branched sections in a radial direction, will develop a form approaching spherical with an equatorial contraction. Various geometries are possible depending on the relationship between growth rate and rate of branching.

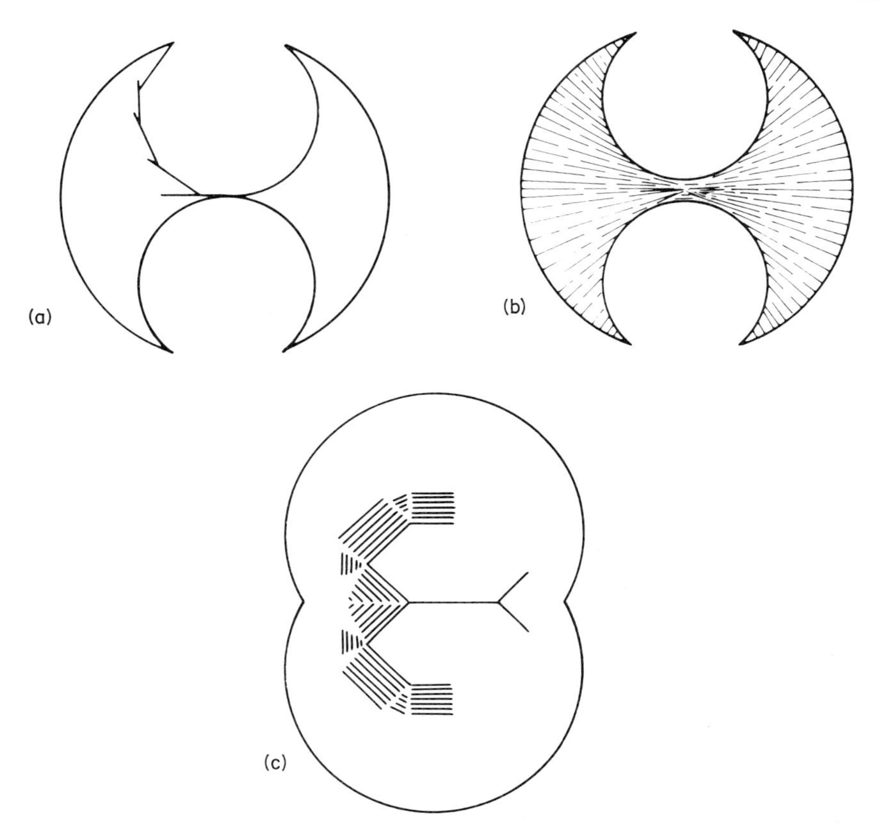

Figure 5-8 (a) Model of spherulite development by branching of an acicular crystal during growth, after Shubnikov.[17] (Reproduced by permission of American Institute of Physics) (b) Shubnikov's idealization of a graphite spherulite by a branching mechanism. (Reproduced by permission of American Institute of Physics) (c) Possibility of spherulitic growth from a graphite platelet by branching during growth. (From Minkoff,[9] based on the model of Shubnikov. Reproduced by permission of The Metals Society)

Shubnikov's idealization of graphite spherulitic growth by a branching mechanism is shown in Figure 5-8(b). This was suggested to be incorrect because of the incorrect orientation of (0001) planes[9] and a possible model on the basis of Shubnikov's ideas in shown in Figure 5-8(c).[9] This figure shows growth of a graphite platelet with branching initially occurring in either direction at the platelet periphery. The (0001) planes were suggested to align radially as determined by the relative growth rates normal to (0001) and in the 0001) plane.

Model of Oldfield, Geering, and Tiller[18]

In this analysis, an attempt was made to show the transformation of a plane

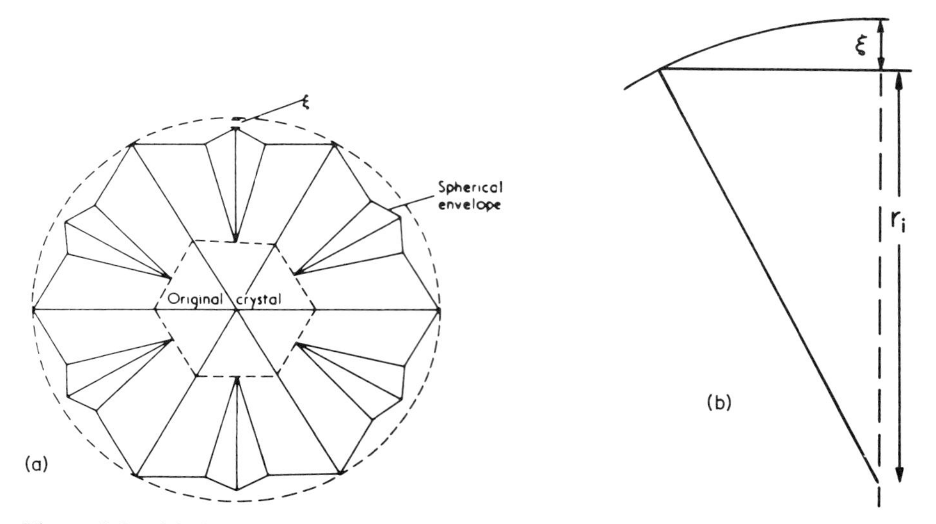

Figure 5-9 (a) Growth of a spherulite by interface breakdown described by Oldfield, Geering, and Tiller.[18] (b) Crystallite develops at centre of face when crystal has attained size r_i. Spherical envelope is distance ξ ahead of growing crystallite.[18] (Reproduced by permission of The Metals Society)

bounding surface of a crystal to an assembly of crystallites oriented approximately radially, and which is spherical in shape.

A polyhedral single crystal is shown in Figure 5-9(a). It is supposed that a new, radially oriented, crystallite develops at the centre of a facet when the crystal has attained size r_i (Figure 5-9b). After a short period of growth, it is suggested that the new crystallite is now growing at the same velocity as the corners of the crystal. It will always lag behind the corners by the distance which separated it from the spherical envelope at initiation. This implies that while growing as fast in the radial direction as the corner of the polyhedron, the spherical envelope is always a distance ξ ahead of the growing crystallite.

When the crystal assembly has reached size R:

$$\frac{\xi}{R} \to 0 \qquad \text{if } \xi \text{ is negligible compared to } R$$

In the theory developed, the idea of interface breakdown is central. The filaments or dendrites resulting from interface breakdown are postulated to form an array which is polyhedral in outline. This would be the result of anisotropy of the liquid–solid surface energy and the growth kinetics. This polyhedral outline subsequently becomes spherical to optimize the requirements of latent heat dissipation and transfer of solute.

Experiments on ice crystals were made using glycerol–water solutions and photographs show interface breakdown and filamentary nature of the spherulic growth for ice.

5-5 Undercooling and Spherulitic Growth

In the following sections, research is described on the manner of unstable growth of graphite as this is influenced during solidification by the different elements present in the cast iron composition. The mechanisms of instability are important as is also the categorization of the effects of the different elements. The role of these elements is now reasonably well understood. One of the objectives in adjusting the composition of an alloy so as to produce a spherulitic structure is to achieve the necessary undercooling for spherulitic growth. This is shown schematically in Figure 5-10. As has been noted in Section 5-1, this undercooling can be achieved by rapid cooling in a Ni–C alloy. In Fe–C–Si alloys changes of graphite form giving crystals of appearance close to spherulites can be observed on very rapid freezing.[19] In castings production, spherulitic growth cannot normally be achieved by undercooling related to rapid freezing; this must be obtained by adjusting the chemical analysis of the melt. Spherulites form within a very specific temperature range and the structure is difficult to control because a spherulite represents one specific type of instability phenomenon appropriate to one undercooling.

The undercooling to achieve spherulitic growth results from two principal effects referred to in Chapter 1. These are the kinetic undercooling ΔT_K and constitutional supercooling ΔT_{CS}. Both are obtained by adjustment of the analysis.

The complications in producing the spherulitic structure are the following:

(a) Spherulitic growth is an unstable growth process and hence is sensitive to small changes in the physical conditions, temperature, composition, etc.
(b) Melt composition is difficult to control, particularly with respect to trace elements which have large effects on growth.
(c) Interactions between elements occur at the graphite growth temperature so that composition is in a dynamic state of change.
(d) Segregation occurs in the casting, in particular macrosegregation towards the middle of a casting and microsegregation in the vicinity of crystallizing phases. Segregation effects may vary the composition of trace elements by an order of magnitude so that large variations of structure can occur due to this alone, particularly in thick section castings.

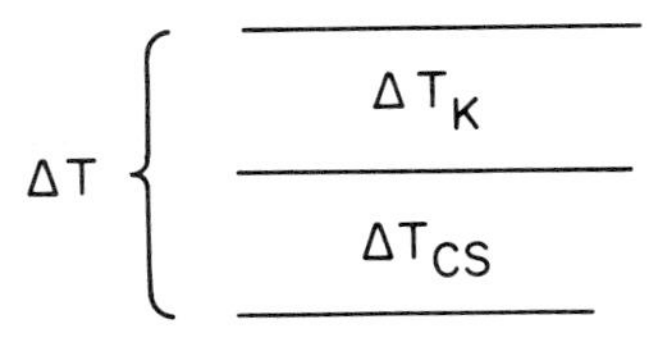

Figure 5-10 Components making up the total interface undercooling: K, kinetic; CS, constitutional supercooling

(e) Part of the undercooling necessary must be achieved by reactive elements, e.g. Mg, Ce. In addition to interacting with graphite these also interact locally with other elements in the vicinity of the crystallizing graphite. They are, furthermore, lost elsewhere by other processes, e.g. by oxidation at the metal–air interface.

The following sections deal with the theory of kinetic undercooling related to reactive elements, e.g. Mg, and the constitutional undercooling effects due to elements like Si. How sulphur influences undercooling is also described.

Kinetic Undercooling; Chernov's Theory of Strong Interactions with a Growing Crystal

Chernov[20] gave a theoretical analysis of the way in which a strongly adsorbing atom at a kink site of a step on a crystal surface slows down the rate of step advance on that surface. Eventually, to achieve nominal growth rates, the liquid must undercool. This is a kinetic effect in which a ΔT_K term is required to drive the attachment of atoms to the step.

Figure 5-11 shows the model. Steps move over a crystal surface by attachment of atoms to the steps at kinks. The rate of advance of a step in this case is a function of the distance atoms must move over the surface to find a kink and hence to the distance between kinks, λ. Any change in λ changes the rate of advance of the step. When the surface is growing from steps moving from a dislocation source, this changes the rate of advance of the surface.

The theory is developed for interactions between impurity atoms and kinks. The impurities compete for kink sites with the atoms of the growing crystal. When a foreign atom is adsorbed at a kink site, it effectively blocks the attachment of a growth atom at that site. This increases the effective spacing between kinks and slows the rate of step advance.

If $\lambda_0(O)$ is the kink spacing in an impurity-free solution, the spacing be-

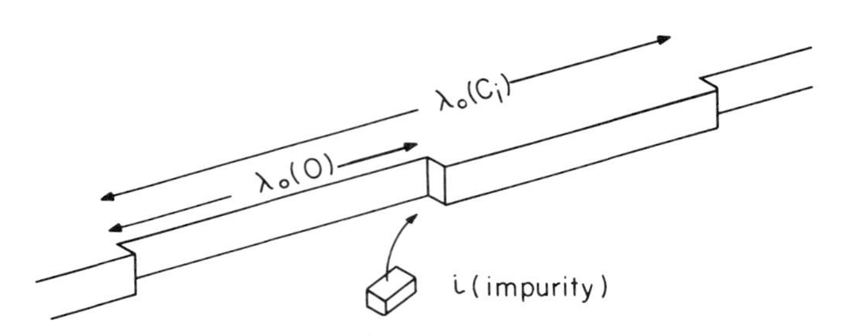

Figure 5-11 Chernov's model[20] for reactive impurity atoms slowing the rate of step advance on a crystal surface. $\lambda_0(O)$ is the distance between kinks in the absence of impurity. Impurity atoms going to kinks block these sites for further atom incorporation in the crystal. The new distance between kinks ($\lambda_0 C_i$) is calculated for different concentrations of active impurity

tween kinks $\lambda_0(C_i)$ for an impurity concentration (C_i) is given by:

$$\lambda_0(C_i) = \lambda_0(O) + \xi_s C_i$$

for

$$\xi_s = \frac{a}{2\Omega\nu^3}\frac{kT^{3/2}}{2\pi m}\exp\left(\frac{W_s + W_1 - W_3}{kT}\right)$$

where a = distance between atoms on the surface
Ω = molecular volume of solution
ν = vibration frequency of impurity in adsorbed state
m = mass of impurity particle
W_s = energy of impurity atom in solution
W_1 = energy of a free kink
W_3 = energy of an impurity-occupied kink

Chernov calculated that for an estimated value of $W_s + W_1 - W_3 = 0.4$ eV $\simeq 10$ kJ mol^{-1} and $\nu = 3 \times 10^{12}$ s^{-1}, the value of $\xi_s = 10^4 a$. Therefore an estimated impurity concentration $(C_i) = 10^{-3}$ changes λ_0 by a factor of two.

Definition of Strongly Adsorbed Atoms

In a further outline[21] strongly adsorbed atoms are considered to form a palisade and hinder growth. A strongly adsorbed atom is classified by its life time on a crystal surface in relation to the time of passage of steps. If l is the interstep distance and V is the rate of advance of a step, a short life time is considered to be when:

$$\tau \ll l/V$$

where τ is life (or dwell) time on the crystal surface. Conversely, long life time is when $\tau \gg l/V$. Chernov calculated that for a face on a crystal inclined 1° to a close-packed face l is approximately 2×10^{-6} cm. If V is approximately 10^{-4} cm s^{-1}, the impurity will have a short life time when $\tau < 10^{-2}$ to 10^{-3} s.

The life time τ is given by an approximate relationship between the vibration frequency ν and the adsorption energy of an atom on a crystal face U:

$$\tau \simeq \nu^{-1}\exp(U/kT)$$

where $\nu \simeq 10^{12}$ s^{-1}. For $U \simeq 20$ J mol^{-1}, $\tau \simeq 10^{-8}$ s at room temperature. This adsorption energy would be considered weak. Chernov calculated a strong interaction to be for $U = 50$–60 J mol^{-1}.

Obviously, for the case of graphite, realistic figures must be taken for adsorption energies and passage of steps. However, the physical model is a reasonable one and may be valid when considering the different elements in solution in cast iron and their interactions with graphite. Since the interaction energies are related to electron configuration in the atom, different interaction energies, e.g. for Mg and La or Ce, must be considered. These have different effects on graphite growth.

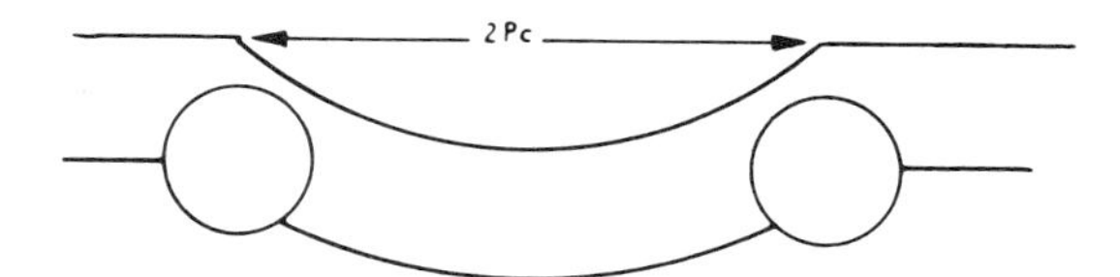

Figure 5-12 Theory of Cabrera[22] for impurity blocking of steps. If distance between impurity atoms is less than $2\rho_c$, where ρ_c is the critical radius for step nucleation at the prevailing temperature, step advance is blocked. The system can undercool to a new temperature and new ρ_c for growth

Kinetic Undercooling; Cabrera's Theory

In the theory of Cabrera,[22] strongly adsorbed atoms attach to the crystal surface and form a network through which the steps must advance (Figure 5-12). The steps will be held up if the spacing between adsorbed atoms is $<2\rho_c$ where ρ_c is the critical radius for nucleation of a step at the growth temperature. Only when the system undercools to a temperature where ρ_c has a smaller value will the advance of the step be allowed.

Cabrera gave the following expression for the velocity of advance of a step:

$$V = V_\infty \sqrt{(1 - 2\rho_c d^{1/2})}$$

where d = average distance between impurities
V_∞ = velocity of a straight step in an impurity-free environment

Thus V is a function of ρ_c and the spacing between impurities.

The Kinetic Undercooling of Graphite Growth by Reactive Impurities in Cast Iron

Reactive impurities in cast iron undergo strong interactions with graphite and also with other elements in the melt, e.g. with S and O.

In order to investigate the mode of interaction of a reactive impurity with graphite, a study was made by Minkoff[23] and Minkoff and Nixon[24] of the interaction between La and graphite in Ni–C alloys. For these experiments, metallic La was added to the melt in quantities insufficient to form spherulites. The microstructure was observed by optical microscopy and on extracted crystals by scanning electron microscopy. The distribution of La between graphite and the melt was determined by analysing for La using paper chromatographic techniques.[25] The chemical results are shown in Table 5.1.

For an experiment in which the alloy had the smallest La addition, the microstructure shown in Figure 5-13(a) demonstrates branching of the graphite and depressions in the crystal surface. Figure 5-13(b) shows a scan-

Table 5-1 Distribution of lanthanum between graphite and melt[26]

At % La in graphite	At % La in melt	Distribution coefficient
0.013	0.054	0.25
0.016	0.09	0.18
0.017	0.09	0.17
0.034	0.38	0.089

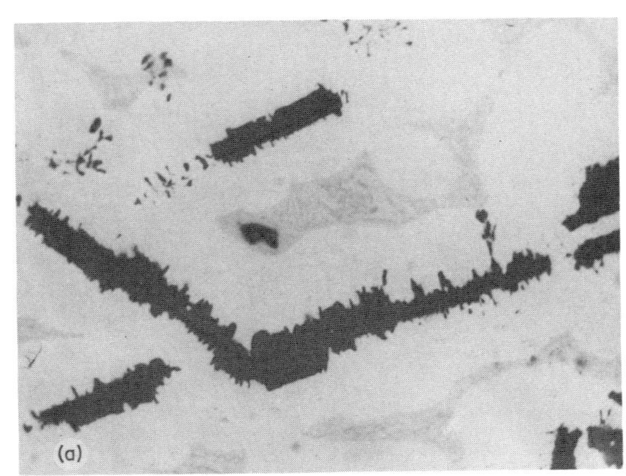

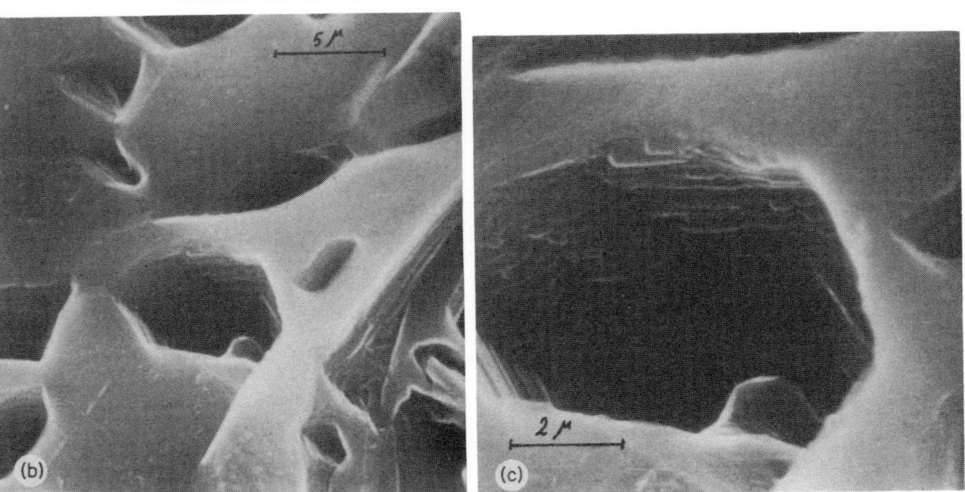

Figure 5-13 (a) Graphite crystal in Ni–C alloy with addition of lanthanum. Branching of the crystal and depressions in the surface. (× 55) (b) Scanning electron micrograph of extracted graphite crystal corresponding to specimen of (a). The holes correspond to the surface depressions seen in the two-dimensional metallographic section. (c) Enlargement of a hole in a crystal where growth is blocked by the adsorption of lanthanum, showing surface steps. (From Minkoff and Nixon.[24] Reproduced by permission of American Institute of Physics)

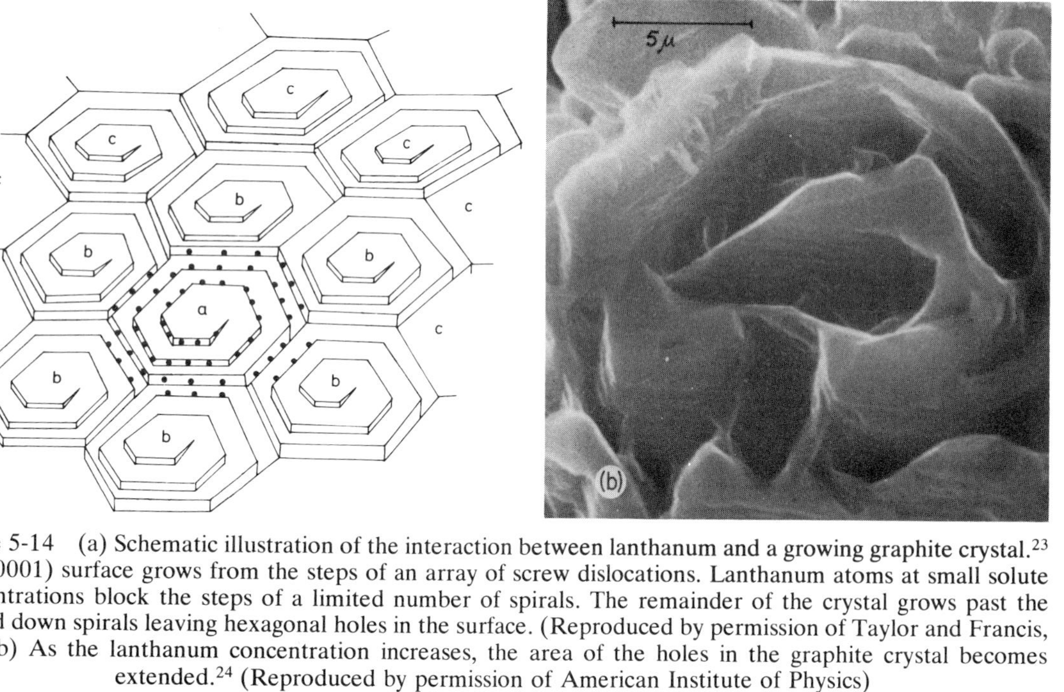

Figure 5-14 (a) Schematic illustration of the interaction between lanthanum and a growing graphite crystal.[23] The (0001) surface grows from the steps of an array of screw dislocations. Lanthanum atoms at small solute concentrations block the steps of a limited number of spirals. The remainder of the crystal grows past the slowed down spirals leaving hexagonal holes in the surface. (Reproduced by permission of Taylor and Francis, Ltd. (b) As the lanthanum concentration increases, the area of the holes in the graphite crystal becomes extended.[24] (Reproduced by permission of American Institute of Physics)

ning electron micrograph of an extracted crystal. The holes correspond to the surface depressions seen in Figure 5-13(a). An enlargement of a hole (Figure 5-13c) shows steps in the hole interior formed by the growth process of the crystal. That the holes are due to a real interaction in growth of the crystal surface with impurities is seen in the micrograph of Figure 5-13(a). They are observed as an irregular surface of the flake and intrusion of metal into the flake structure. Therefore the holes are not formed by the extraction process in acid and are genuinely due to an interaction in growth.

The suggested mode of interaction between the reactive impurities and the growing graphite crystal surface is shown in Figure 5-14.[23] The (0001) graphite surface grows from an array of screw dislocations and the steps traverse the surface. In the initial concentrations of impurity addition, there is insufficient La to completely interact with the total surface and therefore the action is local. Where the local adsorption is sufficient, step advance is impeded so that these parts of the crystal surface advance at a slower rate. They become shielded from the diffusion flux of carbon so that eventually they cease to grow and leave holes, the crystal growing by them.

Figure 5-14(b) shows that, with increasing La content, the area of holes becomes extended. The effect therefore of La is to hinder growth. At a sufficient impurity coverage, undercooling must occur if growth is to proceed further. The growth forms which then appear are appropriate to instability mechanisms operating at the new growth temperature.

Kinetic and Constitutional Undercooling Effects in Graphite Growth

In papers published by Minkoff,[23] Munitz and Minkoff,[16] and Minkoff and Lux,[27] the different ways were studied in which graphite becomes unstable and the growth forms in cast iron metallurgy are created. The two major influences on the undercooling of the graphite crystal in growth are kinetic and constitutional supercooling. The elements added to cast iron for the former influence are reactive, e.g. Mg, Ce. The element mainly affecting constitutional supercooling is Si. Impurity elements like Pb also form distinct boundary layers. The presence of the element S in solution may have pronounced effects. Quantitative information has been presented of the influence of the elements, and a table prepared of growth forms related to undercooling[16] (see Figure 5-16a).

Constitutional supercooling of the graphite surface results from the boundary layer of solute pushed ahead of the crystal as growth proceeds. From observations made[16] of the effect on growth of different solutes in varying concentrations, constitutional supercooling affects graphite growth in the following manner:

(a) The solute boundary layer unstabilizes steps on the graphite (0001) surface.
(b) As the constitutional supercooling increases, the (0001) surfaces become unstable forming pyramidal hillocks.

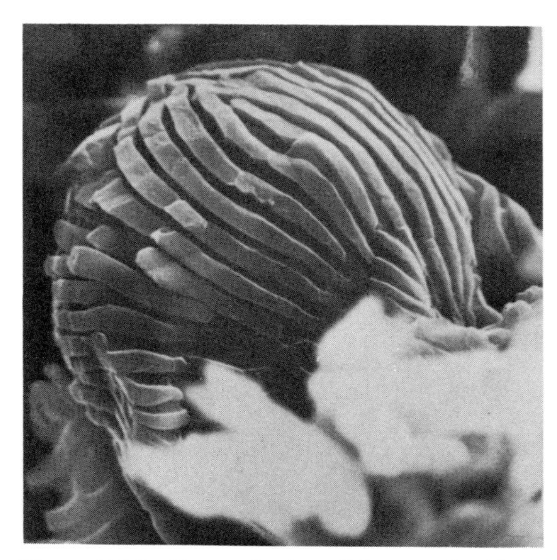

Figure 5-15 Type of spherulitic crystal with distinct ribbed structure resulting from step unstable growth in Ni–C–B alloys (× 1,485). (From Minkoff and Lux[27]) (Reproduced by permission from Pergamon Press Ltd.)

Experiments on the combined effects of constitutional and kinetic undercooling using alloys having Pb and La contents gave a range of growth forms of graphite so that it was possible to correlate the forms with the alloy content. The final forms observed were pyramidal (see Figure 7-9c). Hence graphite forms commence with the plate or flake, the steps become unstable, and both rod and step unstable spherulitic forms appear. With increasing interface undercooling, the range of forms is extended.

The rod or spherulitic forms which develop by steps becoming unstable were referred to in Section 1.4. The rod form when growing in a eutectic manner has been called coral graphite (see Figure 7-2b). The step unstable spherulite (Figure 5-15) is distinguished from normally observed spherulites by the growth pattern of steps on its surface. It has been observed to grow under conditions of a constitutionally supercooled interface, in particular where experiments were conducted in Ni–C systems with high additions of the element boron.[27] Lanthanum additions to melts give the usually observed spherulitic forms while combined additions of elements contributing both kinetic and constitutional components to the undercooling lead to the elongated pyramids (see Figure 1-15g).

Undercooling and Spherulitic Graphite Forms

Calculations were made of the undercooling for obtaining different graphite forms, related to kinetic and constitutional undercooling.[16] Particular interest was attached to the calculated undercooling for obtaining spherulitic forms.

Kinetic Undercooling The kinetic undercooling was calculated from a relationship due to the Cabrera theory,[22] assuming that a network of adsorbed atoms slows the rate of step advance. The spacing between the adsorbed atoms determines the undercooling required for growth since this determines the diameter of the critical nucleus.

The value of the critical radius is then given approximately by:

$$\rho_c = \tfrac{1}{2}\sqrt[3]{C_s}$$

where C_s is the concentration of impurity atoms in the solid and is equal to kC_L, where C_L is the concentration of impurity atoms in the liquid. The value of the distribution coefficient k for lanthanum between graphite and liquid Ni–C alloy is given in Table 5-1 and does not appear to differ greatly from values obtained for Fe–C alloys.[26]

The radius of the critical nucleus is inversely proportional to the undercooling. Knowing the radius of the critical nucleus ρ_1 for one undercooling ΔT_1, the undercooling can be calculated for some other radius ρ_2. This is fixed by the impurity adsorption and is related by the distribution coefficient k to the concentration of impurity in the liquid alloy. The value of ΔT_2 is given by $\Delta T_1\rho_1/\rho_2$. The value of r_1 at $\Delta T = 4$ °C was calculated to be 20 nm.[16]

Constitutional Supercooling A measure of constitutional supercooling at a growing crystal surface may be taken as the maximum value of the difference between the constitutional supercooling gradient G_{CS} and the actual temperature gradient G existing in the liquid.[16] The value of ΔT as a function of distance x from the interface is given by the following:

$$\Delta T = \left\{\frac{mC_0}{k_0}(1 - k_0)[1 - \exp(-Rx/D)]\right\} - Gx \qquad (5\text{-}9)$$

where C_0 is the composition of the original alloy and k_0 is the equilibrium distribution coefficient.

In the experiments reported,[16] m is the slope of the binary liquidus in the systems between Ni or Fe and the added element and R is the growth rate of the graphite–liquid interface. It is assumed that R is the initial growth rate of the planar interface which then becomes unstable, giving the different growth forms. The value of x at ΔT_{max} is given by

$$x = -\frac{D}{R}\ln\left[\frac{Gk_0D}{mC_0(1 - k_0)R}\right] \qquad (5\text{-}10)$$

The value of ΔT_{max} is found by substituting (5-10) in (5-9).

Correlation Between Growth Forms of Graphite and Undercooling

Figure 5-16(a) shows a suggested correlation between the different growth forms observed in graphite and the undercooling ΔT, calculated from kinetic

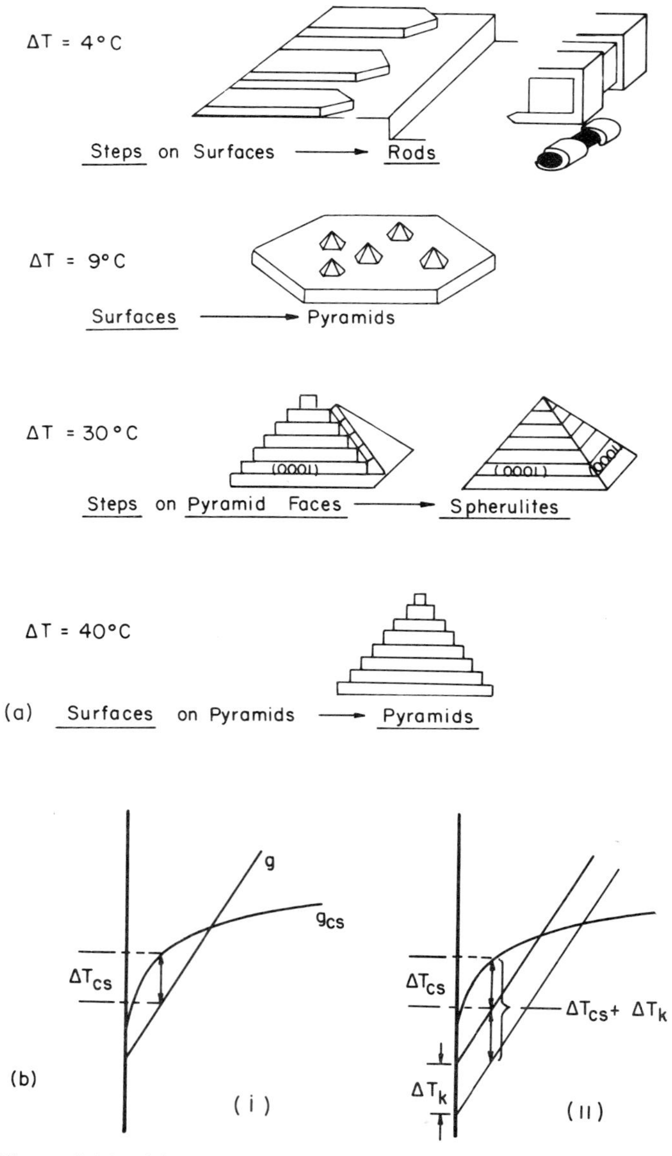

Figure 5-16 (a) Correlation between the different types of instability observed in graphite growth and the growth morphologies. The different instability phenomena are related to different degrees of undercooling and the growth forms useful in cast iron technology are associated with these different growth temperatures. The growth of a spherulite was correlated with the instability of steps on pyramidal faces. (b) Combination of constitutional (i) and kinetic undercooling components to give total undercooling of the graphite phase in growth (ii)

and constitutional undercooling in the experiments described above.[16] Figure 5-16(b) shows the manner in which the constitutional and kinetic undercooling components are combined to provide an estimate of the total undercooling of the graphite phase in growth.

Figure 5-16(a) gives the estimated values of the undercooling for the different growth forms of graphite observed in cast iron. Each form has its own temperature for growth which must be achieved by correct composition adjustment. Graphite flakes have their own temperature for growth and similarly for coral or rod forms. With an increase of the undercooling, pyramidal instability commences growth on the graphite crystal surfaces.

The final growth form noted for graphite in this diagram is a pyramidal one bounded by $(10\bar{1}1)$ faces (see Figure 7-9c). The spherulitic form precedes this, occupying a range of undercooling approximately calculated to be between 29 and 35 °C. This range of spherulitic growth was noted to correspond to the instability of steps on $(10\bar{1}1)$ pyramidal surfaces of flakes.[16] This step unstable growth on $(10\bar{1}1)$ was suggested as a contributing factor in the development of spherulitic form.

The pyramidal crystals correspond to a higher undercooling and are part of the series of imperfect forms noted particularly in thick-walled spherulitic graphite iron castings (see Section 7-7).

Imperfect Spherulitic Forms

Imperfect spherulitic forms are due to incorrect (kinetic) undercooling and imbalance of impurities influencing constitutional supercooling. The spheruli-

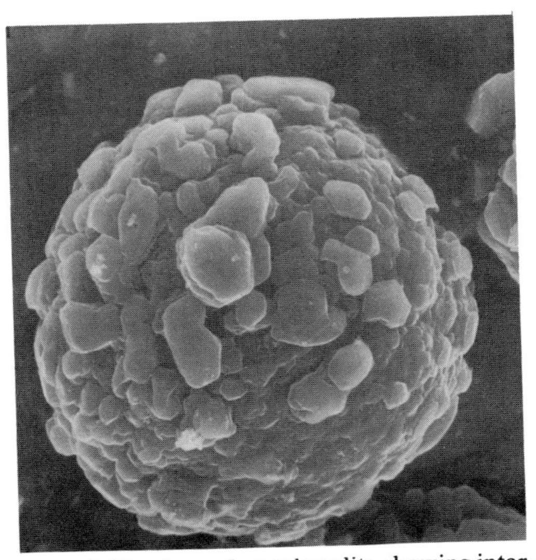

Figure 5-17 Imperfect spherulite showing interface protuberances[16] (× 975)

tic growth range is mainly achieved by the effect of impurities of the kinetic kind. Deficiency of these impurities then leads to intermediate growth (termed vermicular—see Chapter 7) or to growth in which the spherulite is incomplete. Constitutional supercooling impurities at the spherulitic interface perturb the surface and give interface protuberances. An example of this is the effect of Pb seen in Figure 5–17.

The combined effect of kinetic and constitutional impurities, particularly because of segregation, may give elongated pyramidal growth forms.

Segregation and Spherulitic Form; Problem of Section Size

Graphite form is particularly sensitive to the variations of solute content resulting from segregation. The solute variations can be due to microsegregation between γ dendrites or to macrosegregation resulting from the bulk transport of solute to the interior of the casting. The concentration of the element in liquid, C_L, at a fraction of the liquid phase, f_L, can be calculated from the equation of Scheil.[28] If C_0 is the initial concentration of solute and k is the distribution coefficient

$$C_L = C_0 f_L^{k-1}$$

The calculated effects of such segregation of impurity on the graphite form in large castings are described in Chapter 7.

5-6 Spheroidal Graphite Growth Studied by Thermal Measurements

Studies employing cooling curves have been made to examine the solidification sequence of alloys giving spherulitic forms. These have led to quite sophisticated techniques which may be employed as a means of controlling cast structure. These techniques are of importance in commercial production and may enable the melt analysis to be finalized before casting in order to give the desired structure. They are described in Section 5-8.

Studies by Morrogh

In his 1954 paper,[29] Morrogh emphasized concern with the solidification sequence in spherulitic graphite cast iron melts and not with actual mechanisms for formation of a spherulite. Samples of 50 gm of alloy were melted using high purity iron and graphite crucibles. Melting was performed at 1450 °C so that the alloys were hyper-eutectic. The alloy was treated with 1 per cent. NiMg containing 15 per cent Mg to give a spherulitic graphite structure.

The cooling curves were always characterized by an inflection before the eutectic arrest, which occurred over a range of temperature (Figure 5-18). The start of the eutectic arrest seemed to be clearly defined but the end was always indefinite. The eutectic reaction was characterized by Morrogh as the

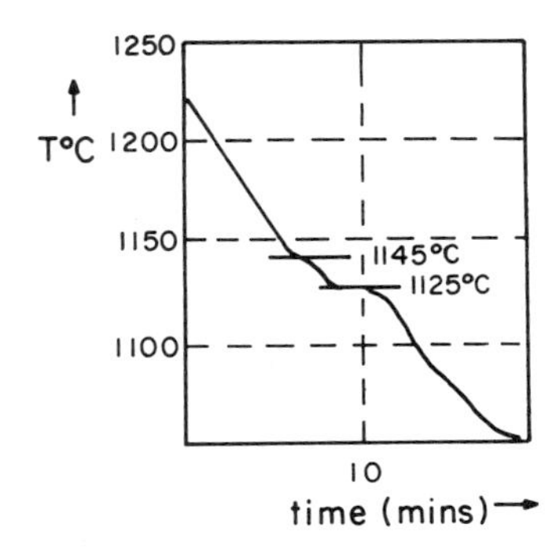

Figure 5-18 Cooling curve for spheroidal graphite cast iron determined by Morrogh.[29] (Reproduced by permission of The Metals Society)

simultaneous growth of preexisting nodules and γ envelopes surrounding the nodules. The nodules had nucleated in the liquid.

At points remote from the hyper-eutectic nodules, or in regions from which these nodules have floated, there is a primary deposition of γ. This continues until the liquid composition becomes sufficiently high in carbon to give a further deposition of spherulites at or near the beginning of the eutectic arrest.

Morrogh suggested that the rate of diffusion of carbon from the liquid through the γ envelopes is the controlling feature of the solidification process. When the cooling rate is so high that the rate of diffusion of carbon in the γ cannot satisfy the transfer required, the remaining liquid will solidify as acicular carbide, which may subsequently decompose. As the γ envelopes increase in thickness and the diffusion distance for carbon increases, the tendency for the remaining liquid to solidify as ledeburite also increases.

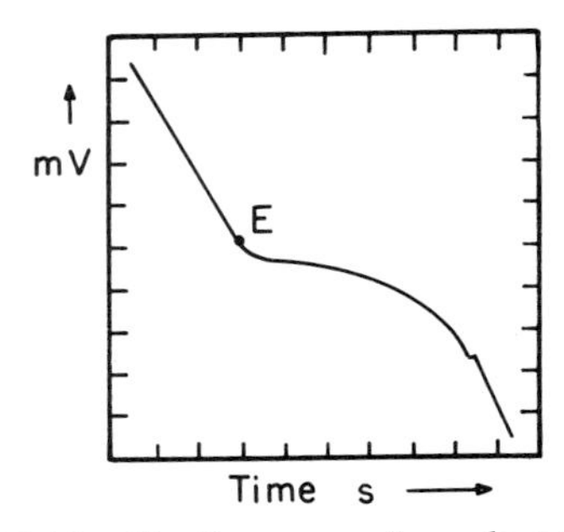

Figure 5-19 Cooling curve for a hyper-eutectic spheroidal graphite cast iron determined by Weterfall, Fredriksson, and Hillert.[30] (Reproduced by permission of the The Metals Society)

Studies by Weterfall, Fredriksson, and Hillert

Weterfall, Fredriksson, and Hillert[30] studied the solidification of three different alloys using carefully controlled techniques of melting and quenching. They determined the start of eutectic solidification by employing microstructural examination of samples quenched at different times. The eutectic start point E was defined by the formation of new graphite nodules. The type of cooling curve for a hyper-eutectic alloy is shown in Figure 5-19. This commences solidification by primary precipitation of graphite and very large graphite nodules are then formed. Below the eutectic temperature, γ is nucleated and forms dendrites, enveloping the primary graphite nodules. New nodules form gradually during the solidification.

Emphasis was placed on studying the formation of the γ shell, the role of which was thought to be incompletely understood in spherulitic graphite growth. According to Weterfall, Fredriksson, and Hillert, the experimental techniques employed removed the uncertainty regarding the formation of this shell. They gave the following scheme of growth in spherulitic graphite solidification:

(a) Growth of spherulite in direct contact with the melt is interrupted when the spherulite gets into contact with a γ dendrite due to flotation.
(b) New spherulites form gradually in the liquid during the whole solidification process. Their growth rate is initially controlled by diffusion of carbon in the liquid phase.
(c) After the formation of a shell of γ, the growth continues to much larger sizes, but with a considerably lower rate.
(d) The growth rate in this stage can be accounted for by diffusion of carbon through the γ shell.

Calculations were made of the growth of a graphite spherulite in the liquid. The equation used was:

$$r^2 = \frac{X^{L/\gamma} - X^{L/Gr}}{X^{Gr} - X^{L/Gr}} \frac{V_m^{Gr}}{V_m^L} D_c^L t$$

The values of composition X have their usual significance and were taken from a calculated Fe–C phase diagram due to Hillert (Chapter 2). The subscripts m refer to molar values and V_m is the molar volume; r is the radius of the spherulite and t is the time for growth.

Calculations were then made of the growth of a graphite spherulite with an austenite shell. The values of the diffusion rate of carbon in the liquid due to Hillert and Lange[31] were employed. It was shown that the growth of a spherulite controlled by diffusion in the melt should be twenty times larger than when controlled by diffusion in γ. The sizes of spherulites achieved in the solidification time intervals conformed more to diffusion in γ than to the liquid state diffusion calculations. This suggested that growth controlled by diffusion through the γ shell was the most important of the two processes.

5-7 Reviews of Growth of Graphite Spherulites from the Liquid

In an extensive review of spherulitic graphite growth, Lux[32,33] presented aspects of the theory of growth of spherulites from the liquid. The emphasis here was placed on the initial formative processes.

A detailed discussion of spherulitic graphite growth from the liquid was given by Schöbel.[34] In this, he dealt at length with the original work on this subject by Scheil and Hutter[35] who had published an analysis in 1953 on spherulitic graphite growth. Pohl, Roos and Scheil[36], had also examined the adsorption of Mg in spherulites. They showed that the magnesium was strongly held in the spherulitic crystal, employing as an experimental technique the distillation of Mg from spherulitic graphite samples and its condensation on a cold finger. Later work by Minkoff and Elroi[25] showed that the quantitative information on adsorption, using this technique, might be erroneous, since metallic pockets rich in Mg are held within the graphite. Schöbel[34] discussed experiments centrifuging spheroidal graphite melts during solidification. The appearance in the microstructure of impinged spherulites suggested growth of the spherulites in the liquid. If this were not the case, say for example they grow within austenite, direct impingement of one spherulite with another could not occur.

5-8 Thermal Measurements as a Means of Control of Spheroidal Graphite Cast Iron Melts

The microstructure of a spheroidal graphite cast iron alloy has two variables which are controlled by processes of nucleation and growth in the liquid.[37] These are:

(a) The departure from spherical geometry of the graphite. This is termed nodularity and is given by the relationship $100 \times (d/D)$, the ratio of dimensions of a nodule.
(b) Nodule count, or number of graphite spherulites per unit area, which is determined by the nucleation rate in post-spheroidizing addition inoculation treatments.

These two variables can be reasonably followed by taking a cooling curve of the iron alloy before casting. The perfection of growth of the spherulite is determined in part by the measurable kinetic undercooling. The nodule count is determined by the rate of nucleation. The form of the cooling curve is determined by the initial undercooling and the recalescence due to the rate of nodule nucleation. This determines the rate of nodule evolution during the transformation of liquid to graphite and γ. Provided the rate of heat extraction from samples is maintained constant over the entire solidification interval, the cooling curves obtained may be compared with standards.

Two different approaches to thermal measurement are given in the following sections.

Solidification Control by Cooling Curves; Research of Loper, Heine, and Chaudhari

The solidification characteristics of hypo- and hyper-eutectic cast irons, investigated using cooling curve methods by Loper, Heine, and Chaudhari[37] were reported to show a strong correlation with microstructure. These techniques demonstrated the possibility of predicting microstructures as cast before pouring into moulds and permitted a corrective action to be taken where necessary.

Figure 5-20 shows a predictive diagram relating undercooling and recalescence with microstructure. Figure 5-20 enables the nodularity and nodule count to be predicted. Nodularity is shown to increase with undercooling but to be followed by a sharp transition to a carbide structure. Recalescence increases with nodule count.

Thermal Analysis Experiments of Bäckerud, Nilsson, and Steen

Bäckerud, Nilsson, and Steen[38] used the rate of recalescence in degrees Celsius per minute and the relative undercooling as a means of discriminating between different graphite morphologies. These quantities are shown in Figure 5-21(a). The rate of recalescence is marked on the cooling curves by dot

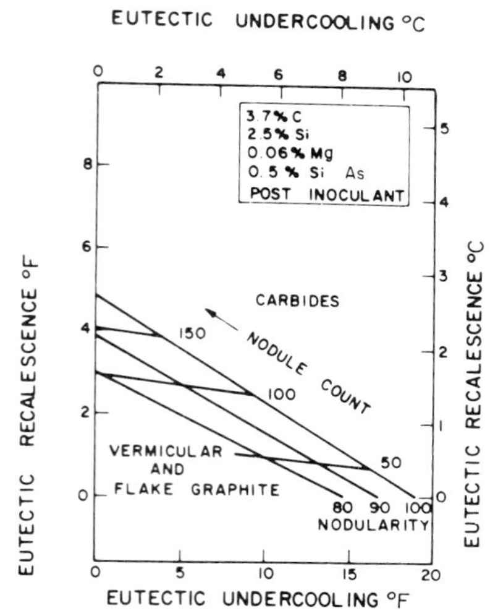

Figure 5.20 Predictive diagram for nodularity and nodule count from Loper Heine, and Chaudhari.[37] The two quantities measured in the melt are eutectic recalescence and eutectic undercooling (Reproduced by permission of Georgi Publishing Co.)

and dashed lines. The characteristic types of cooling curve are noted for flake, spherulite, and intermediate (vermicular) type structures.

The relative undercooling, ΔT is the difference between the maximum undercooling and the peak in the cooling curve (Figure 5-21b). When $dT/d\tau$ was plotted graphically as a function of ΔT, the relationship of Figure 5-21(c) was obtained. The flake and spheroidal graphite structures occurred at the lower part of the curve while vermicular structures occurred towards the upper part. Although the flake and spherulitic values appear close together, it was suggested that spherulitic structures could be predicted with great accuracy. As there was a 15 to 20 °C difference in the absolute growth temperature between these structures there was no difficulty in distinguishing between the morphologies.

5-9 Melt Chemistry and the Periodic System

Spherulitic growth has been presented for the graphite phase as a morphology related to the undercooling. This is determined by the melt chemistry. In the previous sections, the different ideas of growth in the liquid and growth in an austenite envelope were given. Austenite formation is a requirement according to the phase diagram. However, the relationship between chemistry of the melt and how it influences graphite growth from the liquid are by far the most important factors to be considered.

The important subject is melt chemistry and the adjustment of composition necessary to produce the required structures. This must take into account the role of the elements collectively, i.e. their relationship to the undercooling of graphite and individually in their interactions with one another. Section size and segregation enter into melt chemistry.

Other research[32,33] has reviewed the location of elements in the periodic system and their role in graphite growth. A simplified classification is as follows:

(a) Elements of groups I, II, and III, lanthanum, and rare earth elements which are strongly reactive to graphite. This category of elements strongly undercools graphite growth and might be classified by reactivity.

(b) Elements of group IV and boron. These elements form a boundary layer and influence graphite growth by constitutional supercooling.

(c) Elements of group VI. Most of the observations have been made for the element S. In small concentrations, this appears to influence graphite growth by adsorbing at the graphite melt interface. The bonding is by weak van der Waals forces and a decrease of interface energy results. This is predicted to change the temperature of $(10\bar{1}0)$ nucleated growth for a graphite flake, moving it closer to equilibrium. In concentrations larger than required for a mono-layer of adsorbed atoms, the possibility exists of a constitutional supercooling effect. The role of other elements, e.g. Se and Te, is not definite, but Minkoff and Lux[27] observed unusual

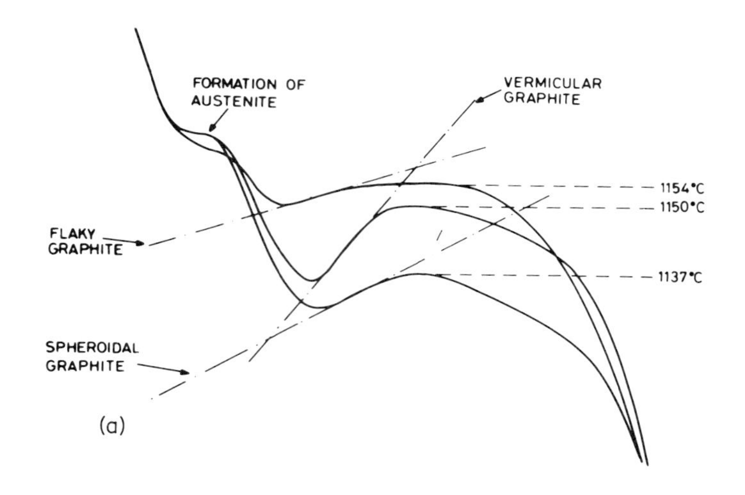
FORMATION OF
AUSTENITE
VERMICULAR
GRAPHITE
1154°C
1150°C
FLAKY
GRAPHITE
1137°C
SPHEROIDAL
GRAPHITE
(a)

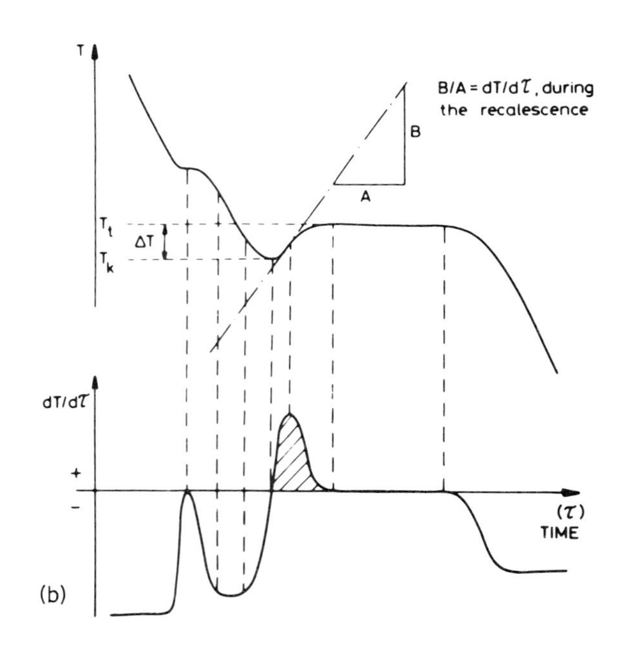
T
B/A = dT/dτ, during
the recalescence
B
A
T_t
ΔT
T_k
dT/dτ
+
-
(τ)
TIME
(b)

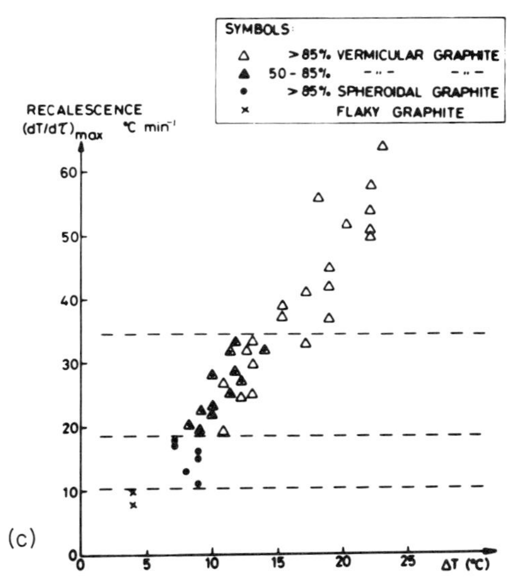

Figure 5-21 (a) Types of cooling curve recorded by Bäckerud, Nilsson, and Steen[38] to show the differences in rate of recalescence (dot and dashed curves) and undercooling for flake graphite, and vermicular and spheroidal graphite cast irons. (b) The rate of recalescence $dT/d\tau$ and the undercooling ΔT. (c) Classification of structures by graphical representation of the rate of recalescence and undercooling ΔT (Reproduced by permission of Georgi Publishing Co)

graphite growth in cast iron containing Se, suggesting an adsorption effect.

'Delta Tee' Method for Structure Control of Cast Iron

This method has been proposed for cast iron by Minkoff[39]. The control of structure in cast iron is difficult because of the sensitivity of graphite growth to small compositional changes. The methods described employing differential thermal analysis seem satisfactory but might only give a 'coarse' control. These methods can tell whether the structure will be in the correct range, but cannot give information on perturbations of structure which have important influences on properties. In making special types of cast iron, including spheroidal graphite iron in large section thicknesses (greater than 100 mm) further problems arise due to segregation. These cannot be sensed by current control methods.

A computational method is proposed for the control of cast iron structure over the entire range of possible graphite forms. The method is based primarily on calculation of the undercooling at the graphite–liquid interface, due to solute influences in solution. For this reason, the method is called 'delta tee' (ΔT) to signify that the structures obtained in cast iron are a direct function of the undercooling at the interface in the conditions under which graphite grows. After analysis of the melt at melt-down, the analytic data are fed into the computer. This processes the data to give corrected analysis of the melt as required for the given structure and properties at the stated casting cross-section. The procedure corrects for melt interactions in solidification and segregation. It provides calculation of any additions to be made, or it may reject the melt-down composition as being unsuited for the type of casting being made. The methods allows control of flake, rod, chunky, and spheroidal graphite cast irons.

References

1. Morrogh, H., and W. J. Williams: *J. Iron Steel Inst. London*, **155**, 321, 1947.
2. Morrogh, H., and W. J. Williams: *J. Iron Steel Inst. London*, **158**, 306, 1948.
3. Herring, C.: *Structure and Properties of Solid Surfaces* (Eds. R. Gomer, R. and C. S. Smith), Univ. of Chicago Press, 1953.
4. Fullman, R. L.: *Acta Metall.*, **5**, 638, 1957.
5. Buttner, F. H., H. F. Taylor, and J. Wulff: *Amer. Foundryman*, **20**, 49, 1951.
6. Keverian, J., H. F. Taylor, and J. Wulff: *Amer. Foundryman,* **23**, 85, 1953.
7. Geilenberg, H.: *Recent Research on Cast Iron* (Ed. H. Merchant), Gordon and Breach, New York, 1968.
8. Hillert, M., and Y. Lindblom: *J. Iron Steel Inst. London*, **176**, 388, 1954.
9. Minkoff, I.: *The Solidification of Metals*, Iron Steel Inst. Publ. 110, London, 1968.
10. Double, D. D., and A. Hellawell: *The Metallurgy of Cast Iron* (Eds. B. Lux, F. Mollard, and I. Minkoff), Georgi Publ. Co., Switzerland, 1975.
11. Double, D. D., and A. Hellawell: *Acta Metall.*, **17**, 1071, 1969.
12. Double, D. D., and A. Hellawell: *Acta Metall.*, **22**, 481, 1974.
13. Tsuchikura, H., T. Kusakawa, and T. Okumoto: *Proc. Int. Conf. Electr. Micr.*, London, 1954.
14. Hunter, M. J., and G. A. Chadwick: *J. Iron Steel Inst. London*, **210**, 707, 1972.
15. Hunter, M. J.: *J. Iron Steel Inst. London*, **211** 85, 1973.
16. Munitz, A., and I. Minkoff: *Int. Foundry Congress*, Budapest Congress Papers 45th, Paper 32, 1978.
17. Shubnikov, A. V.: *Sov. Phys. Crystallog. (Engl. Transl.)*, **2**, 578, 1957.
18. Oldfield, W., G. T. Geering, and W. A. Tiller: *The Solidification of Metals* Iron Steel Inst. Publ. 110, London, 1968.
19. Minkoff, I., and A. Goldis: *Int. Foundry Congress*, Zurich Congress Papers 27th, Paper 21, 1960.
20. Chernov, A. A.: *Sov. Phys. Usp. (Engl. Transl.)*, **4**, 116, 1961.
21. Chernov, A. A.: *Adsorption et Croissance Crystalline*, Centre Nat. Rech. Sci. Paris, No. 152, 265, 1965.
22. Cabrera, N., and D. A. Vermilyea: *Growth and Perfection of Crystals*, p. 393, John Wiley, New York, 1958.
23. Minkoff, I.: *Philos. Mag. London*, **12**, 1083, 1965.
24. Minkoff, I., and W. C. Nixon: *J. Appl. Phys.*, **37**, 4848, 1966.
25. Minkoff, I., and D. Elroi: *Studies of Graphite Growth from Melts*, Technion Res. and Dev. Found. Haifa, June 1964.
26. Oron, M., and I. Minkoff: *Recent Research on Cast Iron* (Ed. H. Merchant), Gordon and Breach, New York, 1968.
27. Minkoff, I., and B. Lux: *Micron* **2**, 282, 1971
28. Flemings, M. C.: *Solidification Processing*, McGraw-Hill, New York, 1974.
29. Morrogh, H.: *J. Iron Steel Inst. London*, **176**, 378, 1954.

30. Weterfall, S. E., H. Fredriksson, and M. Hillert: *J. Iron Steel Inst. London*, **210**, 323, 1972.
31. Hillert, M., and N. Lange: *J. Iron Steel Inst. London*, **203**, 273, 1965.
32. Lux, B.: *Cast Met. Res. J.*, **18**, 25, 1972.
33. Lux, B.: *Cast Met. Res. J.*, **18**, 49, 1972.
34. Schöbel, J. D.: *Recent Research On Cast Iron* (Ed. H. Merchant), Gordon and Breach, New York, 1968.
35. Scheil, E., and L. Hutter: *Arch. Eisenhuettenw.*, **4**, 24, 1953.
36. Pohl, D., E. Roos and E. Scheil. *Int. Foundry Congress Madrid, Congress Papers* 26th, Paper C-14, 1959.
37. Loper, C. R., R. W. Heine, and M. D. Chaudhari: *The Metallurgy of Cast Iron* (Eds. B. Lux, F. Mollard, and I. Minkoff, Georgi Publ. Co., Switzerland, 1975.
38. Bäckerud, L., K. Nilsson, and H. Steen: *The Metallurgy of Cast iron* (Eds. B. Lux, F. Mollard, and I. Minkoff), Georgi Publ. Co., Switzerland, 1975.
39. Minkoff, I.: *Delta Tee*, Technion Res. and Dev. Assoc., 1981.

Bibliography

Doremus, R. H. (Ed.): *Growth and Perfection of Crystals*, John Wiley, 1958.

Karsay, S.: *Ductile Iron*, Vols. I and II, Quebec Iron and Titanium Corp., 1976.

Parker, R. L.: *Crystal Growth Mechanisms. Energetics, Kinetics and Transport*, Vol. 25, Solid State Physics, Academic Press, 1970.

Chapter 6
Liquid Metal Preparation

The subject matter of this chapter is the melting process in liquid iron preparation and the influence of process variables on the chemistry of the alloy. Some of the important studies of the thermodynamics of reactions in cast iron melting are described. These are centred initially on the oxygen equilibrium and the contribution of the separate C–O and Si–O equilibria. Oxygen equilibrium plays an important role in nucleation and, as discussed later in the chapter, in desulphurization in basic line cupolas.

Nitrogen adsorption in melting processes is described and the influence of sulphur on the kinetics of gas dissolution. Variations in nitrogen solubility between iron melted in a cupola and iron melted in an induction furnace are discussed, and relationships given between properties of cast iron melted in these units.

Some treatment is given of carbon solution kinetics in induction melting and in electric arc furnaces. The development of melting units is discussed and also the treatment of liquid metal to prepare material with greater purity for spheroidal graphite iron production.

6-1 Oxygen Adsorption in Iron Melting; The Iron–Oxygen–Carbon–Silicon System

This system has been investigated in different researches[1,2] and has been used in theories of nucleation of graphite.[3]

In the analysis of Östberg[1] the oxygen solubility in cast iron was estimated using information relating to the equilibria between oxygen, carbon, and silicon in molten iron alloys. He used available data on Fe–C–Si, Fe–O–Si, and Fe–O–C, and used activity coefficients based on formulae suggested by Chipman.[4] In the experimental work, the dissolved oxygen content in a malleable iron producing operation was examined at different stages including casting. The iron was obtained from a duplex process melting unit consisting of an acid lined hot blast cupola and air furnace. Samples were taken from the furnace, ladle, and mould. The differences at each stage were examined in relation to reactions with the atmosphere, with the mould, and internally.

Thermodynamic Basis for Oxygen Adsorption in Cast Iron

Östberg took CO or SiO_2 as the saturation phase with the following reactions:

$$C + O = CO \qquad K_{CO} = \frac{a_{CO}}{a_C a_O}$$

$$Si + 2O = SiO_2 \qquad K_{SiO_2} = \frac{a_{Si}}{a_{Si} a_{O_2}}$$

In the calculations, a_{CO} and a_{SiO_2} were assumed to be unity and represent the standard states of CO at atmospheric pressure and pure SiO_2 respectively.

The activity of oxygen a_O can then be expressed:

CO equilibrium: $a_O = \dfrac{1}{K_{CO} a_C}$

SiO_2 equilibrium: $a_O = \dfrac{1}{(K_{SiO_2} a_{Si})^{1/2}}$

Chipman's approximation[4] of the activity coefficient was used:

$$f_O = f_O^O f_O^C f_O^{Si}$$

$$\log f_O = \log f_O^O + \log f_O^C + \log f_O^{Si}$$

where $f_O^O f_O^C f_O^{Si}$ are the partial activity coefficients of oxygen in Fe–O, Fe–C, Fe–Si solutions respectively; f_O^O is considered equal to unity while f_O^C is given as follows:

$$\log f_O^C = e_C^C c_C$$

where c_C = carbon content
e_C^C = an interaction coefficient

To take into account the effect of the third element, Si, on carbon in solution, Östberg used for c_C the carbon content which in the system Fe–O–C would have the same carbon activity as in the system Fe–O–C–Si being considered. This was called the effective carbon content.

Figure 6-1(a) shows the results of calculations for three alloys which are given on the graph as follows:

I: 2% Si–1% C
II: 1% Si–3% C
III: 2% Si–3% C

Line a is the oxygen solubility in Fe–O–Si at 1 per cent silicon. Line b is the oxygen solubility in Fe–O–C at 2–3 per cent carbon. The dashed lines are extensions of the Si–O and C–O equilibria. The full lines show the calculated oxygen solubilities for the different alloys. Above 1350–1400 °C according to Östberg's determinations the CO reaction determines the solubility of oxygen at equilibrium conditions and is not very temperature dependent. However, Neumann and Dötsch[3] have shown that this reaction does not reach equilib-

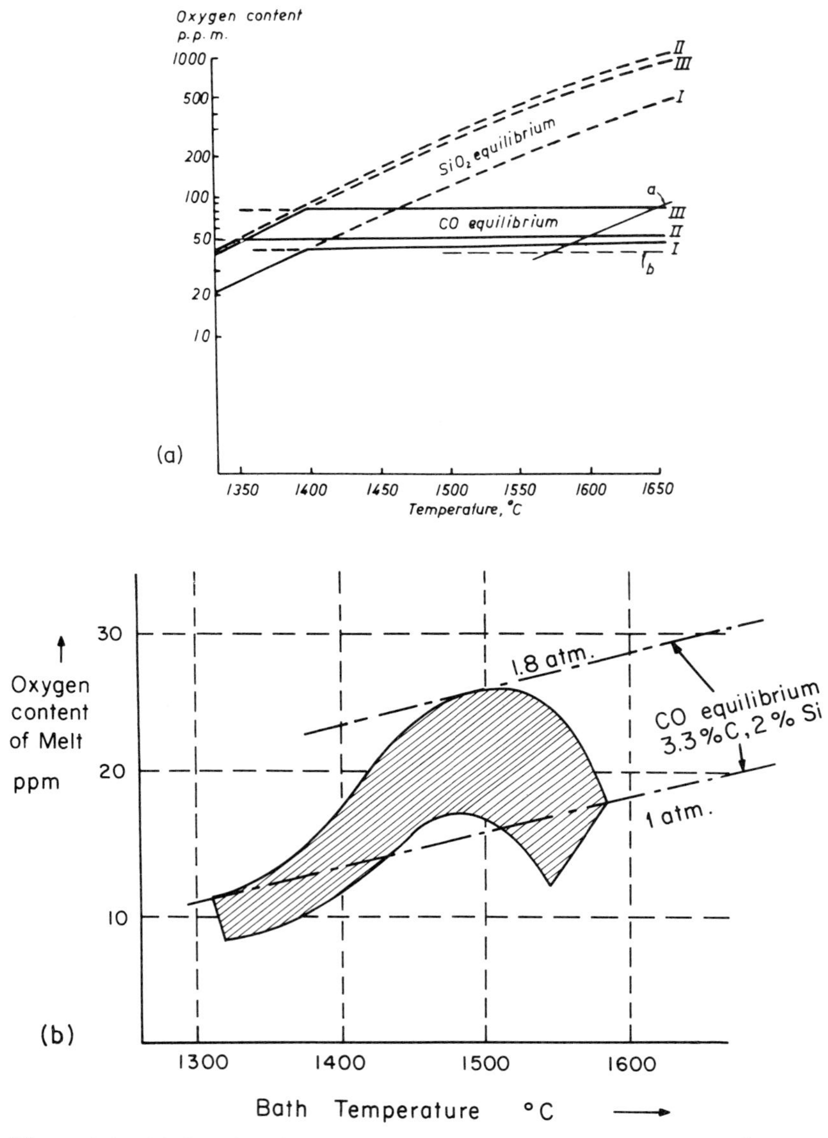

Figure 6-1 (a) Graphs of oxygen solubility in Fe–C–Si alloys due to Östberg.[1] The full lines show the calculated oxygen solubilities. The dashed lines are extensions of the Si–O and C–O equilibria. The calculations were made for three alloys—I: 2% Si, 1% C, II: 1% Si, 3% C, and III: 2% Si, 3% C. Line a is the oxygen solubility in Fe–O–Si at 1% Si. Line b is the oxygen solubility in Fe–O–C at 2–3% C. (Reproduced by permission of The Metallurgical Society of AIME.) (b) Graph of oxygen solubility in Fe–C–Si alloys due to Neumann and Dötsch.[3] Between 1400 and 1450 °C, the Si–O equilibrium is followed. At a higher temperature the C–O equilibrium is once again followed. (Reproduced by permission of Georgi Publishing Co.)

rium, and the Si–O equilibrium determines the oxygen content in the given temperature range. Similar results were shown by Katz and Rezeau[5] and are referred to in Section 6-3.

Oxygen Behaviour in Melting of Cast Iron and the Nucleation of Graphite

Experimental observations of cast iron melting and inoculation have shown that oxide formation plays a role in nucleation of graphite. Neumann and Dötsch[3] made a thermodynamic study of the Fe–C–Si–O system with a view to examining the possible role played by the oxygen content of cast iron in nucleation. Their graph for the oxygen content of liquid cast iron alloys in dependence on temperature is shown in Figure 6-1(b)

The graph of Neumann and Dötsch demonstrates that between 1400 and 1450 °C the Si–O equilibrium is followed and with a corresponding increase of the oxygen content. At a higher temperature, the C–O equilibrium is once again followed so that the melt oxygen becomes lowered.

Neumann and Dötsch were interested in defining temperature regimes which resulted in maximum oxygen absorption by the melt and the variation of this oxygen content as a function of temperature. The understanding of such regimes were important in relating melting practice to nucleation.

6-2 Nitrogen in Cast Iron

Nitrogen is absorbed into iron both in the solid and in the liquid state. The effect of nitrogen on cast iron is marked. It can act as a pearlite stabilizer and lead to complete suppression of the α phase. In heavy sections of gray cast iron, it has been noted[6] that high nitrogen levels tend to promote the formation of a compacted form of graphite. It may possibly act as an influence in promoting white iron structures.

The net effect is that increasing the nitrogen content of gray iron can have a significant effect on the tensile strength (as for example is noted in comparison of properties between cupola and induction melted iron).

Nitrogen Solubility in Carbon Saturated Liquid Iron Alloys

Opravil and Pehlke[7] studied the solution of nitrogen in carbon-saturated iron alloys. As indicated by the small magnitude of the enthalpy of solution, the nitrogen solubility is only slightly dependent on the temperature. For the influence of alloying elements on the solubility of nitrogen, the following interaction coefficient was introduced:

$$e_{\mathrm{N}}^{j} = \frac{\delta/\delta N \log f_{\mathrm{N}}}{\delta j/\delta N}$$

where f_{N} is the activity coefficient of nitrogen in iron and j represents the alloy element addition. A positive value of e_{N}^{j} is indicative of a decrease in nitrogen solubility. Log f_N can be evaluated in alloy systems, when expressed in the form of a Taylor series expansion in the method due to Wagner[8] (Chapter 4-2). It is given in this chapter for the Fe–C–Si system.

Opravil and Pehlke also referred to the method of Chipman and Corrigan.[4] These authors proposed a semi-empirical method for predicting nitrogen solubility in multicomponent alloys of steel. They considered expressions which included the heats of solution as well as dilute solute interactions.

Experimental Results for the Iron–Carbon–Nitrogen System

Opravil and Pehlke[7] used a Sieverts technique and the following relationships were employed:

$$\tfrac{1}{2}N_2 = \underline{N}$$

$$K = \frac{a_{\underline{N}}}{(pN_2)^{1/2}} = \frac{\%\underline{N} f_N}{(pN_2)^{1/2}}$$

where a_N = activity of nitrogen
f_N = nitrogen activity coefficient
$\%\underline{N}$ = weight per cent. of nitrogen dissolved in the liquid alloys
pN_2 = partial pressure of nitrogen gas in equilibrium with the liquid metal expressed in atmospheres

For convenience in handling the data, the standard state for nitrogen in solution was selected as the infinitely dilute solution of nitrogen in iron which is saturated with carbon. Then:

$$K = \%N/(pN_2)^{1/2} = \%N \text{ at one atmosphere of nitrogen pressure}$$

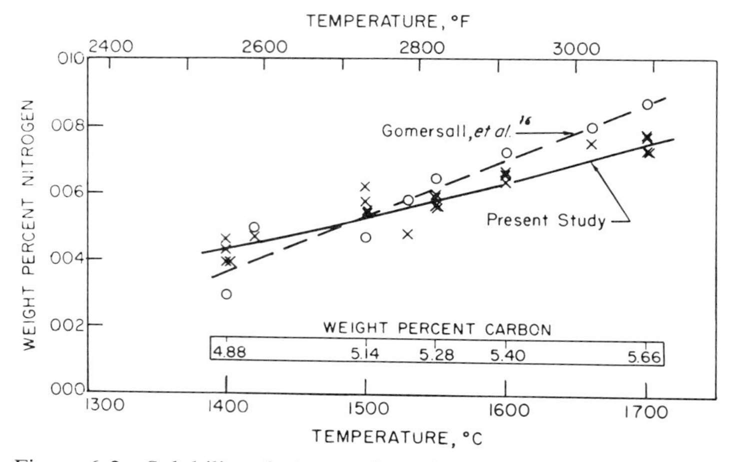

Figure 6-2 Solubility of nitrogen in carbon saturated iron at 0.1 N mm^{-2} pressure. The carbon contents are calculated.[7] (Reproduced by permission of American Foundrymen's Society)

The solubility data determined are given graphically in Figure 6-2. This shows both the increase of carbon solubility and the increase of nitrogen with temperature. The increase of nitrogen solubility is a true temperature effect. It may be shown that increasing carbon actually has the effect of decreasing nitrogen solubility. The graph shows the net effect of the temperature increase and the carbon decrease.

Kinetics of Nitrogen Adsorption in Cast Iron

Figure 6-3 shows the curve obtained for nitrogen solution in saturated Fe–C at 1400 °C. The initial sharp rise is due to the transient on introducing the nitrogen gas. After adjusting the pressure to 1 atmosphere, the pressure was maintained constant and adsorption proceeded until equilibrium was reached.

If C_e is the equilibrium concentration of nitrogen in the melt at the prevailing gas pressure and C is the concentration at any time t, the slope of the solubility line is the negative of the apparent reaction rate constant:

$$\frac{d[\log(C_e - C)]}{dt} = -k'$$

This constant has the dimensions s^{-1} and the concentration is expressed in wt per cent. The relatively high value of k' suggests that N_2 solution in carbon saturated iron is diffusion controlled. In the experiments reported using clean materials, equilibrium was achieved in about 10 minutes.

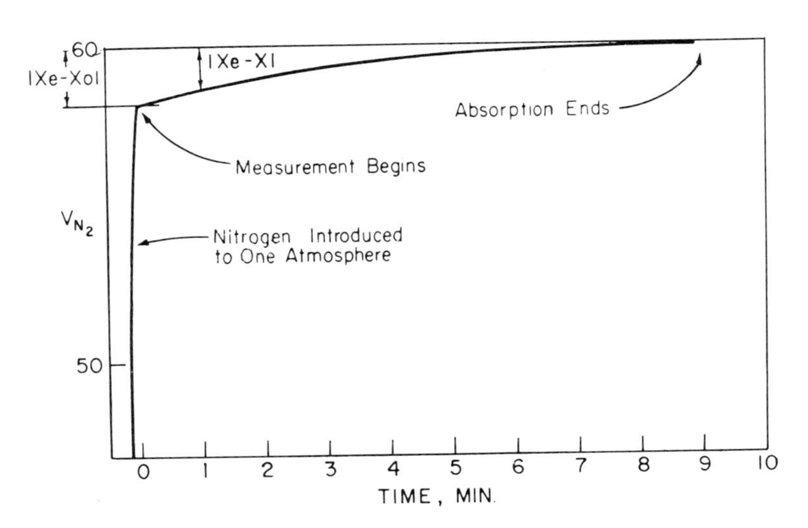

Figure 6-3 Experimental curve for solution of nitrogen in carbon saturated Fe–C alloy at 1673 K. Equilibrium is achieved in about 10 minutes.[7] (Reproduced by permission of American Foundrymen's Society)

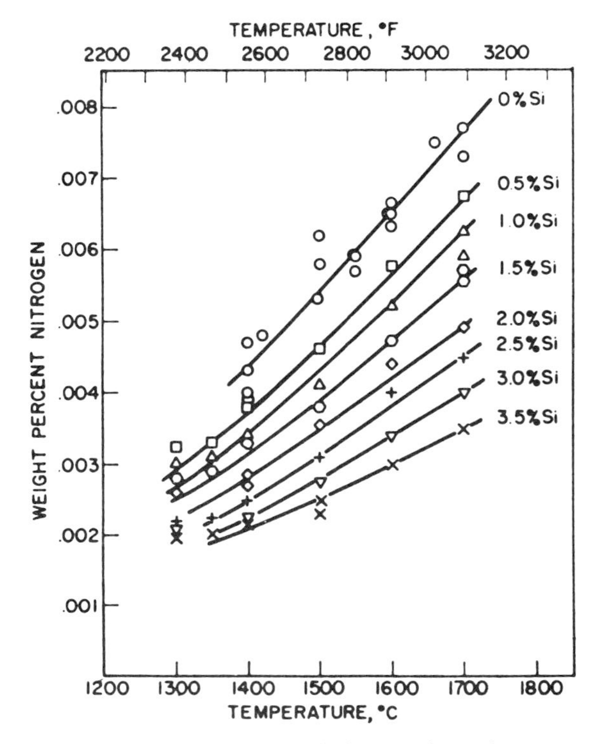

Figure 6-4 The solubility of nitrogen in carbon saturated Fe–Si alloys at 0.1 N mm^{-2} N_2 pressure. Silicon markedly decreases the nitrogen solubility.[7] (Reproduced by permission of American Foundrymen's Society)

The Solubility of Nitrogen in Liquid Iron–Carbon–Silicon Alloys

Figure 6-4 shows data for the solubility of nitrogen in the temperature range 1300–1700 °C for carbon saturated alloys having varying silicon contents.[7] It is noted that silicon markedly decreases the solubility of nitrogen in carbon saturated iron alloys. Since there is also an effect of carbon solubility with temperature tending to decrease the nitrogen solubility, the influence of silicon must be even more pronounced than the graph shows.

The rate coefficient is found to increase very slightly with Si content.

Evaluation of the Solubility of Nitrogen in Iron–Carbon–Silicon

Uda and Pehlke[9] evaluated the nitrogen solubility data for alloys in the range 0–4 weight per cent of silicon and 0–4 weight per cent of carbon. The temperature range explored was 1450–1750 °C.

The equilibrium constant K evaluation for Fe–C–N[7] is independent of pressure and hence f_N is not affected by pressure in the range considered. The choice of an infinitely dilute solution of nitrogen in pure iron as the reference state implied that f_N, the activity coefficient of nitrogen in pure iron, is equal to 1.

The standard free energy of solution of nitrogen in iron is

$$\Delta F^\circ = \mathrm{RT} \ln K = \Delta H^\circ - T\,\Delta S^\circ$$

where ΔH° = standard enthalpy of solution of N_2 in calories per gram-atom

ΔS° = standard entropy of solution of N_2 in calories per gram-atom degree Kelvin at T (degree Kelvin)

K as a function of temperature is then expressed by the Van't Hoff equation

$$\log K = -\frac{\Delta H^\circ}{4.575T} + \frac{\Delta S^\circ}{4.575}$$

Since ΔH and ΔS can be taken as constant over reasonably short intervals of temperature:

$$\log K = \frac{A}{T} + B$$

where $A = -\dfrac{\Delta H^\circ}{4.575T}$ and $B = \dfrac{\Delta S^\circ}{4.575}$

For pure liquid iron, the solubility was measured to be:

$$\log(\text{wt }\%\underline{\text{N}})_{1\text{ atm N}_2} = -(306/T\ ^\circ\text{K}) - 1.20$$

For an alloy of iron with 2 per cent. silicon the nitrogen solubility was:

$$\log(\text{wt }\%\underline{\text{N}})_{1\text{ atm N}_2} = -(572/T\ ^\circ\text{K}) - 1.178$$

For Fe–C–Si it was confirmed that $\log f_N$ was directly proportional to the increase of Si content up to about 4 weight per cent. of silicon.

Therefore the value of the interaction parameter e_N^{Si} could be obtained at various temperatures. The following definitions were made:

$$e_N^{Si} = \frac{\delta \log f_N}{\delta\ \%\text{Si}}\ _{\%\text{Fe}\to 100}$$

$$f_N = \frac{\%\underline{\text{N}}(\text{pure Fe})}{\%\underline{\text{N}}(\text{Fe} - \text{Si})}\ _{p\text{N}_2, T}$$

The temperature dependence of e_N^{Si} is:

$$e_N^{Si} = \left(\frac{171.4}{T}\right) - 0.0309$$

In the work of Uda and Pehlke,[9] e_N^{Si} at 1600 °C was 0.060.

A similar temperature dependence of the interaction parameters in the molten iron alloy had been reported by Wada, Gunji, and Wada.[10] These authors utilized an equation for the interaction parameter from excess quantities between each component on the basis of the quasi-chemical theory.

The expression of Wagner (Section 4-2) for f_N in multicomponent alloys was given in the form of a Taylor series. For terms up to second order, the alloy system Fe–C–Si gives:

$$\log f_N = \frac{\delta \log f_N}{\delta \%N} \%N + \frac{\delta \log f_N}{\delta \%C} \%C + \frac{\delta \log f_N}{\delta \%Si} \%Si$$
$$+ \frac{1}{2} \frac{\delta^2 \log f_N}{\delta(\%C)^2} (\%C)^2 + \frac{1}{2} \frac{\delta^2 \log f_N}{\delta(\%N)^2} (\%N)^2 + \frac{1}{2} \frac{\delta^2 \log f_N}{\delta(\%Si)^2} (\%Si)^2$$
$$+ \frac{\delta^2 \log f_N}{\delta \%C \, \delta \%N} \%C \, \%N + \frac{\delta^2 \log f_N}{\delta \%Si \, \delta \%N} \%Si \, \%N + \frac{\delta^2 \log f_N}{\delta \%Si \, \delta \%C} \%Si \, \%C$$

The following simplifications were made:

(a) The first and fifth terms = 0. This results from the standard state being considered as the infinite dilute solution of nitrogen in liquid pure iron and from Sievert's law.
(b) The seventh and eighth terms are negligible. This results from the small values of the nitrogen contents in the alloys in comparison with the concentrations of C and Si.
(c) The sixth term is neglible. This is because $\log f_N$ in liquid Fe–Si alloys is proportional to the Si content.

The final result of these simplifications is:

$$\log f_N = e_N^C (\%C) + e_N^{Si} (\%Si) + \gamma_N^C (\%C)^2 + e_N^{Si,C} (\%Si)(\%C)$$

where
$$e_N^i = [\delta \log f_N / \delta (\%i)]_{\%Fe \to 100}$$
$$e_N^{i,j} = [\delta^2 \log f_N / \delta (\%i)\, \delta(\%j)]_{\%Fe \to 100}$$
$$\gamma_N^i = [\delta^2 \log f_N / \delta (\%C)^2]_{\%Fe \to 100}$$

Predictions of Nitrogen Solubility in Iron–Carbon–Silicon

Uda and Pehlke,[9] on the basis of the computations in the preceding sections, gave the following equations for predicting nitrogen solubility in Fe–C–Si systems:

$$\log f_N = (\%C)(280/T) - 0.055 + (\%Si)(171/T) - 0.031$$
$$- 0.0050(\%C)^2 + 0.0037(\%C)(\%Si)$$

$$\log(\%N)_{equi.} = (-306/T - 1.201) + \tfrac{1}{2} \log pN_2$$
$$- [(\%C)(280/2 - 0.055) + \%Si(171/T - 0.031)$$
$$+ 0.0050(\%C) + 0.0037(\%C)(\%Si)]$$

It was reported that the difference in nitrogen solubility between experimentally measured values and those calculated were small, with a maximum difference of 10 ppm.

Evaluation of Effects of Oxygen and Sulphur on Nitrogen Solution in Cast Iron

Pehlke and Elliott[11] had shown that surface active elements, in particular oxygen and sulphur, markedly reduced the rate of solution of nitrogen in liquid iron. Other elements have little effect on the rate except when they function as deoxidizers.

At low O and S, the solution of nitrogen in Fe–C–Si alloys is seen to be controlled by diffusion. In the presence of sulphur, the rate of solution is very much reduced and controlled by a surface reaction. In general, these surface active elements slow down the rate of absorption. An increase of 0.05 weight per cent of sulphur at low S levels was reported to reduce the nitrogen solution rate coefficient by approximately 50 per cent. The total nitrogen solubility is reportedly unchanged but the effect on kinetics can be important, particularly where melted material is pure. This applies to induction furnace melting with all steel charges and to cupola melting where large percentages of steel scrap make up the change.

Pehlke indicated that this may be one of the reasons for observed values of high nitrogen content for some melting conditions. Since nitrogen is also very soluble in solid iron in the γ phase, heating into this temperature range has been suggested as a source of additional nitrogen in the finally melted alloy.

6-3 Study of Desulphurization Variables in Cupolas

An extensive study of the operation of a cupola with respect to the variables influencing desulphurization was undertaken by Katz and Rezeau.[5] They analysed the operation from the standpoint of the thermodynamics of the reactions which occur in this particular melting process. The following equilibria of importance were studied:

Fe–O
C–O
Si–O
Mn–O

In the reactions governing desulphurization, exchange reactions between sulphur and slag occur, followed by deoxidation reactions. The following are the reactions in decreasing order of ability to desulphurize iron:

$$\underline{C} + \underline{S} + CaO = CaS + CO \quad (6\text{-}1)$$

$$\tfrac{1}{2}\underline{Si} + \underline{S} + CaO = CaS + \tfrac{1}{2}SiO_2 \quad (6\text{-}2)$$

$$\underline{Mn} + \underline{S} + CaO = CaS + MnO \quad (6\text{-}3)$$

$$\underline{Fe} + \underline{S} + CaO = CaS + FeO \quad (6\text{-}4)$$

The elements underlined are in solution in the metal while the other materials are in the gas phase or appear in the slag.

The sulphur exchange reaction between Fe and slag is common to reactions (6-1) to (6-4):

$$\underline{S} + CaO = CaS + \tfrac{1}{2}O_2 \qquad (6\text{-}5)$$

The different deoxidation reactions involved in Equations (6-1) to (6-4) are:

$$\underline{C} + \tfrac{1}{2}O_2 = CO \qquad (6\text{-}6)$$

$$\underline{Si} + O_2 = SiO_2 \qquad (6\text{-}7)$$

$$\underline{Mn} + \tfrac{1}{2}O_2 = MnO \qquad (6\text{-}8)$$

$$\underline{Fe} + \tfrac{1}{2}O_2 = FeO \qquad (6\text{-}9)$$

In the cupola, large variations occur in conditions of time, temperature, and progress of reactions, over the entire reaction zone. The interest in the studies made by Katz and Rezeau[5] lies in the analysis of these variations. Assumptions generally made about equilibrium were shown to be incorrect. They showed that the reaction of Equation (6-6), which is the most important for adjusting the oxygen level in iron, does not achieve equilibrium. This correlates with similar observations, e.g. by Neumann and Dötsch.[3]

The slags have also been shown to be heterogeneous, i.e. they contain precipitates which result from conditions of operation below their liquidi temperatures. Control by basicity of the slag must then be sought in terms of higher operating temperatures or by chemical additions to lower slag liquidus temperatures.

With reference to the oxygen content in the iron, or its activity, the investigators showed that the equilibrium between carbon and oxygen. does not control the oxygen activity in the metal. This was strongly dependent on the Si and Mn levels in the iron and the Sio_2, MnO, FeO contents of the slag. Hence, desulphurization is sensitive to the quality of the charge materials.

The actual sulphur levels in the cupola could be compared with theoretical predictions and from these it could be shown how far removed from the possible levels the real levels fall in practice.

The variables influencing desulphurization could also be quantitatively compared. It appeared that the most important variable was the oxygen activity in the iron which accounts for a tenfold variation in desulphurization. The other two variables, the constant for the C–O equilibrium and the slag basicity, each affect the desulphurization by roughly a factor of 3. Since the oxygen activity is reduced by its dependence on the Si and Mn levels, desulphurization in the cupola becomes less efficient than that possible.

Thermochemistry Details

Details of the processes occurring in cupolas could be followed by studying the equilibria related to the reactions in Equations (6-1) to (6-4). The activity of oxygen a_{O_2} can be written for each reaction of the series (6-5) to (6-9). This oxygen activity is related to the other activities of the reactants and products of the respective equations as follows:

$$(a_{O_2})_5 = (K_5 a_{CaO} a_S / a_{CaS})^2 \tag{6-10}$$

$$(a_{O_2})_6 = (a_{CO}/K_6 a_{\underline{C}})^2 \tag{6-11}$$

$$(a_{O_2})_7 = (a_{SiO_2}/K_7 a_{\underline{Si}})^2 \tag{6-12}$$

$$(a_{O_2})_8 = (a_{MnO}/K_8 a_{Mn})^2 \tag{6-13}$$

$$(a_{O_2})_9 = (a_{FeO}/K_9 a_{Fe})^2 \tag{6-14}$$

If equilibrium is established for any of the reactions (6-1) to 6-4) then the oxygen activity for the corresponding deoxidation reaction and the oxygen activity for the sulphur exchange reaction will be the same.

An example was given as follows. It could be shown that equilibrium for Equation (6-4) is rapidly established:

$$\underline{Fe} + \underline{S} + \text{CaO} = \text{CaS} + \text{Fe}$$

The the following relationship for oxygen activity should hold for cupola iron and slag:

$$(a_{O_2})_5 = (a_{O_2})_9 \tag{6-15}$$

If equilibrium only proceeded this far, the level of cupola desulphurization would be low compared with the levels theoretically possible for the equilibrium reactions (6-1) to (6-3). To determine whether any of the more efficient reactions reach equilibrium in the cupola, $(a_{O_2})_5$ values from Equation (6-10) need to be compared with the corresponding values for Equations (6-11) to (6-13).

Experimental Determination

Extensive experimental data were collected to establish which of the possible equilibria obtained in the cupola, 250 pairs of slag–iron samples being used in the thermochemical analysis of cupola desulphurization. These samples were obtained from 11 production, water-cooled, front slagging cupolas, representing operation with both acid and basic slags.

Data previously collected by Hatch and Chipman[12] on the sulphur equilibrium between iron blast furnace slags and metal were included for comparison. Laboratory data were also collected on desulphurization.

Examination of Results

Figure 6-5 shows the activity of oxygen from the sulphur transfer data compared with the respective oxygen activities estimated for the C, Si, and Mn deoxidation reactions. The lines drawn on the diagram for each comparison

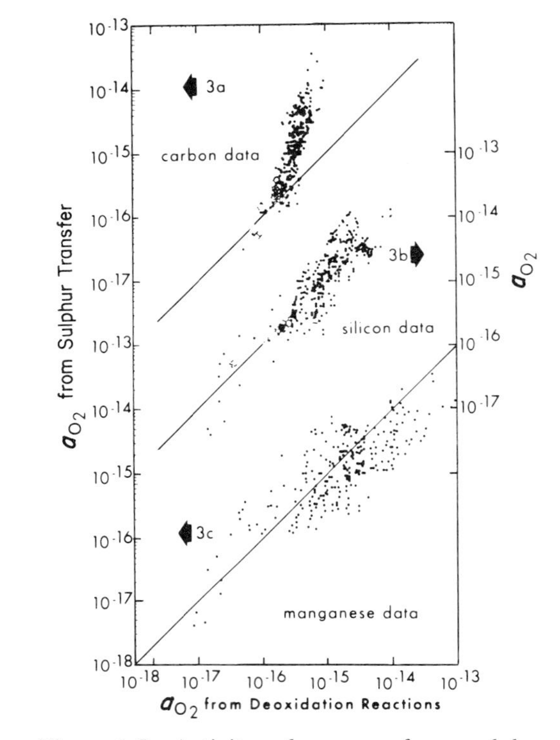

Figure 6-5 Activity of oxygen from sulphur transfer data compared with oxygen activity estimated from C, Si, and Mn deoxidation reactions respectively. The lines represent perfect correlation and show good fit for the Si and Mn reactions but not for C, suggesting a carbon equilibrium did not obtain.[5] (Reproduced by permission of American Foundrymen's Society)

represent perfect correlation. It is observed that good fit exists for the activity of oxygen in iron produced by Si and Mn deoxidation and the activity of oxygen obtained for the respective sulphur distributions.

It is observed, however, that the activity of oxygen from carbon deoxidation did not correspond. This indicated that the reactions (6-2) and (6-3) reached equilibria while the thermodynamically most-favoured reaction involving carbon, (6-1) did not.

Hence it was concluded that apart from the carbon equilibria a general state of equilibria did exist between iron and slag at the cupola taphole.

Attainment of Carbon–Oxygen Equilibrium in Cupolas

The failure to obtain carbon–oxygen equilibrium in the cupola was suggested to be due to the slow rate at which the dissolved oxygen and carbon combine to form CO. In the laboratory investigations, it was noted that for the temperature range equal to that for cupola tapping, a period of 5–10 hours was required for equilibrium. This is very long relative to the residence time of slag and iron in the cupola, which is about one hour. The rates for the other deoxidation reactions are considerably faster.

Observation of Sulphur Levels

In the cupola tapping range, 1450–1550 °C, the sulphur levels in iron for the Si equilibrated system fall between 0.015 and 0.030 per cent sulphur at the low end and 0.01 and 0.02 per cent sulphur at the high end. These figures corresponded to the lower limit of S observed in the systems in practice and was taken as support for the cupola desulphurization model presented. From Figure 6-6 it is observed that depending only on the cupola for desulphurization of irons with $S < 0.01$ per cent would require high Si contents and operating temperatures in excess of 1550 °C.

These observations pointed out the limitations of basic cupola desulphurization, while providing very clear evidence of the extent to which thermochemical reactions proceed under the existing operating conditions. Some

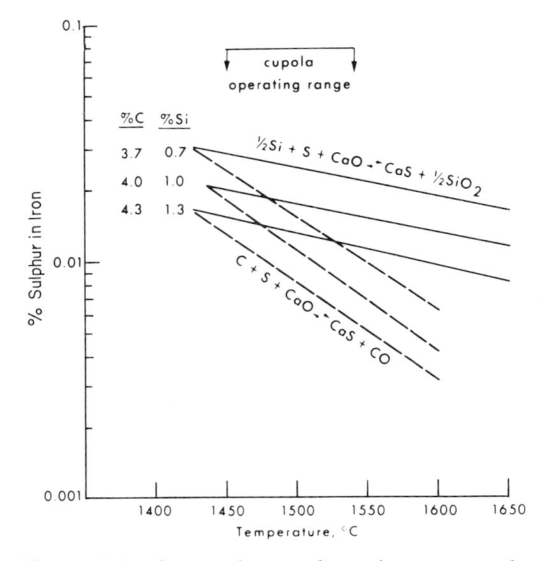

Figure 6-6 Comparison of optimum cupola desulphurization performance based on reactions involving Si + S + CaO and C + S + CaO. Sulphur levels for silicon equilibrated system fall between 0.015 and 0.030 per cent S at high end. Cupola desulphurization would require high silicon contents and high operating temperatures in excess of 1823 K. (Reproduced by permission of American Foundrymen's Society)

available options were proposed, e.g. new and powerful deoxidizers and desulphurizers. It is possible to consider agents that suppress the formation of heterogeneous slags.

If, however, the C–O equilibrium were to be established, the cast iron produced could have <0.01 per cent sulphur. The temperatures required would be slightly above 1500 °C.

6-4 Melting in the Induction Furnace

Induction furnace melting of cast iron offers flexibility for melting different grades of iron. The furnaces which are employed are of the coreless type having mains frequency, medium frequency (3,000 Hz), or high frequency, (30,000 Hz). Channel type furnaces are used for duplexing.

Induction furnace melting offers aspects of melt chemistry which are different from those introduced by cupola melting. In the latter, the basic element of the charge is mostly pig iron or returned foundry scrap, although all steel charges, or high percentages of steel, can be employed. Melting is in contact with coke and is a progressive process taking place in conditions exposed to gases.

Induction furnace melting can be mainly from scrap steel with additions of carburizing materials and ferro-alloys. Charging and pouring sequences of furnaces can be adjusted for melting efficiency and to allow least exposure to the external environment.

When comparison is made of properties of ordinary grades of cast iron melted either in the cupola or in the induction furnace, a number of differences are observed. These relate to composition of the alloy, particularly with respect to trace elements and gas content, especially nitrogen.

Comparison of Properties between Cupola Melted and Induction Melted Cast Iron

Comparative studies have been made of cupola melted and induction melted cast iron alloys. Russian investigations[13] refer to these as synthetic cast irons and assign a synthesis ratio. This is related to the percentage of steel scrap in the charge. Figure 6-7(a) shows that as the percentage of scrap is increased, there is a tendency for an increase of mechanical properties. Part of this increase can be related to the influence of nitrogen (Figure 6-7b) as it affects graphite and matrix and part to the relative influence of elements which, according to Levi,[13] are Ni and O.

Comparison of Cupola and Duplex Arc-Induction Melted Iron

Table 6-1 shows a comparison due to Hallot[14] of statistical data relating properties of duplex arc-induction and cupola melted nodular iron. For compositions which are identical in principle for elements other than for sulphur, the

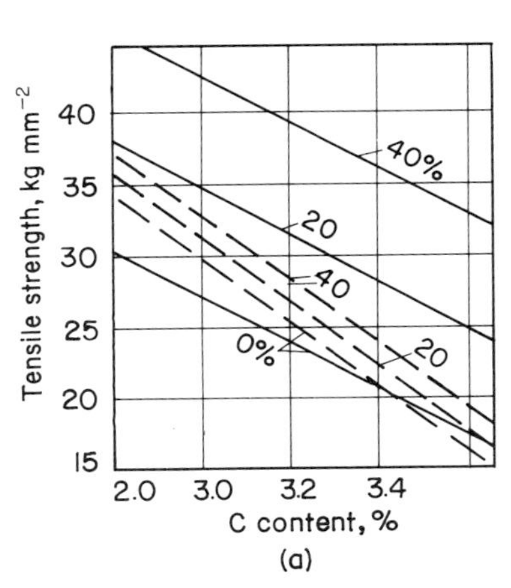

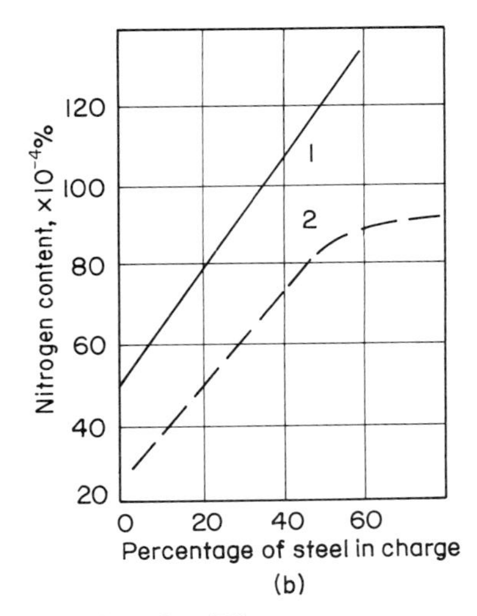

Figure 6-7 (a) Increase in tensile strength of cast iron for different percentages of scrap in charge. Full lines are for induction melted iron. Dashed lines are for cupola melted iron. (b) Nitrogen content of iron as function of steel scrap percentage in charge, full line (1) is for induction furnace; dashed line (2) is for cupola. (From Levi[13])

Table 6-1 Comparison of properties of cupola and Duplex Arc-induction melted irons[14]

Variables	Cupola iron	Duplex Arc-induction melted iron
C	3.6	3.6
Si	2.25	2.25
Mn	0.35	0.35
P	0.03	0.03
S	0.08	0.02
Mg	0.05–0.08	0.05–0.08
Tensile strength N/mm^2	723	750
El %	5.44	6.1
Brinell hardness	255	265

duplex arc-induction melted iron shows higher hardness and tensile strength together with higher elongation.

Carburizing Materials

Brokmeier[15] showed the rates of carburization of a liquid iron bath for some of the materials commonly used as additions (Figure 6-8). Table 6.2 shows the analyses of carbon, sulphur, and ash in these materials. The experiments were performed at 1540–1560 °C. The additions were made as 1 per cent and

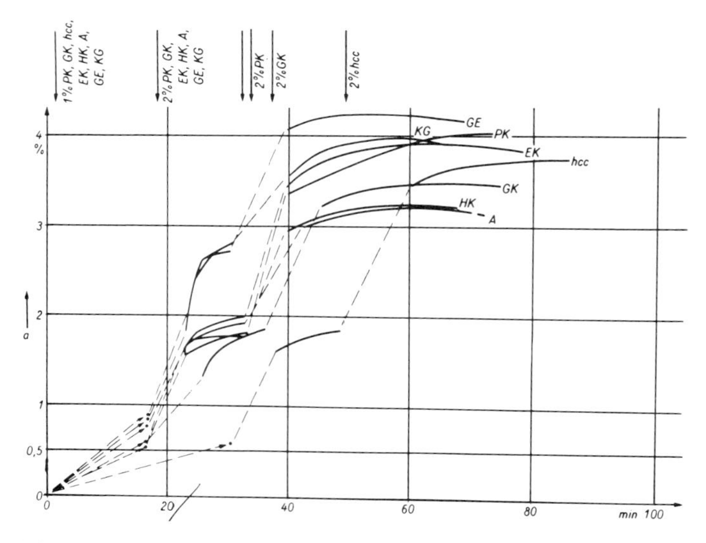

Figure 6-8 Rate of carburization in a liquid iron bath for some of the materials commonly used as additions. Times of addition of extra carburizer are indicated at top of diagram. (After Brokmeier.[15] Reproduced by permission of Brown, Boveri and Cie)

Table 6-2 Analyses of carbon, sulphur, and ash in carburizing materials[15] Reproduced by permission of Brown, Boveri and Company

	Ash %	C %	S %
Petrol coke	0.8	98.5	0.84
Gas coke	11.6	87.5	1.34
High carbon coke	4.6	92.5	1.25
Electrode carbon	1.3	98.8	0.19
Wood charcoal	3.7	84.2	0.19
Anthracite	3.3	88.6	0.18
Electrode graphite	0.2	99.5	0.19
Granulated coal	0.6	98.8	1.01

The letters on the individual graphs of carbon content of melt as a function of time refer to the following:

PK = petrol coke
GK = gas coke
A = anthracite
GE = electrode graphite
KG = granulated coal
hcc = high carbon coke
EK = electrode carbon
HK = wood charcoal

two 2 per cent additions, giving a total of 5 per cent addition. The granulated material was 4–8 mm grain size. The graphs show rate factors in carburizing of baths between grades of material, e.g. graphite electrode material and high carbon coke. Brokmeier has given numerous carburizing graphs showing the influence of composition on rate of carbon solution.

6-5 Solubility of Hydrogen in Iron Base Alloys

The solubility of hydrogen in iron base alloys has been reported in several papers. Some of the results are discussed in the following.

Investigations of Bagshaw, Engledow, and Mitchell

The study[16] was made of hydrogen solubility in a number of binary and complex liquid iron-based alloys. These included Fe–Co, Fe–Mo, Fe–Ni, Fe–Si, and Fe–Cr Interaction parameters were also calculated for complex solutions.

A Sieverts system was employed as described by Humber and Elliott.[17] Discussion was given of variations reported for hydrogen solubility in iron base alloys related to some of the experimental procedures adopted. For high purity iron at 1600 °C and atmospheric pressure a hydrogen solubility was determined of 27.7 ± 0.8 cm^3 per 100 gm.

The solubility of hydrogen estimated in Fe–Ni, Fe–Cr, and Fe–Si is shown in Figure 6-9. Both in Fe–Ni and Fe–Cr, increases of hydrogen solubility are noted for increased concentration of added element. For Fe–Si there is a decrease initially on adding Si and the solubility curve shows a minimum value at 1600 °C of about 3 cm^3 per 100 gm at the Fe–Si composition.

For complex solutions, interaction parameters were calculated according to the procedures of Chipman and Corrigan[4] and Wagner[8] (see Chapter 4). Figure 6-10 shows a comparison between measured and calculated values of hydrogen solubility in Fe–Co–Si and Fe–Ni–Cr. It is noted that the results for Si show that this element affected the correlation between experiment and prediction.

Weinstein and Elliott

This investigation[18] reported hydrogen solubility results for liquid iron alloys having different elements in solution. The results for Fe–C are shown in Figure 6-11(a). The element carbon reduces the hydrogen solubility in iron at 1592 °C from 28.0 cm^3 per 100 gm to less than 16.0 cm^3 per 100 gm at a carbon content of 3.5 per cent.

The results of solubility experiments of hydrogen in liquid iron at 1 atmosphere and 1592 °C are given in Figure 6-11(b). This shows that the alloying elements Al, B, C, Co, Cu, Ge, P, S, and Sn decrease the solubility while Cb, Cr, Mn, and Ni increase it.

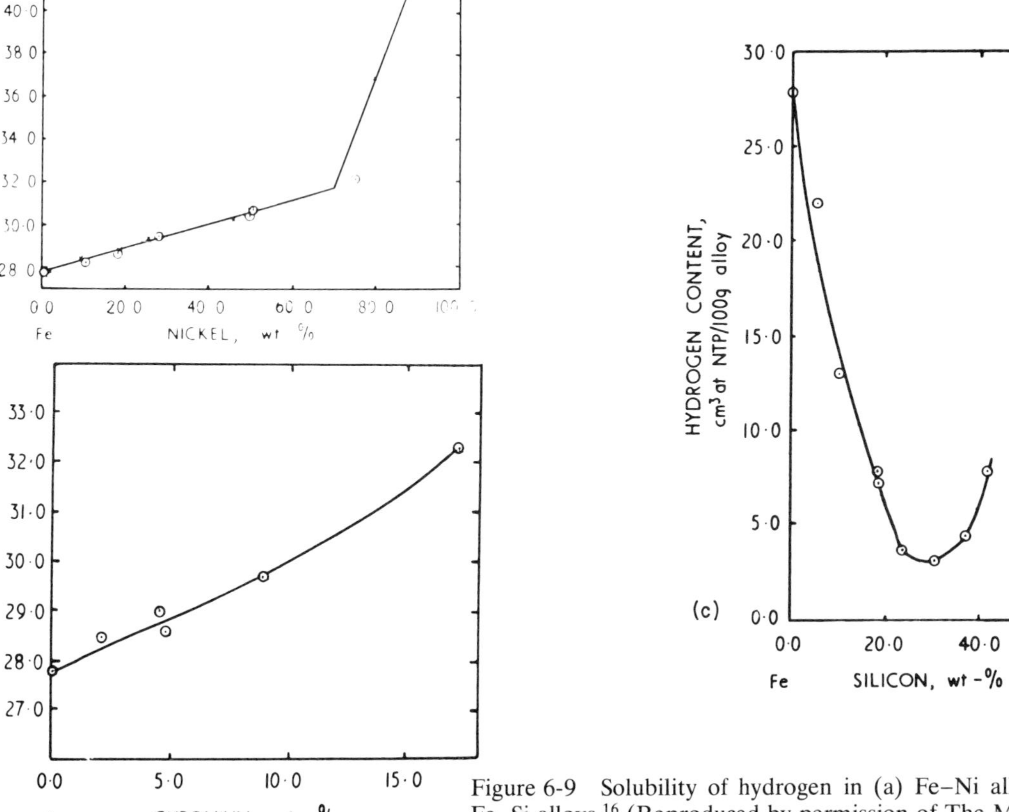

Figure 6-9 Solubility of hydrogen in (a) Fe–Ni alloys, (b) Fe–Cr alloys, and (c) Fe–Si alloys.[16] (Reproduced by permission of The Metals Society)

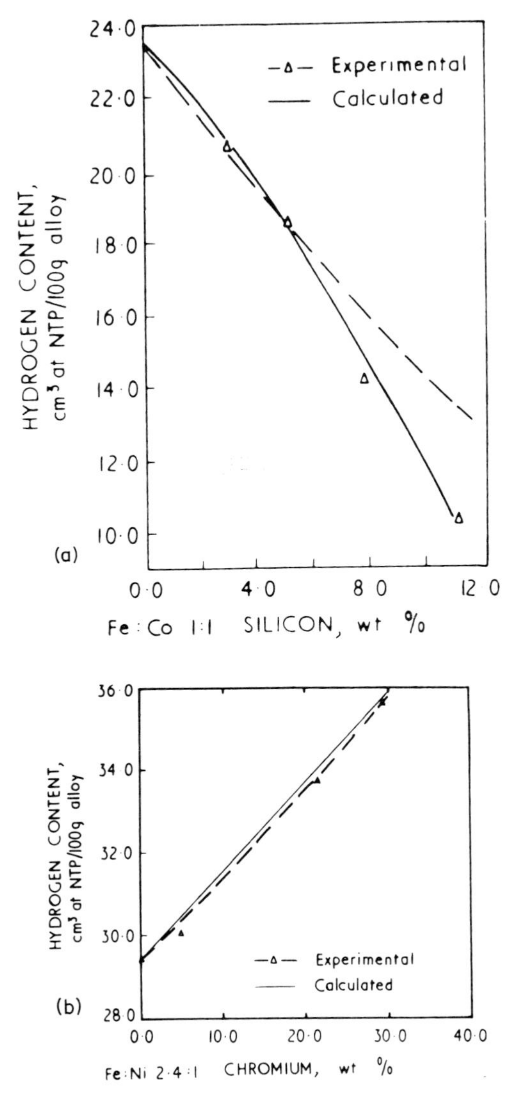

Figure 6-10 Experimental and calculated curves for hydrogen solubility in (a) Fe–Co alloys containing Si and (b) Fe–Ni alloys containing Cr.[16] (Reproduced by permission of The Metals Society)

For the influence of temperature on solubility, it was recommended that for all alloys other than those containing B, P, Ge, and Cr, the temperature coefficient for hydrogen solubility should be taken as equal to that for iron (3.1 ppm per 100 °C). Data were given for the other alloys.

In Figure 6-11(b), it should be noted that 1 ml H_2 (STP) per 100 gm is equal to 0.9 ppm.

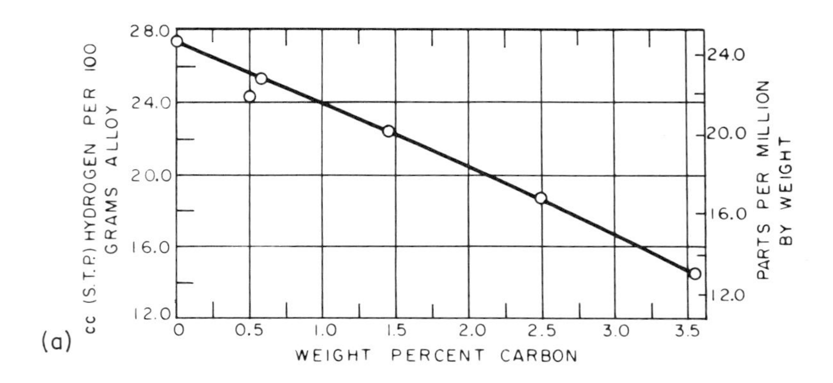

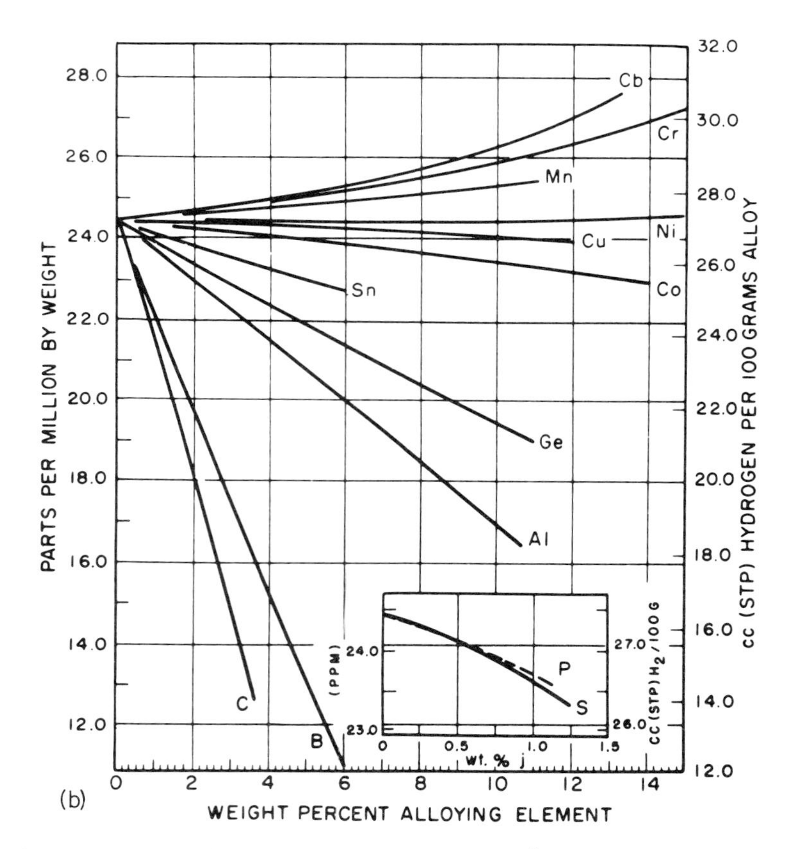

Figure 6-11 Solubility of hydrogen at 0.1 N mm^{-2} pressure at 1865 K in (a) alloys and in (b) liquid iron as a function of weight per cent of alloying element.[18] (Reproduced by permission of The Metallurgical Society of AIME)

6-6 Effect of Dissolved Gases

Hughes[19] gave an overall survey of the influence of gases in cast iron on the structure. The gases of importance ar oxygen, nitrogen, and hydrogen. In the case of oxygen and nitrogen, the main gas content of the cast alloy is the result of melting practice. In the case of hydrogen, considerable changes of gas content may result from the mould as a result of reaction between the liquid metal and moisture.

Hughes estimated that these gases appear in cast iron in the following concentrations:

Oxygen:	0.005–0.01%
Nitrogen:	0.0015–0.01%
Hydrogen:	0.00005–0.000025%

In the case of hydrogen, the adsorption by reaction with water vapour is promoted by reactive minor elements in cast iron, e.g. Al or Mg, and by manganese when present in high concentration.

The influences of these elements on structure as discussed in this text are briefly as follows:

(a) Oxygen probably influences structure through nucleation. Part of the evidence for this centres on the nature of nuclei for graphite where spinel structures have been determined.
(b) Nitrogen is a carbide stabilizer. Pearlite as a transformation product of austenite is one of the contributing factors in enhancing the strength of gray cast iron melted in induction furnaces prepared from high steel scrap charges in cupolas. The influence on graphite growth is also noted.
(c) Hydrogen is surface active for graphite. It is additive in its action with elements like sulphur, selenium, and tellurium so that the combined effect is to reduce the sulphur adsorption on graphite and to enhance the white mode of solidification of cast iron.

6-7 Effect of Elements Dissolved in Cast Iron during the Melting Process

The sensitivity of gray and ductile cast iron to the melting process lies in the influence of small concentrations of impurity on graphite nucleation and subsequent growth. In malleable cast iron, small alloy influences have similar effects on the graphitization process.

In spheroidal graphite cast iron, after reactive elements, and in particular Mg, have combined with some of the impurities including S the remaining impurity elements interact with graphite during the solidification process. Part of the increase of reactivity is related to concentration increases in the liquid due to segregation. In this, dissolved gases also play a role and composition control therefore relates to gases as well as to all the other elements. These are basic features of melting processes.

Hydrogen and the Elements which Lead to Widmanstätten Graphite

The role of hydrogen was discussed in Section 3-2. The observations were reported of experiments on eutectic solidification with increasing amounts of hydrogen in solution. The undercooling for solidification was increased under constant experimental conditions. This indicates that the effect of hydrogen, as an element surface-active to graphite, is to replace the other surface-active element sulphur which is normally present. The effect is equivalent to solidication in nearly sulphur-free melts. Eventually white structures may crystallize.

Particular attention[20] has been directed to the combined effects of hydrogen with selenium and lead. In large castings with the attendant problems of segregation, Widmanstätten graphite can appear. The experimental observations are limited. The elements involved are all influential in promoting undercooling and the mechanism might be sought in an instability in the region between gray and white solidification, possibly in supersaturated austenite.

6-8 Liquid Metal Melting and Inoculation

Moore[21] discussed the melting process for cast iron and its influence on solidification. The structure of cast iron melted in a cupola and after inoculation is related to the carbon equivalent. Low carbon equivalent irons can be inoculated to give a very fine structure of graphite with a high cell count, and high strength values can be obtained. The percentage increase of cell count due to inoculation decreases as the carbon equivalent increases.

Using the standard wedge chill test, the influence of carbon equivalent on chill depth can be noted. As carbon equivalent decreases, the chill depth increases. The interpretation of Moore for the inoculation effect is that these analyses of cast iron undercool markedly, as shown by the chill test, and inoculation leads to a large increase of nuclei at this large undercooling. However, it is difficult to accept this point of view since a large undercooling could not occur without having at the same time austenite phase growth.

The most likely mechanism must be the increase of nuclei related to composition. As the carbon concentration decreases, the equilibrium oxygen content of the melt increases. The first growth of austenite, with its associated increase of solute concentration at the dendrite boundary, must immediately give a remarkably large increase of eutectic cell nucleation due to the increased presence of heterogeneous nuclei. This effect must be associated with an enhanced oxygen content of the melt in lower carbon content alloys.

Moore also gave a synopsis of structure variation in cast iron due to different melting units. In electric arc and induction furnaces, considerable variation of melting conditions can be made, in particular temperature. The order of charging can be varied, e.g. the order of addition of carbon. This paper discusses these effects in detail, showing cast structure coarsening with overheating. It describes the influence of the order of melt additions on nucleation. Moore suggests that these effects are related to the dissolution of graphite nuclei with temperature or the increased presence of graphite nuclei with a late addition of melt carburizer.

While there is no doubt that a particle of graphite in the melt could act as a nucleus for the graphite eutectic, no overall theory exists relating nucleation of the eutectic to graphite. Present experimental and thermodynamic evidence, and also the study of nuclei, suggest that these effects may be better related to the variation of oxygen levels in the melt.

6-9 Liquid Metal Melting and Refining

Improvement of the cast iron melting process can lead to improved melting efficiency and also to better compositional control. For more efficient melting, particularly of borings and turnings, a plasma arc furnace has been described.[22]

Several possibilities exist for processes leading to the refining of composition, in particular the removal of elements which are harmful to structure. The most common of such processes is the removal of sulphur using CaC_2 or Na_2CO_3. In the Gazal process,[23] gas is passed through porous corundum plugs at a pressure of 1 to 5 atmospheres, the stirring leading to a highly efficient contact between liquid and additions. Approximately 80 per cent desulphurization is reported. This increases with temperature. In the Mg treatment of ductile iron, the efficiency is increased with the stirring introduced by the gas stream.

The availability of direct reduced iron makes possible the melting of charges for ductile iron production with the minimum introduction of harmful impurities. Some great difficulties in melting sponge iron are encountered,[24] including large slag production. The use of direct reduced iron would appear to be advantageous if its purity is greater than oxygen-blown pig iron or high quality steel scrap.

References

1. Östberg, G.: *Trans. AIME*, **212**, 678, 1958.
2. Patterson, W., and W. Standke: *Giesserei Techn. Wiss. Beih.*, **15**, 1, 1963.
3. Neumann, F., and E. Dötsch: *The Metallurgy of Cast Iron* (Eds. B. Lux, F. Mollard and I. Minkoff), Georgi Publ. Co., *Switzerland, 1975.*
4. Chipman, J., and D. A. Corrigan: *Trans. Metall. Soc. AIME*, **233**, 1249, 1965.
5. Katz, S., and H. C. Rezeau: *Trans. Am. Foundrymen's Soc.*, **87**, 367, 1979.
6. Morrogh, H.: *Foundry Trade J.*, **143**, 302, 1977.
7. Opravil, O., and R. D. Pehlke: *Trans. Am. Foundrymen's Soc.*, **77**, 415, 1969.
8. Wagner, C.: *Thermodynamics of Alloys*, Addison-Wesley, 1952.
9. Uda, M., and R. D. Pehlke: *Cast Met. Res. J.*, **10**, 30, 1974.
10. Wada, H., K. Gunji and T. Wada: *Trans. ISI Japan*, **8**, 329, 1968.
11. Pehlke, R. D., and J. F. Elliott: *Trans. Metall. Soc. AIME*, **218**, 1088, 1960.
12. Hatch, G. G., and J. Chipman: *Trans. AIME*, **185**, 274, 1949.

13. Levi, L. I.: *Russian Castings Prod.* (Engl. Transl.), p. 295, BCIRA, Alvechurch, England, July 1972.
14. Hallot, L.: *Metall. Constr. Méc.*, **100(1)**, 7, 1968.
15. Brokmeier, K.-H.: *Induktives Schmelzen*, Brown, Boveri & Cie, Mannheim, 1966.
16. Bagshaw, T., D. Engledow, and A. Mitchell: *J. Iron Steel Inst. London*, **203**, 160, 1965.
17. Humbert, J. C., and J. F. Elliott: *Trans. AIME*, **218**, 1076, 1960.
18. Weinstein, M., and J. F. Elliott: *Trans. AIME*, **227**, 382, 1963.
19. Hughes, I. C. H.: *Trans. Am. Foundrymen's Soc.*, **77**, 121, 1969.
20. Dawson, J. F.: *Trans. Am. Foundrymen's Soc.*, **77**, 113, 1969.
21. Moore, W. H.: Tech. Report No. 7313, Am Foundrymen's Soc., July 1973.
22. Shulhoff, W. P., and H. L. Colthurst: *Int. Cast Metals Res. J.*, **5**, 8, 1980.
23. Galey, J., J. Foulard, N. Lutgen, and W. Toman: *Int. Cast Metals Res. J.*, **5**, 16, 1980.
24. Mayer, H.: Personal Communication 1979.

Bibliography

Cupola Handbook, 4th ed., Am. Foundrymen's Soc., 1975.
Neumann, F.: *Metallurgische Schmelzführung*, Brown, Boveri & Cie, Mannheim, 1972.

Chapter 7
Production of Cast Iron with Intermediate Structures

Appropriate treatment of liquid Fe–C alloys leads to structures between flake and spheroid. In this chapter, the principle types of intermediate structure are described. In one, a rod-like form of graphite is produced by adjustment of the Fe–C–Si composition alone, in the absence of sulphur. This type of graphite was referred to as coral, after its appearance in constructed models. Its structure and mode of growth are described based on electron diffraction, transmission, and scanning electron microscopy. The properties of the alloys are indicated.

The second type of intermediate structure is a graphite form which is determined by unstable growth in the range of undercooling close to that required for spherulites. It has been termed vermicular or compacted graphite. It is produced commercially by additions to the liquid metal of elements or combinations of elements in amounts less than the addition made for spherulitic graphite growth. It is possible to produce these structures by magnesium additions or magnesium with other elements.

A description is given of some experimental investigations and a restatement is made of the theory of interaction between elements in solution and graphite to produce the required undercoolings for the different growth forms. Effects of combinations of elements are discussed. Of importance is the effect of a combination of reactive elements, e.g. Mg and rare earth metals. Only part of the required kinetic undercooling is then associated with the magnesium. The remainder of the undercooling is obtained by the rare earth metals. Because of their smaller reactivity, less variation is achieved of undercooling and hence less variation of the structure.

Finally, segregation effects in solidification lead to marked effects on the graphite form and are discussed as intermediate structures. Large increases in concentration due to segregation can be calculated for some elements while important elements can disappear by reaction. This can be serious towards the centre of large castings. The Scheil segregation equation can be used to estimate some of these effects.

7-1 Intermediate Structures

Up to this point, the subject of cast iron structure, as it is related to graphite in alloys, has been focused on two modes of graphite growth in the liquid. The two types of structure which have been characterized are:

(a) The flake. Growth is by a defect mechanism. Instability occurs on the $(10\bar{1}0)$ face of graphite, leading to a branched crystal essentially of a planar character. This is the basic growth type of the flake graphite cast irons. The study of its formation requires a knowledge of branching mechanisms, nucleation, and impurity effects, particularly that of sulphur. Its growth as a eutectic has been discussed.

    Flake types were the basic alloys of cast iron in structural engineering applications until 1948. Their properties were related to the duplex system of a plate-like phase distribution in a matrix. The plate-like phase was of limited ductility. The matrix structure and properties could be varied by alloying and heat treatment.

(b) The second type which has been characterized is the spherulite. This grows from the liquid by an instability mechanism determined by the undercooling. In this range of temperature, the normal branched unstable forms have been superseded. Spherulitic growth is observed to have its origin in a pyramidal instability. At even greater undercooling beyond that of the spherulite, another form has been observed; and this is the superpyramid.

The forms which occur at an undercooling smaller than that of the spherulite are relatively easy to describe, and this is done in the present chapter. They grow at an undercooling which is intermediate between flake and spherulite. There are different morphologies and different names have been given. These include chunky graphite, vermicular graphite, and coral graphite. They are given the general title in this chapter of intermediate forms.

These forms are relatively easy to describe because they can be correlated with the manner of growth of graphite in the flake eutectic. An instability in growth occurs on the graphite (0001) faces, either by steps on these faces becoming unstable or by the faces becoming unstable.

In commercial practice, these cast structures must also be obtained by control of the undercooling. Whereas for spheroidal crystals the tendency has been to employ Mg because of the greater kinetic effect produced, at the lesser undercooling required by vermicular or chunky graphite, Ce is a preferred additive. The interaction is now modified by Ti.

When cerium and rare earth elements are added to cast iron, the kinetic undercooling effect is less than that for Mg. This relates to differences in reactivity and adsorption on graphite. The structures obtained with cerium treatment of cast iron were noted by Morrogh and Williams[1] to have a significant shape variation. Microstructures looking more like flake than spherulites were termed quasi-flake.

A rod-like morphology is observed in sulphur-free Fe–C–Si alloys. In this type of growth, silicon acts to promote constitutional supercooling of the interface.

7-2 Observations of Intermediate Structures in Iron–Carbon–Silicon Alloys

Lux and Grages[2] observed the occurrence of a rod-like unstable growth modification of graphite in cast iron alloys, made from pure Fe–C–Si and free of sulphur. The analyses were approximately 3 per cent carbon and 2.3 per cent silicon. These alloys were prepared from pure carbonyl iron powder, pure silicon powder (99.9 per cent) and pure carbon (99.99 per cent). The powders of the pure materials were mixed, pressed, and vacuum sintered for three hours at 800–950 °C and 0.13 N m^{-2}. The electrode thus obtained was cast in a copper mold of a vacuum arc furnace. Specimens were taken for testing and further specimens for examination were melted using the electrode stock in a vacuum medium frequency furnace and cast in test molds.

In the first observations[3] the spatial configuration of the graphite was examined and a model of the graphite was constructed in Wood's metal. This was made in a mold of Perspex sheets cut to form, following microstructural sections taken at 0.5 μm thicknesses.

This model gave an appearance of coral. The graphite formed an elongated highly branched structure which was distinctly non-lamellar

Electron Microscopy and Diffraction

Electron diffraction of graphite samples,[4] extracted from the experimental test pieces and examined close to the edge, pointed to the structural model of Figure 7-1(a). In the elongated crystals, the graphite layers run parallel to the specimen edge so that the $\langle 0001 \rangle$ direction is normal to the surface.

Figure 7-1(b) shows a model suggested by both scanning and transmission electron microscopy.[3] In observing replicas of etched graphite sections the sharp changes of orientation of structure could be observed across the crystal. Because of the (0001) orientation of the outer layers, Lux, Bollmann, and Grages[4] suggested that coral graphite crystals were intermediate in form between lamellar and spherulitic graphite.

Models of Growth of Rod-like Graphite

Further studies of the cast Fe–C–Si alloys were made by Minkoff and Lux[6] and Lux, Minkoff, Mollard and Thury[7] employing scanning electron microscopy on extracted graphite crystals. Figure 7-2(a) shows the general configuration of graphite crystals in an alloy, typical of a fine structure eutectic.[7] Figures 7-2(b) and 7-2(c) show enhanced details of individual crystal growth within the eutectic. In Figure 7-2(b), it is noted that the typical lamellar growth of graphite is changed to a rod-like form in which lamellae wind in layers round the crystal.[5]

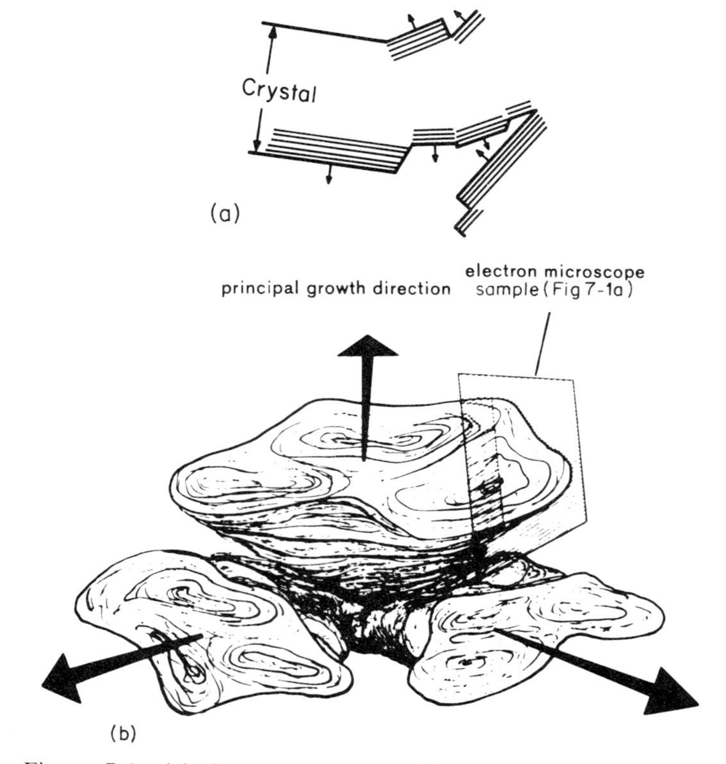

Figure 7-1 (a) Orientation of (0001) planes in coral graphite crystals as determined by electron diffraction. The (0001) direction is normal to the surface.[4] (b) Model of coral graphite crystals suggested by scanning and transmission electron microscopy.[3] (Reproduced by permission of Dr. Riederer-Verlag GmbH, Postfach 447, D-7000 Stuttgart 1)

In Figure 7-2(c), it is noted that the leading edge of the crystal still has an elongated hexagonal form, while further along the crystal, growth has transformed the shape into a rod-like geometry.[6]

Experiments of Minkoff and Lux on Coral Graphite Forms

Further experiments[8,9] were performed to ascertain the mechanism of this type of growth. As a growth model for Fe–C–Si, experiments were made on the Ni–C–B system. Alloys were melted from pure Ni shot and boron. Melting was performed in graphite crucibles in a vacuum furnace and solidified at a rate of cooling equal to 45 °C per minute. Crystals of graphite were extracted from the alloy and examined in the scanning electron microscope. The alloy analysis was approximately 2.5 per cent boron and 2.4 per cent carbon.

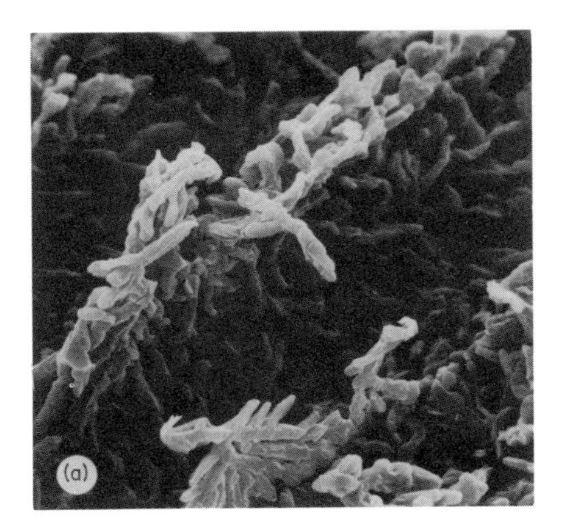

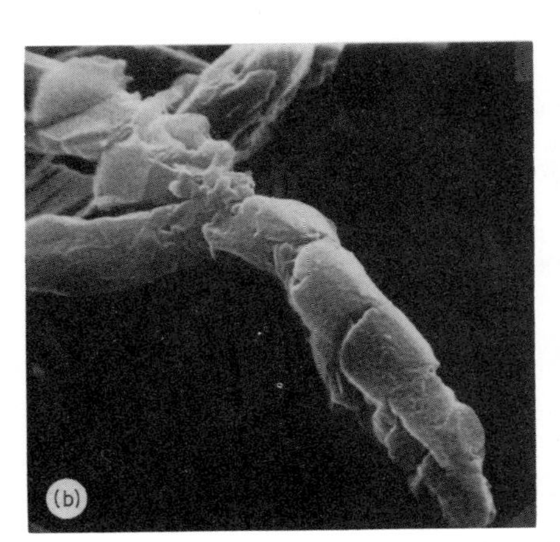

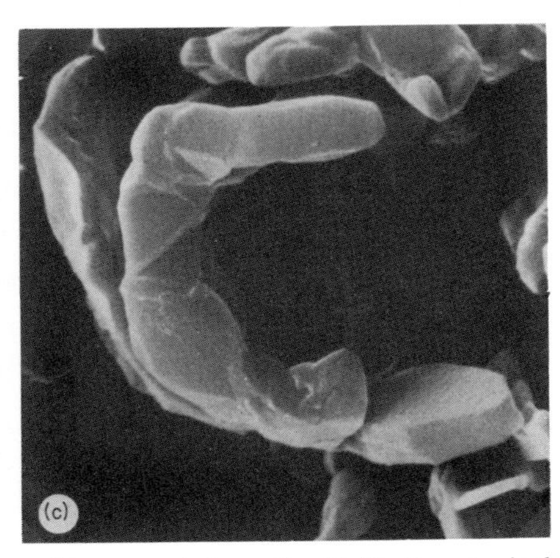

Figure 7-2 (a) General configuration of graphite crystals in an Fe–C–Si alloy, typical of a fine structure coral graphite eutectic[7] (× 990). (Reproduced by permission of Georgi Publ. Co., Switzerland.) (b) Detail of an individual crystal of coral graphite from the structure shown in (a). Note that the lamellae wind in layers round the crystal (× 3,025).[5] (Reprinted with permission from Micron, Vol. 2, p. 286, Fig. 4, by I. Minkoff and B. Lux, Copyright © 1971, Pergamon Press, Ltd.) (c) Coral graphite crystal from Fe–C–Si alloy. Note that the leading edge of the crystal, at the bottom of the illustration, still has an elongated hexagonal form. Further along the crystal, growth has transformed the shape into a rod-like geometry[6] (× 3575). (Reproduced by permission of Georgi Publ. Co., Switzerland)

Figure 7-3 shows the step structure observed on the graphite crystal surface. The normal step arrangement on graphite is replaced by elongated promontories which are the result of step instability. Figure 7-3 shows that the promontories traverse the edge of the crystal to the second face. The mechanism of growth of rod graphite is shown in Figure 7-4(a) and (b).

In ordinary flake growth of graphite, the crystal is bounded by (0001) faces which thicken from the steps of screw dislocations. The actual growth in the

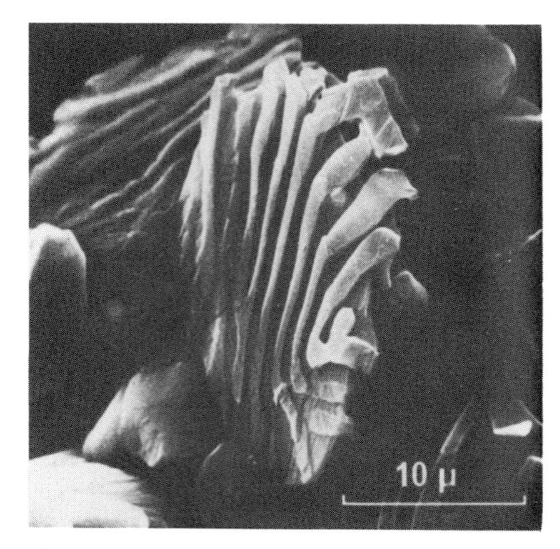

Figure 7-3 Hexagonal graphite crystal in Ni–C–B alloy experiments of Minkoff and Lux.[9] Normal step arrangement on a graphite crystal surface is replaced by elongated promontories (× 1,375). (Reproduced by permission of Dr. Reiderer-Verlag GmbH, Postfach 447, D-7000 Stuttgart 1)

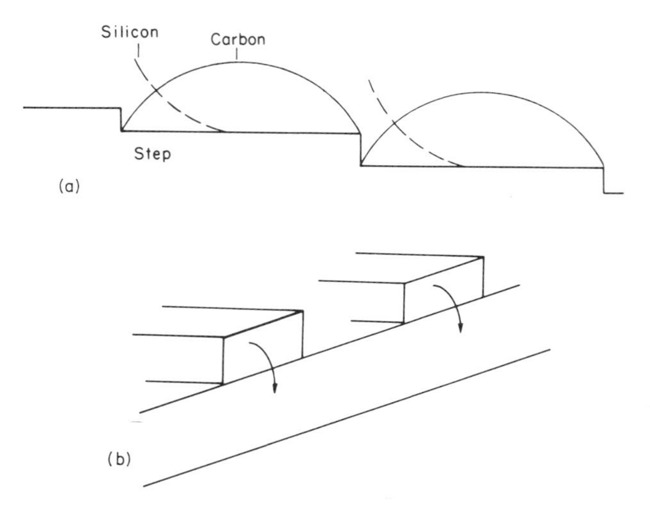

Figure 7-4 Mechanism of growth of rod-like forms of graphite. (a) Distribution of solute ahead of steps. Full line is the carbon distribution. Dashed line is silicon distribution. (b) Movement of steps over crystal edge due to high supersaturation

edge direction is by a nucleation mechanism on steps entering $\{10\bar{1}0\}$ faces generated by twist boundaries. On the (0001) faces of the crystal, steps move from spiral sources and push solute ahead of them. The carbon distribution will be as shown by the full line in Figure 7-4(a). An element like Si which is not incorporated in the structure will have a distribution as shown by the dotted lines. In pure Fe–C–Si alloys, the constitutional supercooling effect of the silicon is sufficient to unstabilize the steps so that these became elongated promontories.

Step instability was calculated to occur at an interface undercooling of 7 K[10] (see Chapter 5).

The driving force for the step promontory to traverse the crystal edge (Figure 7-4b) is the large solute supersaturation existing there. This type of growth is not observed in ordinary cast iron alloys of basic Fe–C–Si composition because of the presence of S.

That this effect is not observed in commercial flake graphite cast irons may be understood by the following reasoning. Small concentrations of sulphur have an effect on the growth temperature at the interface. According to Munitz and Minkoff[10] sulphur adsorbed in small concentrations at the graphite interface decreases the interface undercooling. The experiments on Ni–C–B alloys suggest that the influence of Si in Fe–C–Si alloys is similar to the effect of B in Ni–C, i.e. to increase the interface undercooling and unstabilize the step. These two influences are cancelled in commercial cast iron since the undercooling effects are additive. In this manner, the influence of Si in increasing ΔT is offset by the effect of S in decreasing it.

Mechanical Properties of Coral Type Graphite Irons

Mechanical properties of these types of cast iron were reported by Lux.[11] The results of Lux show that the modulus of elasticity is close to that for spheroidal graphite cast iron, about 137–156 G N m^{-2}, and is independent of the silicon content. Both the strength and the hardness vary as a function of the silicon content. The elongation varied between 1.5 and 2.0 per cent.

Lux examined the influence of cooling rate on the structure and mechanical properties of coral graphite cast iron of composition 3 per cent carbon and 4 per cent silicon. For a sand casting, the properties were tensile strength 362 N mm^{-2} and total elongation 1 per cent. For a casting made in an iron mould, the figures were tensile strength 509 N mm^{-2} and total elongation 2 per cent.

7-3 Cast Iron with Vermicular Graphite (Compacted Graphite)

The second type of cast iron to be discussed in this chapter having an intermediate morphology between flake and spherulite has been given different names. The following terminology was reported.[12]

Country	Terminology		
Japan	Quasi-flake	Upgraded chunk	Semi-ductile
England	Compacted		
Germany, U.S.A.	Vermicular (term also used for Coral in England)		

This form of graphite retains an interconnected arrangement as in a eutectic cell but with a growth form which is non-flake and non-rod-like. It has characteristics close to those of a spherulite and is the result of a melt treatment with less than the quantity of addition normally necessary for a spheroidizing treatment.

The actual addition to the melt for obtaining the structures vary. At present, the different melt treatments include magnesium or magnesium with titanium, or cerium with rare earth elements. Small additions may be made of other elements, e.g. copper. Sissener *et al.*[13] reported that the useful employment of cast iron with vermicular graphite was probably suggested for the first time by Estes and Scheidewind[14] and by Schelleng.[15]

Properties

Two properties of interest are the tensile strength and the thermal conductivity. While the former approaches that of spheroidal graphite cast iron alloys, the thermal conductivity is closer to that of gray iron. Some mechanical properties and thermal conductivity of compacted graphite cast iron in comparison with gray and spheroidal graphite cast iron are given in Table 7-1.

Table 7-1 Comparative mechanical and thermal properties of types of cast iron. (a) Mechanical properties.[13] (Reproduced by permission of American Foundrymen's Society

	Flake*	Flake†	Vermicular	Spheroidal graphite
Composition %C	3.10	3.61	3.6	3.56
%Si	2.10	2.49	2.54	2.72
%Mn	0.60	0.05	0.05	0.05
CE	3.80	4.44	4.46	4.47
Matrix structure	Pearlite	Pearlite	Ferrite	Ferrite
Tensile strength, N mm^{-2}	319	107	336	438
%El	Not determined	Not determined	6.7	25.3

*30 mm diameter test bar.
†Test bar with 25 mm wall thickness.

(b) Thermal conductivity.[16] (Reproduced by permission of American Foundrymen's Society)

	Thermal conductivity, $W(m\ K)^{-1}$	
Type of iron	Medium–light section	Heavy section
Gray, ASTM A48-74 class 25	52.3	50.2–56.5
Gray, ASTM A48-74 class 45	47.3	—
Compact graphite	49.4	44–50.2
Spheroidal graphite	33.5	37.7

7-4 Experiments by Thury on the Form of Graphite

Thury[17] made experiments on the formation of graphite in cast iron, making additions of calcium and rare earth metals in the form of Mischmetal. The influence of copper and titanium on the graphite morphology of these magnesium-free irons was studied.

Thury classified the graphite form according to a scheme described by an Iron and Steel Examination Chart.[18] In this, spherulitic graphite is called type K, an incompletely spherulitic form is called type L, and vermicular graphite is termed type P (Figure 7-5)

The analysis of the metal treated was 4.09 per cent carbon, 0.45 per cent silicon, 0.01 per cent manganese, <0.025 per cent phosphorus, and 0.016 per

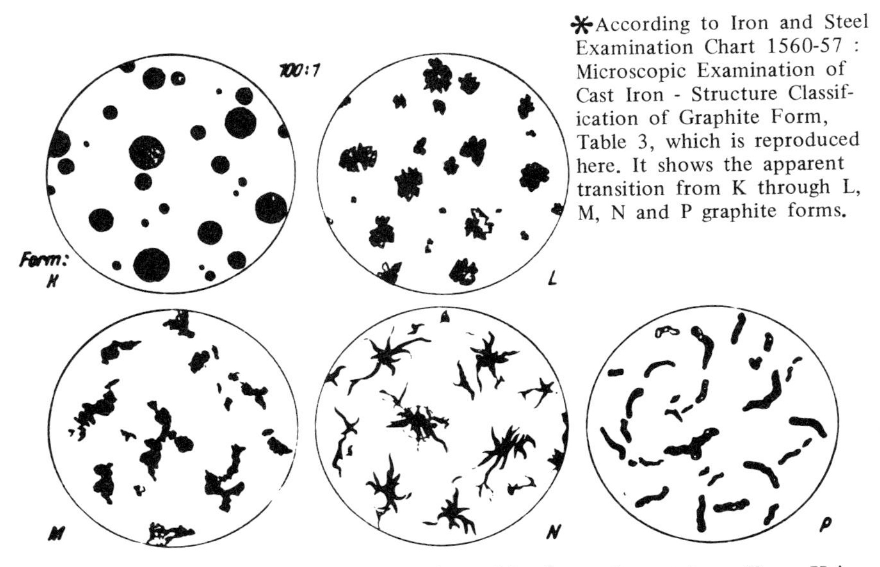

Figure 7-5 Scheme of classification of graphite forms in cast iron. Form K is spherulitic. Form P is vermicular graphite. (According to Iron and Steel Examination Chart 1560–57, shown by Thury.[17] (Reproduced by permission of American Foundrymen's Society)

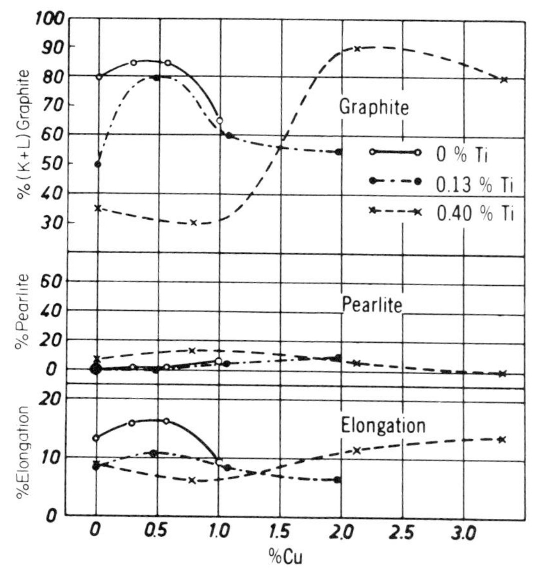

Figure 7-6 Forms of graphite obtained in cast iron adding calcium and rare earth metals in the presence of copper and titanium. The classification of structure is according to percentage K and L types shown in Figure 7-5.[17] Adding Ti initially prevents spherulitic growth. The percentage of spherulitic structure is restored as Ti is added together with approximately 2 per cent Cu. (Reproduced by permission of American Foundrymen's Society)

cent sulphur. The metal treatment commenced with desulphurization to less than 0.002 per cent sulphur. This was achieved by the stepwise addition of finely crushed CaC_2 containing 10 per cent fluorspar. About 0.05 per cent Mischmetal was then added, followed by ferrosilicon inoculation and casting.

In Thury's investigations, completely spherulitic graphite structures were not observed. The forms observed were approximately 90 per cent type K and 10 per cent type L. Figure 7-6 shows the results for structures when Ti and Cu were added to the iron.

Thury's results showed that the influence of Ti in small concentrations is to prevent spherulitic growth. This is seen by the fall in per cent of (K + L) graphite as Ti is increased from 0 to 0.40 per cent. Thereafter, the increasing percentage of graphite in spherulitic form was attributed to the increase in copper, a maximum being achieved when the copper present is approximately five times the amount of titanium. With more copper the percentage of spherulitic graphite again decreases.

Thury concluded that it was not possible to obtain good spheroidal graphite cast iron without the employment of Mg, but in its absence and using different element combinations, other interesting structures could be achieved.

7-5 Combinations of Elements Producing Intermediate Graphite Structures

Reactive elements interact to some degree with other elements in solution. The observations of Thury suggest an interaction between Ce and Ti. This reduces the reactive element effectiveness for spherulite production.

Munitz and Minkoff[10] calculated element effectiveness in solution with respect to effect on the undercooling of graphite. An element may partly interact with Mg or Ce and partly act to undercool the graphite as a constituent of a constitutionally undercooled boundary layer. Any excess Ti and also Cu appear to act in this latter manner, so that increasing copper acts to increase the percentage of spheroidal graphite.

The effect on graphite growth of the reactive elements is additive. Hence Thury noted that differences in the uncontrollable concentrations of free calcium and free rare earth metals gave rise to structures that were not reproducible.

Research of Evans, Dawson, and Lalich

The comprehensive research of these experimenters[16] is reported in the following section. Its end result was to produce a single alloy addition for a cast iron melt capable of giving compacted (U.S.A. vermicular) graphite structures. The principle elements employed were Mg, Ti, and traces of Ce.

The observations made by Evans, Dawson, and Lalich on the use of Mg alone in producing structural changes of graphite in cast iron should be noted. Figure 7-7 shows the correlation between the residual Mg content of cast iron and the microstructure. Curve (a) is for an addition free of Ti and Ce. Small variations of the Mg content lead to large changes of microstructure. Too little Mg does not produce a fully compacted structure while overtreatment produces graphite in spherulitic form.

The differences noted between under- and overtreatment could be as little as 0.005 per cent magnesium. Combinations of elements were able to extend the range of compositions capable of producing vermicular graphite structures. Each added element contributes individually to the undercooling, thus leaving only part of the necessary undercooling to be provided by Mg.

The combination of Mg and Ce in vermicular graphite production is particularly interesting since both elements contribute to the kinetic undercooling. The reactivity of Ce with graphite appears to be less than that of Mg. Titanium has a mixed role, initially reacting with some of the Mg and then becoming a constitutional supercooling constituent. The net result of a complex addition is then to extend the range of composition of the added elements which can produce a required structure. This means that Mg by itself is too reactive and hence its composition range of effectiveness is very small. Cerium is insufficiently reactive and depends for its action on the carbon content. The combination of Mg and Ce allows control within a broader composition range of added elements. The addition of Ti acts both to limit the Mg interaction with graphite and also, it appears, to provide some constitu-

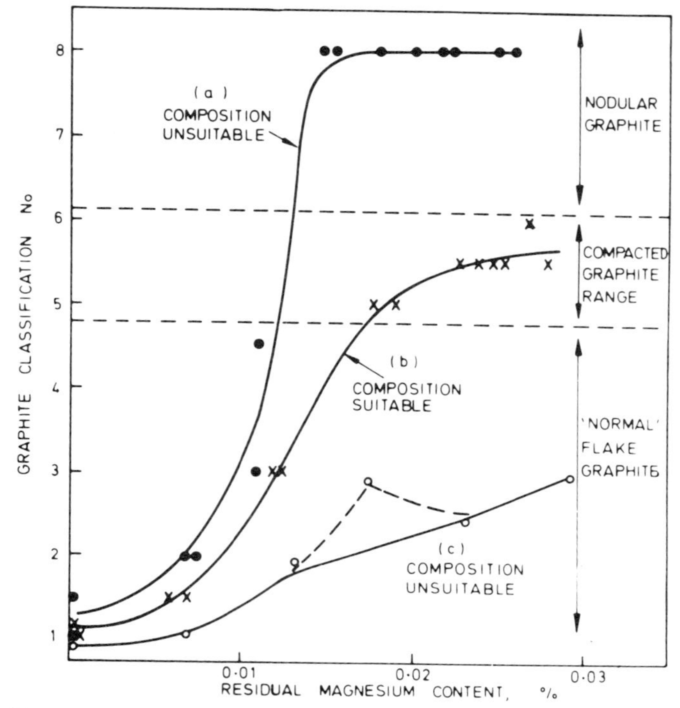

Figure 7-7 Correlation between residual Mg content of cast iron and microstructure. Small variations of Mg content lead to significant changes of microstructure. Note especially curve (a) which is for an addition free of Ti and Ce.[16] (Reproduced by permission of American Foundrymen's Society)

tional supercooling required to promote the vermicular structure. In the experimental programme, these authors showed the effect on microstructure of a high-purity cast iron when Mg additions were varied. A compacted graphite structure was obtained with a content of 0.016 per cent magnesium while spheroidal structures were obtained with 0.033 per cent magnesium.

Using combinations of Mg, Ce, and Ti, the curve of Figure 7-7 was obtained. This showed that the required structures could be obtained with residual Mg contents up to 0.03 per cent, an amount of Mg which normally would give a spheroidal graphite structure.

Curve (c) was obtained for treatments of alloys in which Ti and/or Pb was present, but the treatment did not include Ce.

The recommended alloy addition for the production of vermicular structures gave contents of elements as follows:

Mg	0.015–0.035%
Ti	0.06–0.13%
Ce	Trace

When 0.1 and 0.5 per cent cerium bearing Mg–Fe–Si alloys were used, the results were always similar. It was suggested that this indicated the presence of Ce to be essential but that the actual amount required was extremely small.

The alloy developed for treatment was a 5 per cent Mg–Fe–Si containing about 7 pe cent titanium and 0.3 per cent cerium. With this it was found possible to operate on a wide range of S contents. However, it appeared that compacted graphite structures were easier to produce at compositions approximately eutectic.

7-6 Review of Theory with Respect to Morphological Change

The theory outlined in Chapter 5 for morphological change of graphite in growth is concerned with progressive changes of shape with progressively increasing undercooling. The undercooling is achieved by adjusting the composition of the liquid so that the crystallizing graphite grows with an interface temperature determined by the solutes present. With respect to the growth of intermediate forms it is convenient here to restate the different solute effects as noted:

(a) An effect related to strongly adsorbed solutes. These are reactive in solution, examples being magnesium and rare earth elements. The effect is termed kinetic since it influences the attachment of atoms to the interface, slowing down the growth kinetics and producing an interface undercooling.
(b) An effect related to a layer at the interface enriched in solute, leading to constitutional supercooling. This interface undercooling is influenced by the type of solute and its concentration. The solute type is categorized by its k value (i.e. distribution coefficient—the ratio of solute in solid to solute in liquid) and the slope of the liquidus. Elements like P, Pb, Sn, and Cu can have a marked interface undercooling effect.
(c) An effect related to surface active elements which adsorb with weak, van der Waals forces at the interface and lower the interface energy. This has an effect of changing the growth temperature for graphite, bringing it closer to the equilibrium value.

The influence of solutes in solution is additional but interactions occur, particularly with solutes which behave in a strongly reactive manner. For solutes of the same type, the addition principle appears to hold, i.e. Mg or Ce may be added together so that a part of the kinetic effect is contributed by both elements. In this way, the strong influence of Mg can be lessened. These effects play an important role in obtaining intermediate vermicular structures and also spheroidal graphite structures.[20]

The actual morphology of the graphite crystals can vary between coral type crystals and spherulites, but for the compacted graphite type the preservation of an interconnected character is a requirement. Figure 7-8 shows one type of crystal observed, its form being neither flake nor spherulite.[21]

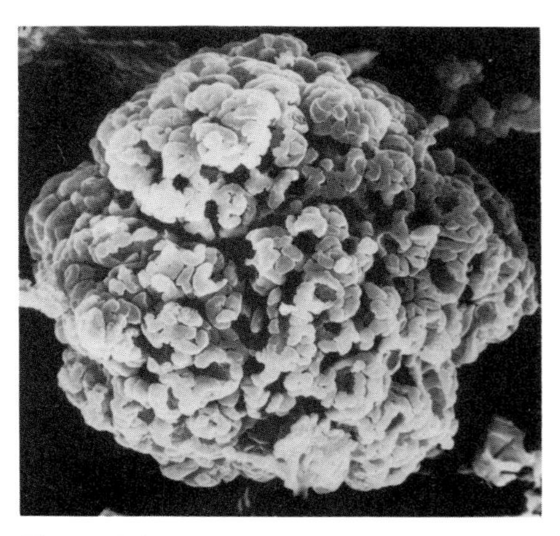

Figure 7-8 Type of crystal obtained by a melt addition which is insufficient to produce a spherulite[21] (× 780)

7-7 Scheil Segregation Equation[19]

It is important to take into account the possibility of marked segregation of impurity elements in solidification. Any change of solute at the boundary may change the mode of growth of the graphite.

The Scheil equation can be written:

$$C_S = kC_0(1 - f_S)^{k-1}$$

or

$$C_L = C_0 f_L^{k-1}$$

where C_L = solute concentration of the liquid
C_o = solute concentration of the original solution
f_L = fraction of liquid
k = distribution coefficient

For elements like Bi, Pb, and Sn which have a k value of approximately 0.1, the interdendritic concentration of solute in the liquid areas between freezing solid will be as follows for the case of approximately 10 per cent. of remaining liquid:

$$\frac{C_L}{C_0} = (0.1)^{-0.9}$$

This has a value of approximately 8. Hence, if initially the total impurity concentration of the elements Bi, Pb, and Sn is approximately 30 ppm, the segregated value may be approximately 240 ppm. Munitz and Minkoff[10] calculated an interface undercooling for graphite in Ni–C due to Bi, Pb, or Sn of about 3.5 K for every 100 ppm of solute. Therefore the effect of segregation would be to increase this figure to approximately 8.5 K.

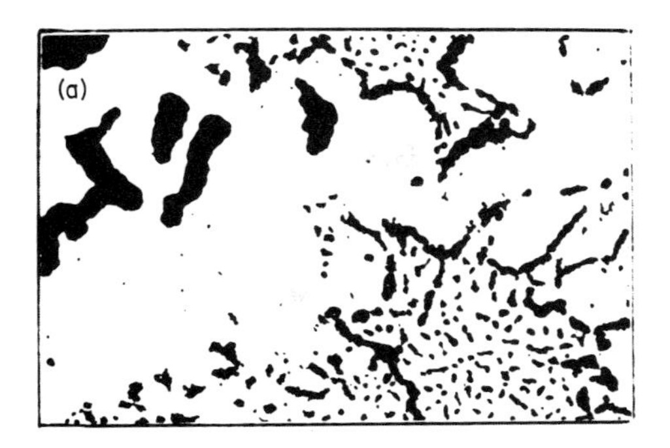

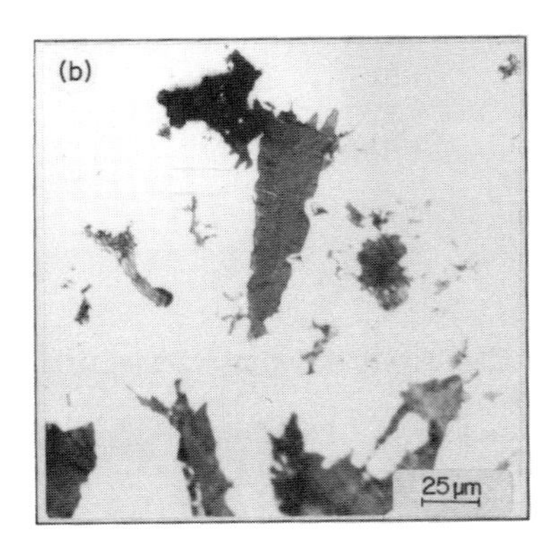

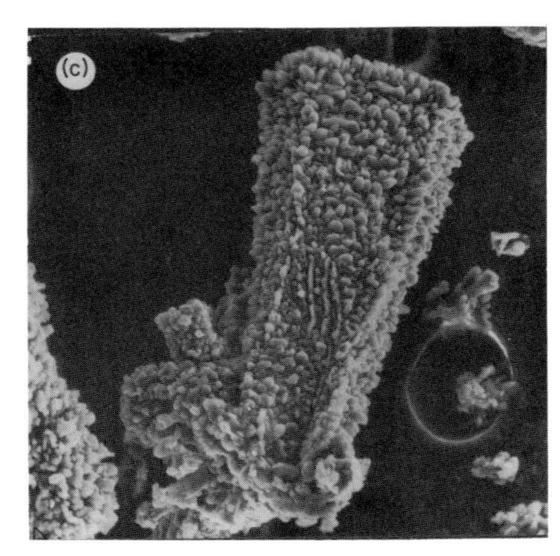

Figure 7-9 (a) Structures in cast iron, shown by Sissener *et al.*[13] Note particularly the elongated forms in the upper left-hand corner. (Reproduced by permission of American Foundrymen's Society) (b) Forms observed by Munitz[21] in Ni–C alloys with high Mg and high Pb contents (× 220). (c) Extracted graphite pyramidal crystal conforming to the type of metallographic section observed in (b). The pyramid is bounded by $(11\bar{2}1)$ faces[10] (× 550)

The two main features of segregation in spheroidal graphite or intermediate structure irons are the concentration of elements like Bi, Pb, and Sn in some areas and the removal of elements like Mg in others. The latter may occur principally by interaction with sulphur. Initially, the sulphur content of the melt is determined by the reactive element addition, i.e. Mg, Ce, etc. This brings the sulphur level down to low values. However, in solidification, sulphur segregates, as does oxygen and other elements which interact with the magnesium. In consequence, many different structures are observed, particularly in thick-wall castings where segregation is problematic.

Figure 7-9(a) shows a microstructure from the research of Sissener *et al.*[13] It shows crystals resembling the pyramidal forms in metallographic section. A typical microstructure of pyramidal form is shown in Figure 7-9(b), while an extracted pyramidal growth form is shown in Figure 7-9(c). This was observed by Munitz and Minkoff[10] in Ni–C alloys with Mg and high Pb contents.

The manner of dealing with these effects was discussed by Mayer and Hämmerli.[22] They suggested careful selection of charge materials to control segregation. In the method suggested by Minkoff[23] it is possible to make calculations of segregation and define limiting values of impurities to avoid harmful structures.

References

1. Morrogh, H. and W. J. Williams: *J. Iron Steel Inst. London*, **158**, 306, 1948.
2. Lux, B., and M. Grages: *Pract. Metallogr.*, **5**, 123 1968.
3. Lux, B., M. Grages, and D. Sapey: *Pract. Metallogr.*, **5**, 587, 1968.
4. Lux, B., W. Bollman, and M. Grages: *Pract. Metallog.*, **6**, 530, 1969.
5. Minkoff, I., and B. Lux: *Micron*, **2**, 282, 1971.
6. Minkoff, I. and B. Lux. *The Metallurgy of Cast Iron* (Eds. B. Lux, F. Mollard and I. Minkoff), Georgi Publ. Co., Switzerland, 1975.
7. Lux. B., I. Minkoff, F. Mollard and E. Thury. *The Metallurgy of Cast Iron* (Eds. B. Lux, F. Mollard and I. Minkoff), Georgi Publ. Co., Switzerland, 1975.
8. Minkoff, I., and B. Lux: *Nature*, **225**, 540, 1970.
9. Minkoff, I., and B. Lux: *Pract. Metallogr.*, **8**, 69, 1971.
10. Munitz, A., and I. Minkoff: *Int. Foundry Congress*, Budapest Congress Papers 45th, Paper 32, 1978.
11. Lux, B.: *Giessereiforsch.*, **19**, 141, 1967.
12. CIATF Commission 7.6: Compacted Graphite Cast Irons, Minutes of Meeting, Detroit, April 24, 1978, C. R. Loper Chairman.
13. Sissener, J., W. Thury, R. Hummer, and E. Nechtelberger: *AFS Cast Metals Res. J.*, **8**, 178, 1972.
14. Estes, J. W., and R. Scheidewind, *Trans. Am. Foundrymen's Soc.*, **63**, 541, 1955.
15. Schelleng, R. D.: *Trans. Am. Foundrymen's Soc.*, **74**, 700, 1966.
16. Evans, E. R., J. V. Dawson, and M. J. Lalich: *Trans. Am. Foundrymen's Soc.*, **84**, 215, 1976.
17. Thury, W.: *AFS Cast Metals Res. J..*, **6**, 163, 1970.
18. Iron and Steel Examination Chart 1560–57. (See Ref. 15.)
19. Scheil, E.: *Z. Metallk*, **34**, 70, 1942.
20. Lo-Kan: *Fonderie*, **367**, 167, 1977.
21. Munitz, A. D.Sc. Thesis. Technion, Haifa, 1977.
22. Mayer, H., and F. Hämmerl: *Sulzer Technical Review*, **1**, 1, 1972.
23. Minkoff, I.: *Delta Tee*, Technion Res. Dev. Assoc. 1981.

Chapter 8
Alloy Cast Iron Systems

The addition of alloying elements to cast iron considerably extends the range of structures obtainable on casting and on subsequent heat treatment. A short presentation is made in this chapter of some of the fundamental features of the binary and ternary systems for the different alloy cast irons. The treatment is given along the following lines:

1. White cast irons with chromium; the Fe–Cr–C system; martensitic Cr–Mo white cast irons.
2. Gray or white solidification in alloy cast irons.
3. Nickel in cast iron; nickel chromium martensitic white cast iron (Ni-hard); austenitic cast irons (Ni-resist), gray and spheroidal graphite types.
4. Low alloy gray and spheroidal graphite cast irons; pearlitic irons; bainitic or acicular cast irons; martensitic gray or spheroidal graphite cast iron.
5. Aluminium cast irons; the Fe–C–Al system.
6. Ferritic cast iron; silicon and silicon molydenum alloys.
7. Copper in cast iron; tin in cast iron.

8-1 Alloy White Cast Irons

Alloy white cast irons fall into the following three categories:

(a) High chromium, high carbon irons, possibly with small additions of other alloying elements, e.g. Ni, Mo, etc.
(b) Martensitic low alloy white irons (Ni-hard type).
(c) Low alloy white irons, mainly with small Cr additions.

The Fe–Cr–C ternary diagram provides a basis for a general understanding of alloys in this system and will be considered first.

The Iron–Chromium–Carbon System

Figure 8-1 shows the Fe–Cr binary equilibrium diagram[1] which is characterized by a closed γ loop and a region of σ phase. The Fe–C phase diagram has an extended γ region.

An understanding of cast structures and phase relationships in the

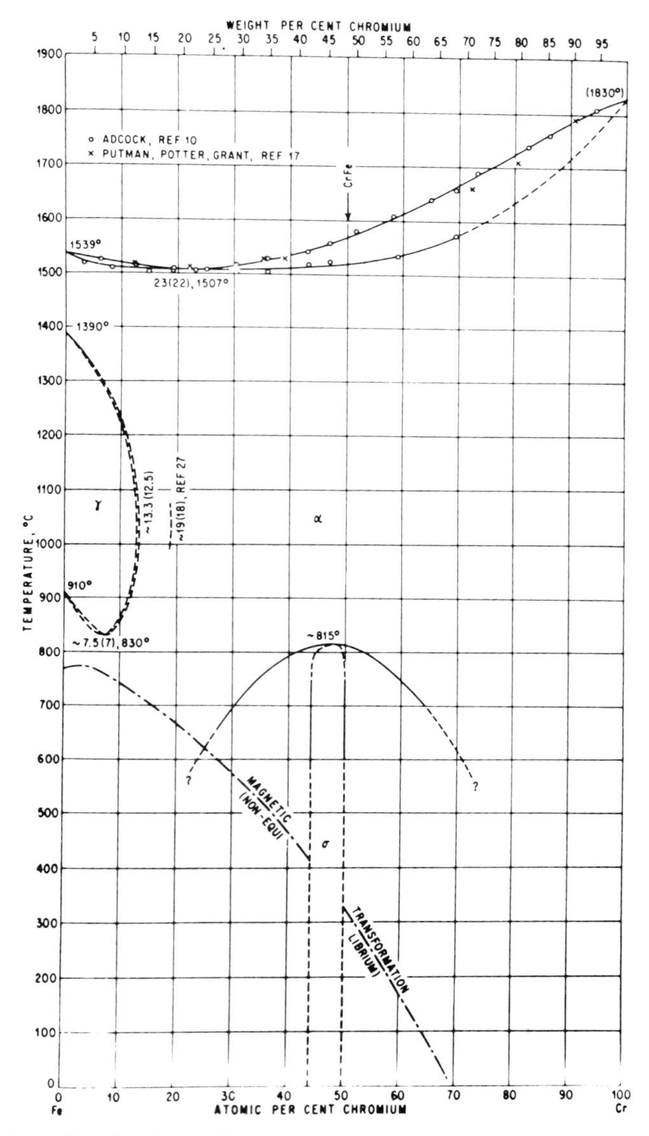

Figure 8-1 Fe–Cr binary diagram.[1] (From *Constitution of Binary Alloys* by M. Hansen. Copyright © 1958 McGraw-Hill Book Company. Used with the permission of McGraw-Hill Book Company)

Fe–Cr–C system can be obtained from the ternary diagrams of Figures 8-2 and 8-3. Figure 8-2(a) shows that in high chromium alloys, the structure is distinguished by the presence of the K_2 (M_7C_3) phase. At low chromium contents, the K_c phase (M_3C) characterizes the structure. These two carbide

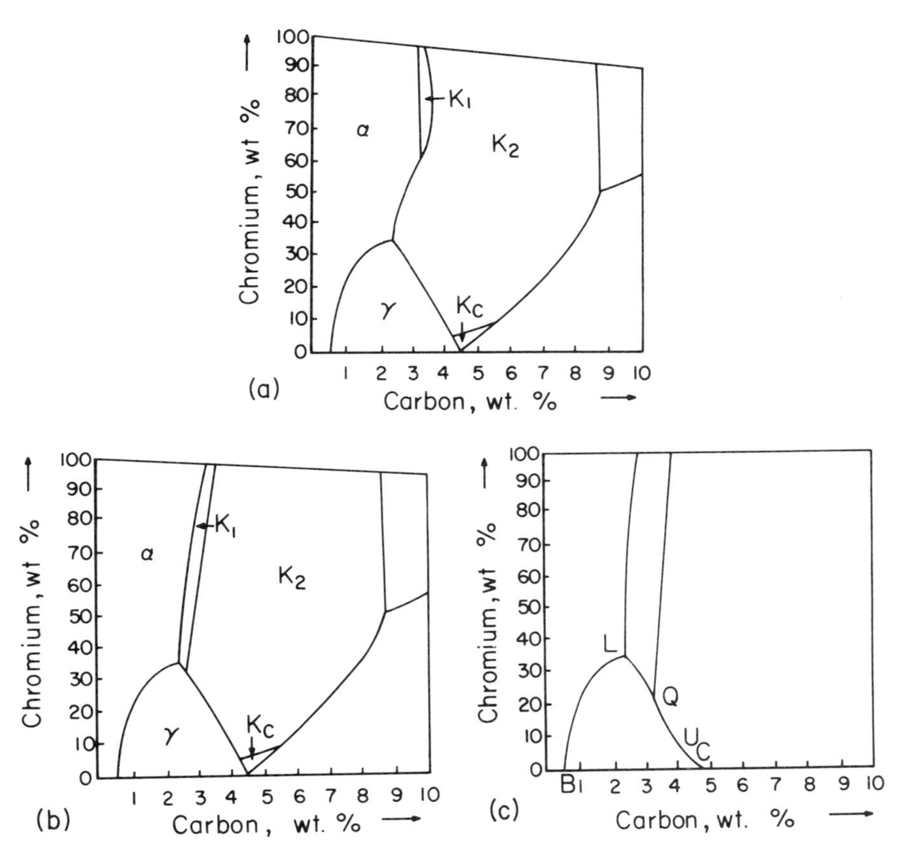

Figure 8-2 Two possible forms of the liquidus surface for the ternary Fe–Cr–C system up to 10 per cent C.[2,3] The two diagrams differ in defining the extent of the K_1 field. In (a), the K_1 field terminates on the α liquidus, while in (b), the K_1 field terminates on the γ liquidus. (Reproduced by permission of *The British Foundryman.*) (c) Regions of the liquidus surface given by Fredriksson and Remaeus[10]

phases form eutectics with the γ phase. Depending on composition, the ternary alloy can be hypo- or hyper-eutectic. If the former is the case, primary γ crystallizes first, followed by eutectic of a carbide phase and γ. If the alloy is hyper-eutectic, the carbide phase will crystallize in a characteristic manner, followed by eutectic. The coarseness of the structure will depend on the freezing rate and section size.

From the slope of the right-hand side of the triangle of the γ liquidus, an approximate relationship between chromium and carbon content to establish a completely eutectic structure can be obtained (Table 8-1). The total percentage of carbides in the structure is roughly given by the following relationship:

$$\text{Total \% carbides} = (12.33 \times \%\text{C}) + (0.55 \times \%\text{Cr}) - 15.2$$

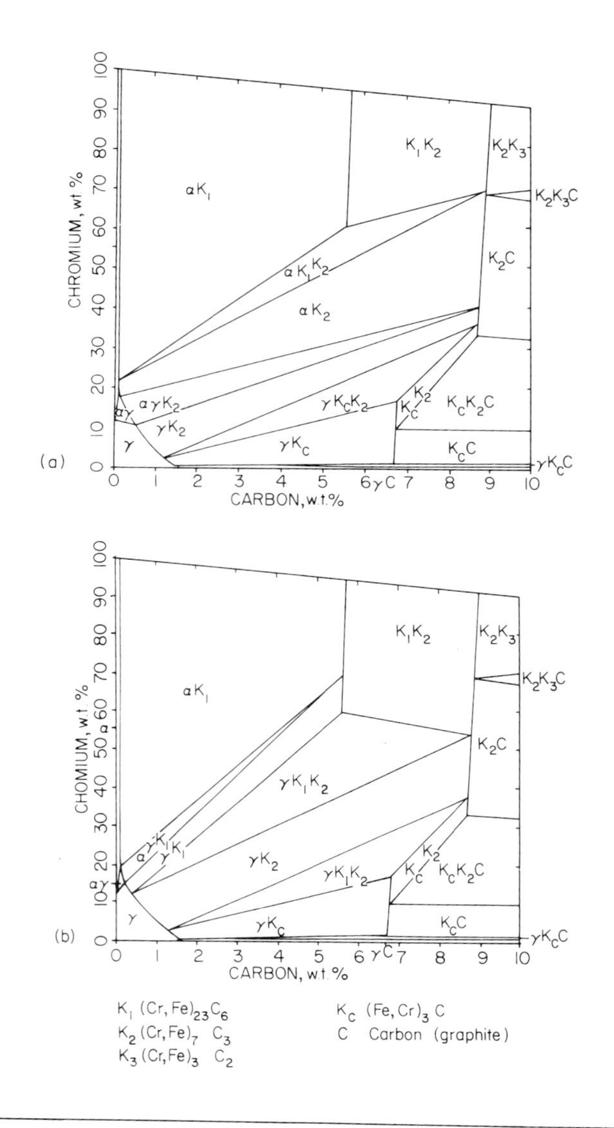

Figures 8-2(a) and (b) show two possible forms of the liquidus surface for the ternary Fe–Cr–C system up to 10 per cent carbon.[3] It can be seen that there are five primary liquidus surfaces for the α, γ, K_1, K_2, and K_c phases where the carbide phases are as follows:

$$K_1 = (Cr, Fe)_{23}C_6$$

$$K_2 = (Cr, Fe)_7C_3$$

$$K_c = (Cr, Fe)_3C$$

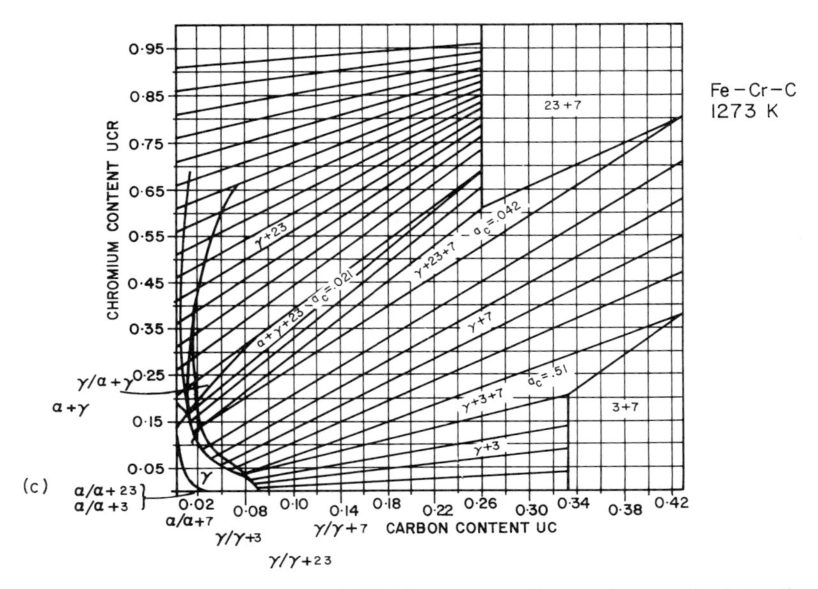

Figure 8-3 Ternary sections at 1273 K corresponding to the two liquidus diagrams of Figures 8-2(a) and (b). From (a) the K_2 phase precipitates from γ in the solid state, but in (b) the carbide phase which precipitates would be K_1.[3] (Reproduced by permission of *The British Foundryman.*) (c) Computer-calculated isothermal section of the Fe–Cr–C system at 1273 K[5]. (Reproduced by permission from Pergamon Press Ltd)

The two suggested diagrams for the liquidi surfaces differ in defining the extent of the K_1 field. Some research has terminated this field on the α liquidus (Figure 8-2a)[3] while other published research[4] terminated the K_1 field on the γ liquidus (Figure 8-2b). These two suggested diagrams would give different ternary sections at 1273 K. Jackson[3] showed the two ternary sections (Figure 8-3a and b) corresponding to the two liquidus diagrams.

Lundberg, Waldenström, and Uhrenius[5] published computer calculated isothermal sections of the Fe–Cr–C system in the temperature range of 873–1373 K, from which Figure 8-3(c) is reproduced. This shows the isothermal section of the Fe–Cr–C system at 1273 K which compares with the section given by Jackson in Figure 8-3(b). This suggests the liquidus corresponding to Figure 8-2(b) as the correct one.

Table 8-1 Cr–C relationships for a completely eutectic structure

%Cr	%C
15	3.6
20	3.2
25	3.0

In Figure 8-3(c), the concentrations are represented by the mole fraction of carbon, related to the ordinary mole fraction X as follows

$$U_C = \frac{X_C}{1 - X_C} = \frac{X_C}{X_{Fe} + X_{Cr}}$$

The $M_{23}C_6$ and M_7C_3 carbides are represented by 23 and 7 respectively and cementite M_3C is denoted by 3.

Liquid State Reaction

The five liquid surfaces are separated by grooves, all sloping down to the γ/Fe_3C eutectic. The composition limits and structures of the carbide phases are given in Table 8-2, which is taken from the data of Jackson.[2]

Fredriksson and Remaeus,[10] using a liquidus surface conforming to Figure 8-2(b), have labelled the important regions as follows and have defined the peritectic reactions of Table 8-3:

BLQUC	γ primary liquidus
BL	Peritectic between liquidus surfaces bordering α and γ phases
L	Peritectic line meets a eutectic line between the liquid, γ, and $(Cr, Fe)_{23}C_6$ phases

The binary eutectic which crystallizes from high carbon–high chromium alloys is γ-$(Cr, Fe)_7C_3$.

Table 8-2 The composition limits and structure of the carbides in Fe–Cr–C (up to 9 per cent carbon)

Carbide	Structure	Composition limits
$K_c(Cr, Fe)_3C$	Orthorhombic	6.67%C. Can dissolve up to 20% Cr by weight[6]
$K_1(Cr, Fe)_{23}C_6$	Cubic	5.6%C. Can dissolve up to 59.0% Cr by weight[7]
$K_2(Cr, Fe)_7C_3$	Hexagonal	9%C. In equilibrium with α contains 26.6%Cr (min.) and 70% Cr (max.)[8,9]

Table 8-3 Peritectic reactions in Fe–Cr–C according to Fredriksson and Remaeus[10]

Point in diagram	Reaction	Temperature, °C	%Cr	%C
L	$L + \alpha \rightarrow \gamma + M_{23}C_6$	1275	34	2.4
Q	$L + M_{23}C_6 \rightarrow \gamma + M_7C_3$	1255	23	3.5
U	$L + M_7C_3 \rightarrow \gamma + M_3C$	1175	8	3.8

The σ Phase

The σ phase in the Fe–Cr–C system is sluggish in its transformation kinetics and may not normally form. However, it can appear in alloys used for extended periods at high temperature. Figure 8-4 shows a vertical section of the Fe–Cr–C ternary diagram at 1.5 per cent carbon.[11] In alloys containing 15 to 40 per cent chromium or more, depending on the carbon content, the α solid solution may be gradually converted to the hard brittle σ phase by holding it within the temperature range of 550–820 °C for long periods.

Carbon Content of the γ Phase; Hardness of Martensite

The ternary section at 1000 °C of Figure 8-3(b) shows the composition relationships between carbon in the γ phase and the K_2 carbide of the eutectic (region γK_2). For alloys of chromium content in excess of 15 per cent, the austenite is low in carbon, being approximately 0.5 per cent carbon, while the carbon content increases to 0.8 per cent for decreasing chromium.

The hardness of martensite is a function of carbon content. The hardness of a martensitic matrix in Fe–Cr–C alloys is therefore related to the chromium content; this is a factor of influence in selecting compositions for abrasion resistance.[12]

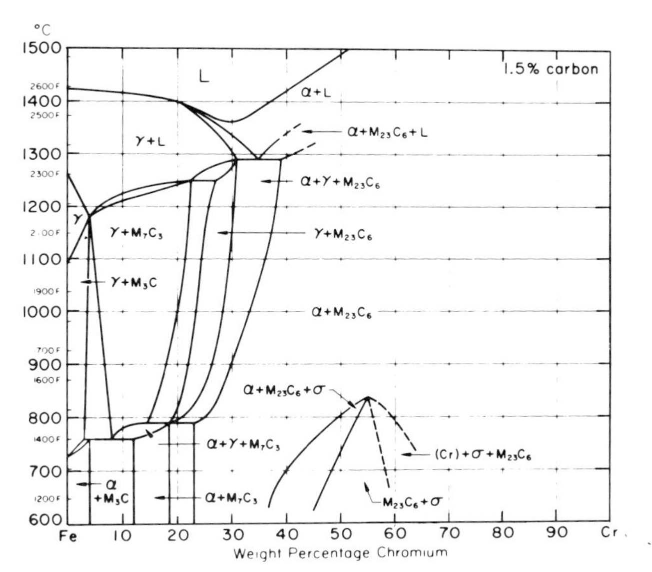

Figure 8-4 Vertical section of Fe–Cr–C ternary diagram at 1.5 per cent C.[11] (Reproduced from W. D. Forgeng and W. D. Forgeng, Jr. "C–Cr–Fe" Metals Handbook: Metallography, Vol. 8, 8th ed. Editor, Taylor Lyman, by permission of American Society for Metals © 1973, p. 403)

Observations of Solidification and Transformation in 26 per cent Chromium, 2.5 per cent Carbon Iron

Fredriksson and Ramaeus[10] investigated the solidification and subsequent properties on heat treatment of a high Cr–high C white iron. The suggested optimum alloy content is that indicated in the previous section as giving a eutectic composition. The influence on cast structure of too high a carbon content is the presence of large primary carbides. These drastically reduce the wear resistance. With too low a carbon content, a primary precipitation of γ occurs. This could result in a soft matrix with low wear resistance.

With these guiding factors, the following alloy was investigated:

2.54 %C 1.10 %Si 0.58 %Mn 0.010 %P 0.027 %S 26.2 %Cr

The influence of increasing the cooling rate was to favour the growth of primary γ. The structure in this case consisted of the γ and M_7C_3 phases. The γ phase was depleted in Cr and C close to the carbide phase. This gave martensite in these areas. At a slow cooling rate, pearlite was formed at 700 °C, consisting of α and $M_{23}C_6$ carbide.

The results of heat treatment showed that this did not increase the hardness of the product. At a high austenitizing temperature, only a low percentage of martensite formed and at a low austenitizing temperature, the matrix was soft. This could be remedied by casting with lower Cr contents which, as noted, give higher C contents in the γ and higher hardness.

General Classification of Chromium White Cast Irons

The chromium cast irons can be classified, according to the characteristics of structure determined by the ternary diagram, in the following three categories:

(a) 33 per cent chromium irons, mainly for high temperature properties.
(b) 14–28 per cent chromium cast irons (containing molybdenum) with a martensitic matrix, employed principally for abrasion resistance.
(c) Low chromium white cast irons. These offer advantages in stability of the carbide phase over the unalloyed iron carbon white alloys.

(a) *33 per cent Chromium White Cast Irons* These alloys have been described in a number of papers (see, for example, Colton[13]). The structure appears to be one of austenite and carbides, but ferrite is also reported. The applications are given as mainly in conditions requiring oxidation resistance up to 1050 °C or in mildly corrosive conditions.

(b) *14–28 per cent Chromium Cast Irons* The as-cast structure consists of a discontinuous $(Cr\ Fe)_7C_3$ phase and γ phase which may be completely or partially transformed to martensite during cooling. For abrasion resistance, the optimum properties are obtained from the cast iron when the matrix is martensite. This is normally achieved by heat treatment, although it is possible to leave the austenite to transform to martensite with mechanical loading in abrasion conditions.

(c) *Cast Irons with Low Chromium Contents* In the solidification process of the Fe–Cr–C alloys, the chromium goes entirely into the carbide phase. With small chromium compositions, this phase will be the Fe_3C type. An effect of Cr is to stabilize the carbide phase against decomposition, i.e. graphite formation, such as in roll manufacture.

High Chromium–Molybdenum Alloys; White Cast Iron Metallurgy

An extensive survey of high chromium–molybdenum irons and the metallurgy of white cast irons within this system was given by Dodd and Parkes.[14] The research was directed to thick-section castings. It reviewed phase relationships in the Fe–Cr–C system, the microstructure of the cast and heat-treated alloys, and the influence of alloying elements on hardenability. The authors pointed out the complex relationships which exist between alloying elements, heat treatment variables, the M_S temperature, and retained austenite.

The limitations in application of the Fe–Cr–C ternary equilibrium diagram to casting were pointed out. Equilibrium is not normally obtained in solidification and other alloying elements change the phase relationships. However, the Fe–Cr–C diagram itself is important in showing the composition of phases at equilibrium for a given temperature, e.g. the carbon and chromium contents in austenite.

Molybdenum and also manganese are soluble in M_7C_3 carbides. In addition to the carbides noted in the Fe–Cr–C system (see Table 8-2), molybdenum-rich M_2C eutectic carbides and also M_6C carbides have been observed in high chromium–molybdenum cast irons. Table 8-4 shows the hardness values of typical microconstituents.

Table 8-4 Typical hardness values of various microconstituents in Cr–Mo cast irons.[14] (Reproduced by permission of American Foundrymen's Society)

Carbides	Vickers hardness
M_3C	840–1,100
M_7C_3	1,200–1,800
Mo_2C	1,500
Matrix	**Vickers hardness**
Ferrite	70–200
Pearlite	300–460
Austenite (high Cr iron)	300–600
Martensite	500–1,000

Different compositions of the molybdenum-containing alloys are selected for different applications. In general, the alloys employed are hypo-eutectic and fall within the γ field of the phase diagram. The hypo alloys solidify by γ crystallization followed by the γ–M_7C_3 eutectic. The hyper-eutectic alloys solidify first by crystallization of the M_7C_3 phase, followed by the eutectic. The hyper-eutectic alloys are used for hammer mill impact bars and are 26 per cent chromium, 2 per cent molybdenum in composition. The hypo-eutectic alloys are normally employed for thick sections and were reported to have compositions of 2.4–3.0 per cent carbon and 18–22 per cent chromium. On cooling, the austenite phase may transform to structures of α + carbide, or to martensite, or it may be retained untransformed. The different transformation products are discussed in Chapter 9. The martensitic structure in these alloys is obtained by heat treatment involving at first the precipitation of secondary carbides at high temperatures (destabilization) followed by transformation on cooling. The composition of the alloy, the time and temperature of the destabilization treatment, and the cooling rate are interrelated. For some casting configurations a recommendation is to obtain a pearlitic structure in the as-cast condition. This facilitates machining and avoids transformation-induced cracking. Subsequent heat treatment is performed to produce a martensitic structure (Chapter 9).

8-2 Gray or White Solidification in Alloy Cast Iron

Gray or white solidification was dealt with for Fe–C–Si alloys in Chapters 3 and 4, where it was shown to be dependent on graphite nucleation and temperature of growth. In alloyed cast irons, composition is a determining factor. It is illustrated in the following for the case of a Ni–Cr alloy cast iron. In Ni–Cr alloy cast irons, either carbide or graphite phases may be present in the structure. In addition to growth from the liquid, graphite can form by solid state decomposition of cementite in cast iron, and alloying elements can accelerate the transformation. Fredriksson and Hillert[15] examined the formation of graphite in a cast iron alloy having a nickel and chromium alloying content. The nickel by itself was sufficient to produce gray solidification and the chromium content by itself was sufficient to produce white solidification. The alloy had the following composition:

3.35 %C, 0.8 %Si, 1.85 %Cr, 4.80 %Ni, 0.35 %Mo, 0.10 %P, 0.30 %S

The nickel content was sufficiently high to prevent a phase transformation of austenite during cooling. The chromium content was sufficiently high to balance the graphitizing effect of Ni and Si. The alloy could therefore commence solidification in a white mode for a sufficiently high rate of cooling. In a mottled cast iron, graphite eutectic colonies generally have a spherical form due to undisturbed growth in the melt. The white eutectic solidification should follow the gray as explained on kinetic grounds (Section 3-4). The spherical colonies in these materials can be used as a means of identifying the graphite formed directly from the melt and distinguishing it from graphite formed by solid state decomposition of Fe_3C. Graphite formed in the solid state is related in morphology to the temperature of formation below the eutectic. Usually the forms are compact in type, but less compact aggregates form at higher temperatures. Flake-like graphite can grow from the solid, reportedly close to the eutectic temperature.[16] Hence it is not always possible to use the shape of graphite flakes to distinguish between graphite grown in the liquid and graphite grown in the solid.

From these studies of graphite formation in the alloy cast iron used for the experiments, it was found that the white eutectic apparently solidified first. All the chromium goes into the carbide phase but nickel segregates to the eutectic cell boundaries. Graphite eutectic solidification then occurs in the Ni-enriched liquid, and when it occurs is confined to the spaces between the white cells and does not have a characteristic cell shape.

In addition to this mode of growth of graphite in these types of alloyed iron, it was reported that some graphite could form by solid state decomposition of cementite. This would take place in the finally solidified region and be related to segregation.

8-3 Nickel in Cast Iron[17]

Nickel influences the structure of cast iron both as a graphitizer and in its effect on the stability or decomposition of austenite. In low alloy irons, it has a pronounced effect on austenite transformation. With higher alloy contents it is the main influence in promoting the stable austenitic phase.

A binary Fe–Ni phase diagram is shown in Figure 8-5[18] The effect of Ni on the γ phase region of Fe is similar to that of C. Nickel expands the region of γ. At 30 per cent nickel, the $\gamma \rightarrow \alpha$ transformation is completely suppressed at room temperature.

In the ternary Fe–C–Ni alloy, a beneficial effect of Ni on solidification is noted by the reduction of the carbon content of the eutectic. The interval between the γ–graphite and γ–Fe_3C eutectic transformation is increased. This latter influence is a factor in suppressing white iron formation and therefore Ni in cast iron acts to prevent a chill or mottling tendency.

In solid state transformations, the influence of the nickel is to move the pearlite knee to higher time intervals and the temperature of the start of the martensite transformation is suppressed. In white cast irons alloyed for heat treatment to promote martensite and containing Ni, additional Cr is generally required to compensate for any tendency of Ni to promote graphite formation in solidification.

Austenite Transformation

The influence of small amounts of Ni in moving the knee of a transformation diagram for cast iron to higher time intervals is shown in Figure 8-6. The

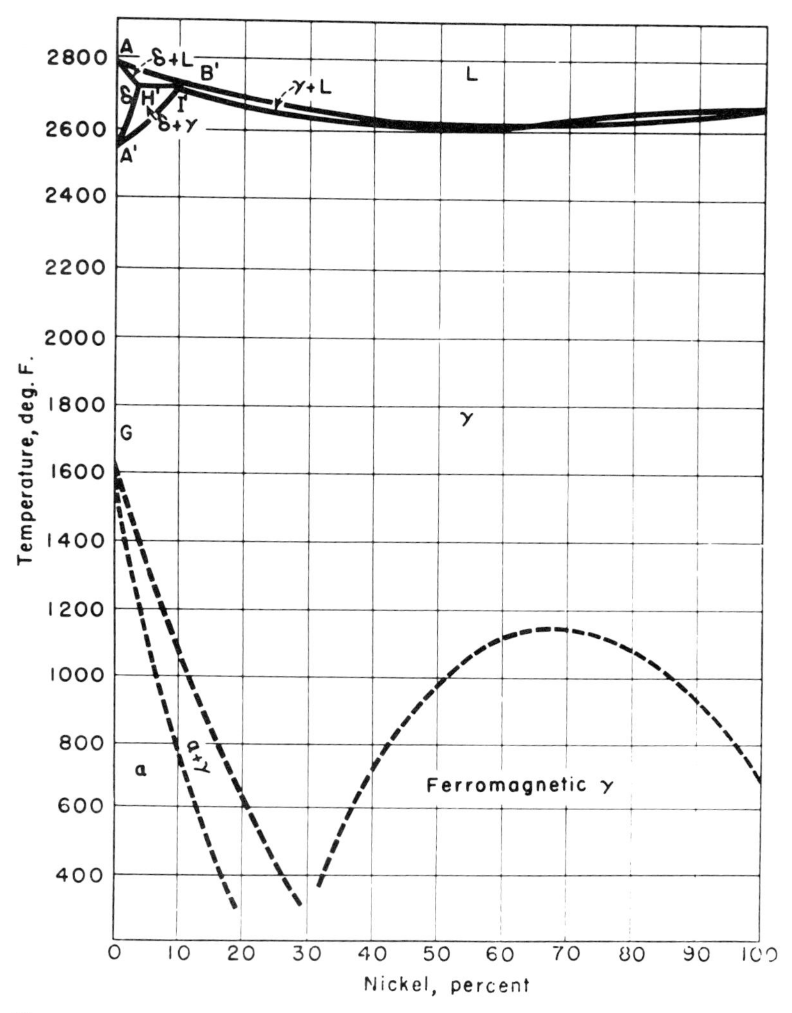

Figure 8-5 The Fe–Ni binary diagram.[18] (From *Alloys of Iron and Nickel* by J. Marsh. Copyright © 1977 McGraw-Hill Book Company. Used with the permission of McGraw-Hill Book Company)

transformation is compared of a nickel-free cast iron with a cast iron containing 2.03 per cent nickel.[19] One influence of nickel on structure is to refine the pearlite. At about 6–8 per cent nickel, the eutectoid transformation is suppressed and in cooling γ transforms directly to martensite. At about 12–14 per cent nickel, the M_S temperature becomes less than room temperature so that γ becomes the stable phase. This gives the following division in nickel cast irons:

(a) Pearlitic cast iron—up to 3 per cent nickel.

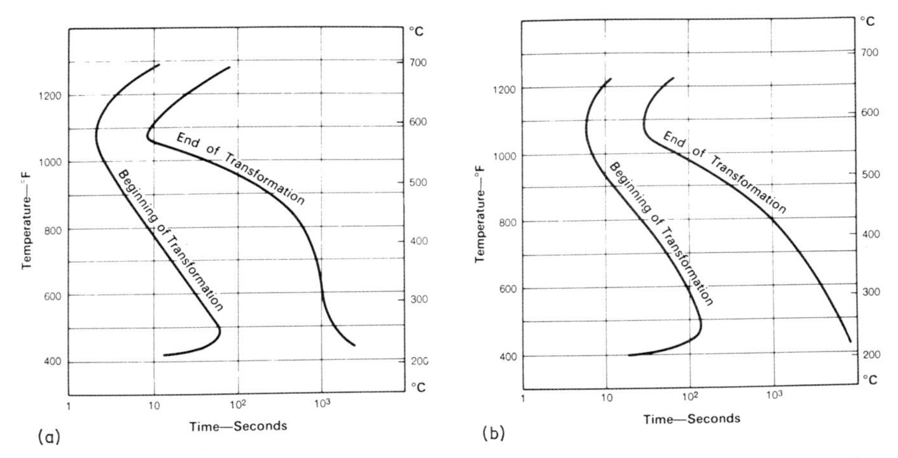

Figure 8-6 Influence of small amounts of Ni in moving knee of transformation diagram for cast iron to higher time intervals. (a) Nickel-free cast iron and (b) cast iron with 2.03 per cent Ni.[19] (Reproduced by permission of Iron Castings Society)

(b) Martensitic cast iron—between 4 and 8 per cent nickel. The structure would depend on the cooling rate.
(c) Austenitic cast iron—between 14 and 20 per cent nickel.

The austenitic irons have an advantage in temperature cycling in the range 670 °C and above which is the range for the $\gamma \rightarrow \alpha$ transformation. The volume changes associated with this transformation lead to warping or cracking of castings. Since the structure in these irons remains austenitic, the alloys resist warping and associated failure. However, it is important to heat treat the alloys to achieve the pre-precipitation of carbide. Otherwise these separate in thermal cycling, leading to volume changes, warping, and cracking.

8-4 Nickel–Chromium Alloy Martensitic White Cast Irons (Ni-Hard Cast Irons)

Cox[20] and Durman[21] have discussed the applications of either the high chromium type of cast iron or the nickel alloy (Ni-hard) cast iron types in wear applications. The Ni-hard alloy cast irons containing approximately 4.5 per cent nickel and 1.5–3.0 per cent chromium are used in fields of application comparable with the high chromium cast irons. These alloys are used for high abrasion resistance. They may be used either in the as-cast state without heat treatment or alternatively after heat treatment at subcritical temperatures.

The difference in structure between the Ni-hard type alloys and the high chromium alloys lies in the nature of the carbides of the eutectic and in the character of the martensitic matrix. The eutectic carbides in Ni-hard are of

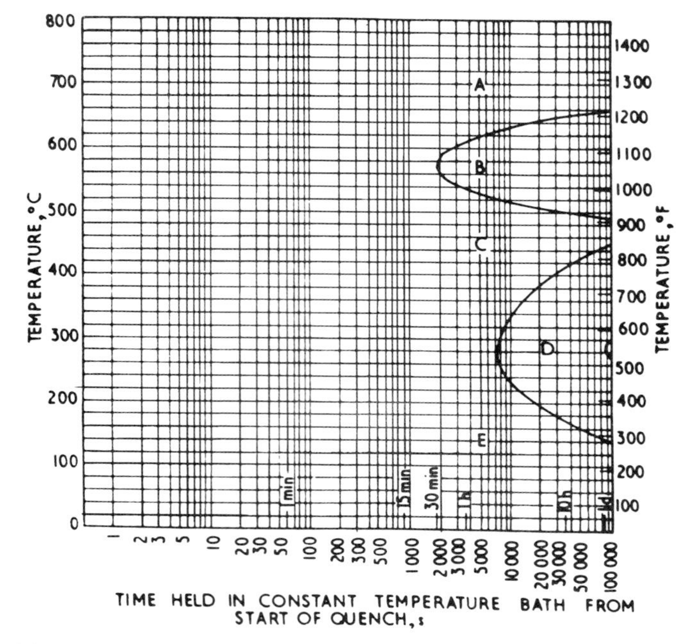

Figure 8-7 Approximate isothermal transformation curve for cast iron containing 3.0–3.5 per cent Ni, 1.5–2.5 per cent Cr, 2.8–3.5 per cent C.[22] Bainite nose 250 °C. (Reproduced by permission of *The British Foundryman*)

the M_3C type. The matrix is predominantly a high carbon twinned martensite together with residual austenite.

Figure 8-7 shows an approximate isothermal transformation curve given by Harrison and Dixon.[22] This is for an alloy having the following composition:

2.8–3.5 %C, 0.2–0.8 %Si, 0.2–0.8 %Mn, 3.0–3.5 %Ni, 1.5–2.5 %Cr, <0.15 %S, 0.4 % max P

Table 8-5 gives the composition and physical properties of two types of Ni-hard (type 1 and type 2). These differ in C content and thus in the amount of carbide. This gives a higher hardness to type 1. Type 2 will have a higher tensile strength and somewhat higher impact properties.

8-5 Austenitic Cast Irons

Increasing the Ni content of cast iron leads to austenitic structures and produces a series of alloys which are used both for corrosion resistance applications and also for high temperatures. Commercial alloys of this character may have varying concentrations of Si, Cr, Al, and Mn in addition to Ni. Both flake and spheroidal graphite grades exist.

Table 8-5 Ni-hard type 1 and type 2 compositions.[17] (Reproduced by permission of American Foundrymen's Society)

	Ni-hard type 1	Ni-hard type 2
C	3.5–3.6	2.90 max.
Ni	4.0–4.75	4.0–4.75
Cr	1.40–3.50	1.50–3.50
Si	0.40–0.70	0.40–0.70
Mn	0.40–0.70	0.40–0.70
Hardness	R_c 53	R_c 52
Hardness rapidly chilled	R_c 56	R_c 55
Tensile strength	275–345 N mm^{-2} (40–50,000 psi)	310–380 N mm^{-2} (45–55,000 psi)
Tensile strength chilled	345–413 N mm^{-2} (50–60,000 psi)	413–483 N mm^{-2} (60–70,000 psi)

A problem of application at intermediate temperatures was pointed out by Cox[23] in the austenitic spheroidal graphite cast irons containing 5 per cent silicon. This was the transformation of austenite to pearlite at 450–600 °C, occurring in relatively short times. If the Si content is reduced the scaling resistance is impaired. To overcome this problem, an alloy of 30 per cent nickel, 5 per cent silicon, and 3 per cent chromium was recommended.

Ni-Resist: Flake Graphite Type

Five basic types of Ni-resist are listed.[24] The most commonly used are types 1 and 2. Type 1 is low in nickel content and has a high copper content to increase the corrosion resistance. Type 1b is similar to type 1 but has an increased chromium content. Ductile counterparts exist for all the types other than types 1 and 1b where the presence of copper does not allow effective treatment with a spheroidizing addition.

The types 1, 1b, and 2 covered by ASTM A436 are given in Table 8-6. Types 1 and 2 are particularly suited to heavy metal-to-metal wear service. Type 1b has superior corrosion resistance compared with type 1, and is harder and stronger. Type 2 is the most commonly used in corrosive environments for heat and oxidation resistance to 700 °C.

Table 8-6 Ni-resist types 1, 1b, and 2 (flake graphite)[24] (Reproduced by permission of INCO Europe Ltd.)

	TC	Si	Mn	Ni	Cu	Cr	S	Brinell	TS
ype 1	3.0 max.	1–2.8	0.5–1.5	13.5–17.5	5.5–7.5	1.5–2.5	0.12	131–183	172 MPa
ype 1b	3.0 max.	1–2.8	0.5–1.5	13.5–17.5	5.5–7.5	2.5–3.5	0.12	149–212	205 MPa
ype 2	3.0 max.	1–2.8	0.5–1.5	18.0–22.0	0.5 max.	1.5–2.5	0.12	118–174	172 MPa

Table 8-7 Ni-resist types 3, 4, and 5 (flake graphite)[24] (Reproduced by permission of INCO Europe Ltd.)

	TC	Si	Mn	Ni	Cu	Cr	S	Brinell	TS
Type 3	2.60 max.	1–2.0	0.5–1.5	28–32	0.5	2.5–3.5	0.12	118–159	172 MPa
Type 4	2.60 max.	5–6.0	0.5–1.5	29–32	0.5	4.5–5.5	0.12	149–212	172 MPa
Type 5	2.40 max.	1–2.0	0.5–1.5	34–36	0.5	0.10	0.12	99–124	172 MPa

Types 3, 4, and 5 are given in Table 8-7. Type 3 is applicable for severe shock resistance up to 200 °C, and erosion in wet steam and corrosive slime. Type 4 is superior to the other Ni-resist types in resisting erosive, corrosive, and oxidizing environments. Type 5 has minimal thermal expansion, and thus has dimensional stability. It has superior shock resistance up to 430 °C

Ni-Resist: Ductile Types[24]

These are ductile iron counterparts to types 2, 3, 4, and 5 of the flake graphite alloys. The presence of graphite in spheroidal form results in higher strength and ductility so that tensile strengths can be achieved between 390 and 550 Mpa with elongations of 4–40 per cent. Several modifications exist in the type 2 to type 5 alloys. As the alloy content of chromium and molybdenum increases in these types, carbides exist in the microstructure. These increase the tensile properties while lowering ductility and toughness. The analyses of the ductile Ni-resist alloys are roughly equivalent to those of the flake graphite alloys given according to types in Tables 8-6 and 8-7.

8-6 Low Alloy Gray and Spheroidal Graphite Cast Irons

Small additions of alloying elements can be employed to increase the mechanical properties of both gray and spheroidal graphite cast irons. Gagnebin[25] showed a table in which the normal unalloyed classes of flake graphite cast iron, extending up to 310 N mm^{-2} (45,000 psi) in mechanical properties, are improved by small alloying additions to achieve a tensile strength of 413 N mm^{-2} (60,000 psi). With lower carbon contents and after heat treatment, flake graphite cast irons can achieve tensile strengths of 550–620 N mm^{-2} (80,000–90,000 psi).

Acicular Structures

Low alloy gray cast irons containing approximately 1 per cent nickel and 1 per cent molybdenum as alloy additions may be treated to have matrix structures of bainite. In cast irons containing these elements, upper bainite shows a marked growth of light-etching α needles, in a structure termed acicular. To

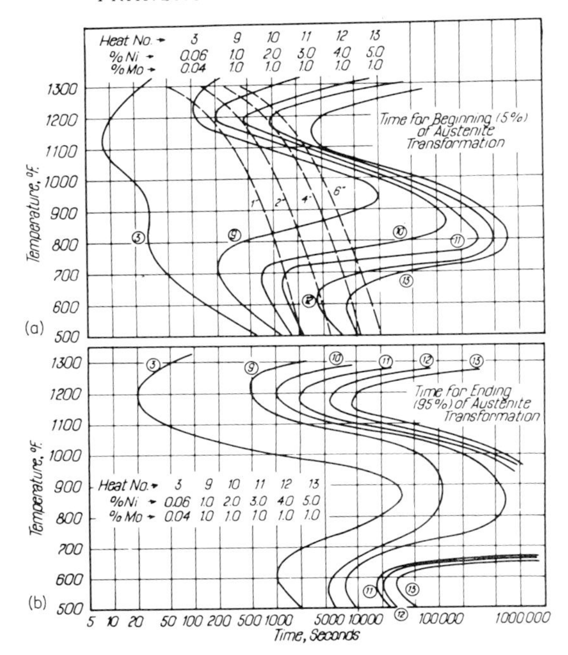

Figure 8-8 Transformation curves of Flinn, Cohen, and Chipman[26] for a series of Ni–Mo alloys having distinct pearlitic and bainitic ranges separated by a deep trough. The influence of Ni and Mo is to deepen the trough. (Reproduced by permission of American Society for Metals)

obtain these structures, the alloy analysis must be propertly adjusted to the section thickness being cast in order to correlate the cooling rate corretly with the isothermal transformation curve.

Figure 8-8(a) shows the start and Figure 8-8(b) shows the end of isothermal transformation curves for a series of Ni–Mo alloys studied by Flinn, Cohen, and Chipman.[26] These curves show the typical isothermal transformation characteristics of the γ phase leading to either pearlite or bainite, separated by a deep trough. Both Ni and Mo widen the bainite field.

Figure 8-9 shows the transformation behaviour related to section thickness for cast irons of varying Ni analysis at a fixed molybdenum content of 1 per cent. From the data of Figure 8-9 it is observed that low alloy contents and large section sizes will promote pearlite. A high alloy content and small section size will tend to yield martensite. Adjustment of the alloy content to the section size being cast yields bainite.

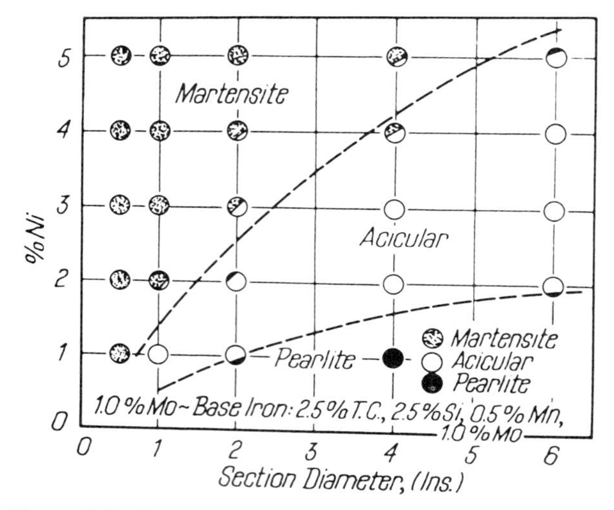

Figure 8-9 Transformation behaviour related to section diameter for cast irons of fixed molybdenum content (1 per cent) and varying nickel content.[26] (Reproduced by permission of American Society for Metals)

Bainite Transformation in Cast Iron

As the temperature of the γ phase of eutectoid composition decreases in the range below the equilibrium eutectoid temperature, the first transformation product possible is pearlite, which is nucleated by Fe_3C.

The next possible transformation is that of bainite which is nucleated by the α phase. Two possibilities have been given for the growth of bainite in cast iron. In the upper temperature range of growth (upper bainite), the α phase grows as a needle and carbon diffuses ahead of the interface. Some precipitation of carbon as carbide can occur. The concentration of carbon in the γ phase makes transformation sluggish and the γ may finally transform to a fine α–carbide aggregate which is dark-etching. The α (ferrite) needles in upper bainite are light-etching.

In the lower temperature range (lower bainite), α needles grow with limited C diffusion so that carbides may precipitate in the α phase. The structure is typically dark-etching and angular, corresponding to bainite found in isothermal transformation in steels.

The light-etching constituent is not common in steel. Table 8-8 shows typical properties of cast irons with pearlite, fine pearlite, and acicular (bainitic) structures.

The strength of bainitic structures in cast iron was reported[27] to be raised by tempering at 315 °C for 5 hours, giving a 40 per cent increase in properties (Table 8-8). This again is not noted for bainitic structures in steel.

Table 8-8 Mechanical properties of low alloy flake graphite cast irons with pearlitic and acicular structures.[26] (Reproduced by permission of American Society for Metals)

Structure	TC	Si	Mn	Ni	Mo	Tensile strength	Brinell
Pearlite	2.49	2.56	0.82	0.06	0.04	419 N mm^{-2} (60,800 psi)	255
						414 N mm^{-2} *(60,100 psi)	255
Fine pearlite	2.26	2.27	0.85	1.09	0.33	490 N mm^{-2} (71,000 psi)	286
						496 N mm^{-2} *(72,000 psi)	286
Acicular	2.30	2.31	0.85	1.03	1.36	490 N mm^{-2} (71,800 psi)	321
						675 N mm^{-2} *(98,000 psi)	340

*Tempered for 5 hours at 315 °C.

8-7 Bainitic and Martensitic Spheroidal Graphite Cast Irons

Gerlach[28] discussed problems of production of bainitic or martensitic spheroidal graphite cast irons related to section size and heat treatment. He recommended the control of as-cast structures by varying the nickel content of castings when they contained a maximum of approximately 1.0 per cent molybdenum. Figure 8-10(a) was reproduced showing different transformation characteristics for cast iron with varying nickel and molybdenum contents.

The problem of controlling the matrix structure is important and in particular with spheroidal graphite cast irons. These have marked segregation patterns. A uniformly transformed structure is dependent on uniformity of composition. In practice mixed structures may be produced. These may include bainite, martensite, and retained austenite. Tempering at 260–270 °C was recommended to transform retained austenite.

Schelleng[29] gave similar information on Ni–Mo bainitic spheroidal graphite cast iron. Tempering of the castings at 315 °C was recommended. Schelleng suggested Ni contents between 2.7 and 5.5 per cent to take into account variations of section size between 2.5 and 15 cm.

Dodd,[30] reviewing spheroidal graphite acicular cast irons, emphasized some of the outstanding properties which may be achieved in relation to those which may be obtained on conventional quenched and tempered martensitic structures. Figure 8-10(b) shows a comparison of properties of the austem-

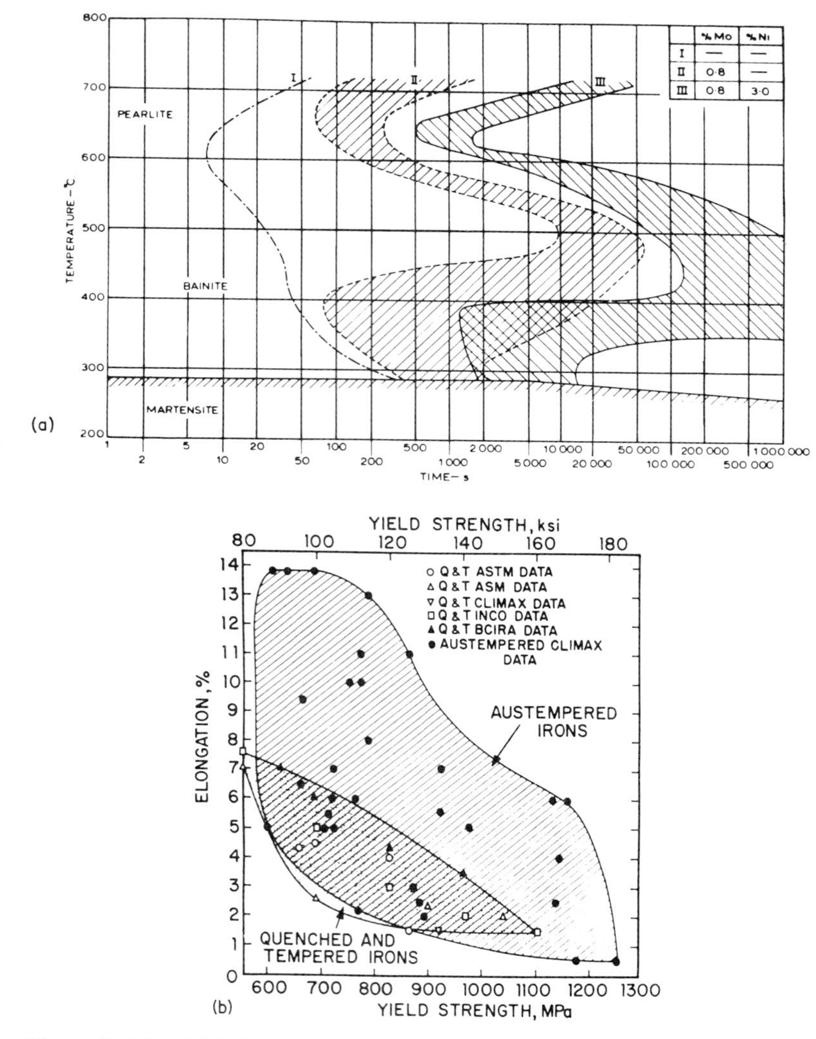

Figure 8–10 (a) Influence of Ni and Mo on the transformation characteristics of cast iron. (After Gerlach.[28] Reproduced by permission of *Metallurgia* Fuel and Metallurgical Journals Ltd.) (b) A comparison of yield strength and elongation for austempered irons in relation to quenched and tempered irons.[30] (Reproduced by permission of Fuel and Metallurgical Journals Ltd.)

pered alloys in relation to alloys which have structures of tempered martensite. This shows that austempered alloys have a more enhanced range of strengths and elongation than quenched and tempered martensite.

The problems of casting and securing properties in these types of alloy cast iron were discussed. Castings of large size and weight are problematic. Dodd recommended the as-cast approach to obtaining properties in the bainitic alloys where problems of cracking in quenching might be encountered. The cost of Ni and Mo to produce the strength required might be high but the properties obtained are impressive.

8-8 Aluminium Alloyed Cast Iron

The addition of aluminium to iron carbon alloys results in an alloy system in which three distinct structures are noted:[31]

(a) Between 0 and 6 per cent aluminium. Structures of flake graphite with mixed ferritic–pearlitic matrix.
(b) Between 6 and 20 per cent aluminium. White cast iron. The carbide is Fe_3AlC_x (x = 0.65 approximately).
(c) Between 20 and 30 per cent aluminium. Flake graphite microstructures are again noted but the matrix is ferritic.

In addition to the gray and white iron alloys, aluminium alloyed ductile iron and aluminium alloyed intermediate graphite structure cast iron are in use.

In the flake graphite materials the present interest in the above series lies mainly in the gray iron alloys of 2–5 per cent aluminium content and in the alloys containing more than 20 per cent aluminium. In the former, high strength values are obtainable combined with good machinability. In the latter, high scaling resistance is theoretically possible at high temperatures.[31] A comparison of strength values of Fe–C–Al alloys with conventional gray cast iron is given in Table 8-9[32]

Table 8-9 Comparison of properties of Fe–C–Al alloys with conventional gray iron.[32] (Reproduced by permission of American Foundrymen's Society)

	UTS at temperatures		
Alloy	27 °C	427 °C	538 °C
Fe–C–Si gray iron	294 MPa	252 MPa	170 MPa
Fe–C–2.4 wt % Al	348 MPa	200 MPa	175 MPa
Fe–C–4.3 wt % Al	324 MPa	213 MPa	193 MPa
Fe–C–6.0 wt % Al	279 MPa	192 MPa	168 MPa
Fe–C–24.0 wt % Al	124 MPa	123 MPa	102 MPa

The Iron–Aluminium Diagram

The Fe–Al diagram[32] is shown in Figure 8-11. As in the Fe–Cr system, the Fe–Al equilibrium diagram is characterized by a closed γ loop. Two bcc solid solutions exist, α–Fe up to 12 per cent aluminium at room temperature and Fe–Al from 20 per cent aluminium. The Fe_3Al phase exists at 15 per cent aluminium.

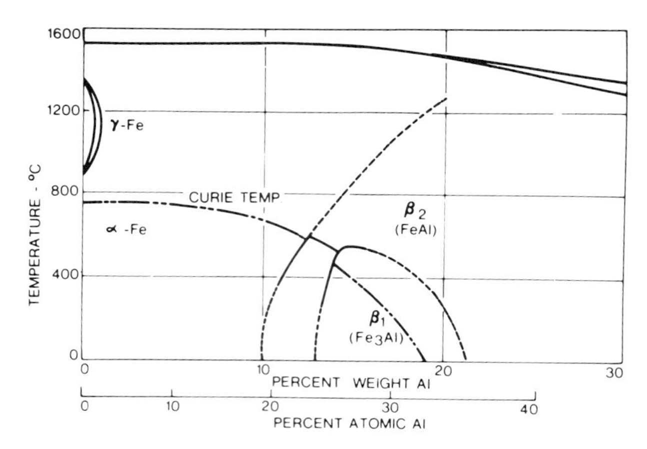

Figure 8-11 The Fe–Al binary phase diagram.[32] (Reproduced by permission of American Foundrymen's Society)

The Iron–Carbon–Aluminium Ternary System

The Fe–C–Al phase diagram was investigated by Morral[33] in 1934 and by Löhberg and Schmidt[34] in 1937. The following details of this ternary system are taken from the publication of Löhberg and Ueberschaer[35] in 1969.

Figure 8-12 shows the Fe corner of the Fe–C–Al system. This region has three ternary invariant equilibria. The phases marked in the diagram have the following lettering:

Σ_i = melt
M_α = α solid solution of C and Al in Fe
M_γ = γ solid solution of C and Al in Fe
K = ternary carbide phase (Fe_3AlC_x)
k = graphite phase

The three ternary invariant equilibria represented are as follows:

1. $\Sigma_i + M_{\alpha_1} = M_{\gamma_1} + K$ (1297 °C)
2. $\Sigma_2 + K = M_{\gamma_2}$ + graphite (k) (1280 °C)
3. $M_{\alpha 3} + M_{\gamma 3} = K$ + graphite (k) (730 °C)

The points $M_{\alpha 3}$ and $M_{\gamma 3}$ coincide approximately with the letters P and S of the Fe–C boundary.

Figure 8-13 shows an isothermal section at 1000 °C taken from Morral.[33] This shows the extent of the γ region, the location of the γ + C (graphite) region on the Fe–C side and the $\alpha + \gamma$ region on the Fe–Al side.

Two three-phase equilibria border the α + K (carbide) phase region. These equilibria are between γ, carbide, and graphite on the carbon-rich side, and between γ, α, and carbide on the aluminium rich side. With increasing Al, a

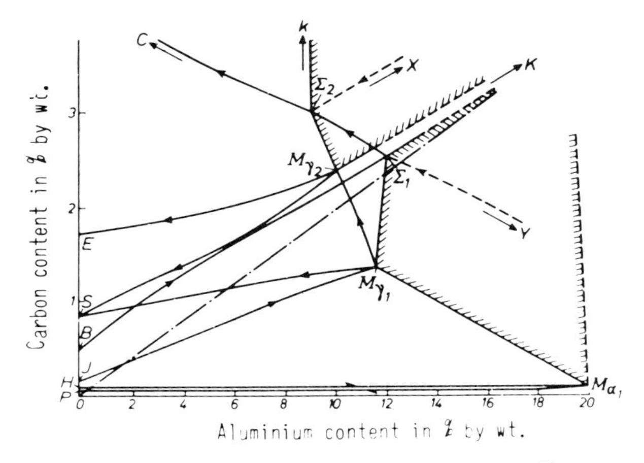

Figure 8-12 Fe corner of Fe–C–Al system.[35] Two ternary invariant equilibria are shown shaded. Phases $M_{\alpha 3}$ and $M_{\gamma 3}$ of third invariant equilibrium correspond with points P and S on the Fe–C boundary. The line marked $C\Sigma_2Y$ is a projection of a eutectic trough in the liquidus. (Reproduced by permission of Giesserei-Verlag)

region of α + carbide occurs followed by a three-phase α + carbide + graphite region and the two-phase α + graphite area.

In Figure 8-12 a line $C\Sigma_2Y$ is shown which is the projection of a eutectic trough in the liquidus. A vertical section through this trough is given in Figure 8-14. This eutectic trough passes through a maximum in temperature at about

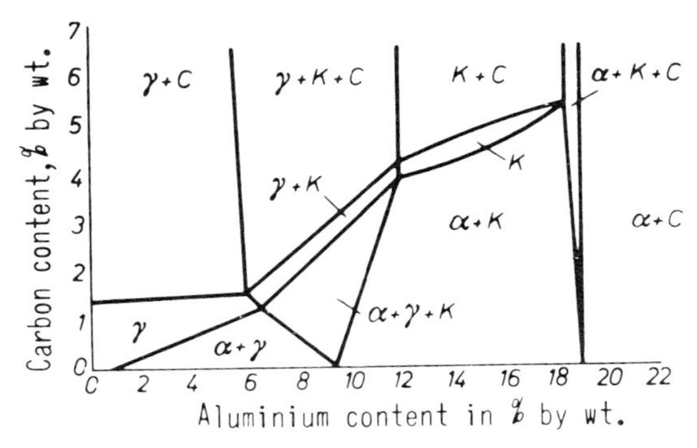

Figure 8–13 Isothermal section in Fe–C–Al system at 1273 K after Morral.[33] In this diagram C is the graphite phase and K is carbide. The effect of increasing Al is to give a region of α + carbide, α + carbide + graphite, and then α + graphite, from.[35] (Reproduced by permission of Giesserei-Verlag)

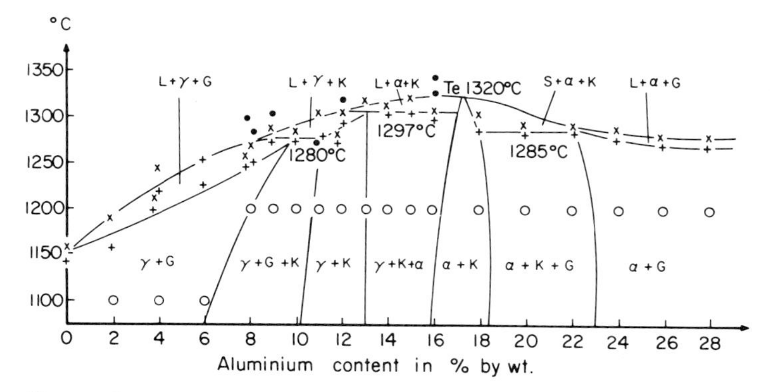

Figure 8–14. Vertical section through eutectic trough marked $C\Sigma_2Y$ in Figure 8–12.[35] (Reproduced by permission of Giesserei-Verlag)

14 per cent aluminium and 2.3 per cent carbon, after commencing on the γ–graphite eutectic at 1150 °C (extreme left of Figure 8-14).

Figure 8-14 also shows a four-phase equilibrium between liquid (marked S), α, carbide, and graphite phases at 1285 °C. This suggested the following reaction:

$$\text{Liquid} + K = \alpha + \text{graphite}$$

The liquid composition for this invariant is approximately 24 per cent aluminium. In support of the phases shown by the equilibrium diagrams, Löhberg and Ueberschaer[35] demonstrated metallographic sections of Fe–C–Al alloys with structures noted in Table 8-10.

Table 8-10 Microstructures of Fe–C–Al alloys[35]

Alloy	Structure
20 %Al – Annealed 1200 °C WQ	α + carbide + graphite
18 %Al – Annealed 1200 °C WQ	α + carbide
24 %Al – As cast	α + graphite

Applications of Aluminium Cast Irons

In the review of this material by Walson[32] potential applications were related to the high strength of the alloys at room temperature and their high resistance to scaling.

Comprehensive test figures for strength in the temperature range 649–982 °C were given by Yaker *et al.*[31] These tests were performed on flake and spherulitic graphite iron castings. The flake graphite iron had compositions of 5.9 and 20.5 per cent aluminium. The spherulitic graphite iron had compositions of 1.25 and 3.8 per cent aluminium.

8-9 Silicon Cast Irons

The Fe–Si binary system (Figure 8-15)[36] is characterized by a closed γ loop. The silicon cast irons are ferritic in matrix structure and can be made in flake graphite or spheroidal graphite forms. The characteristics of these alloys are

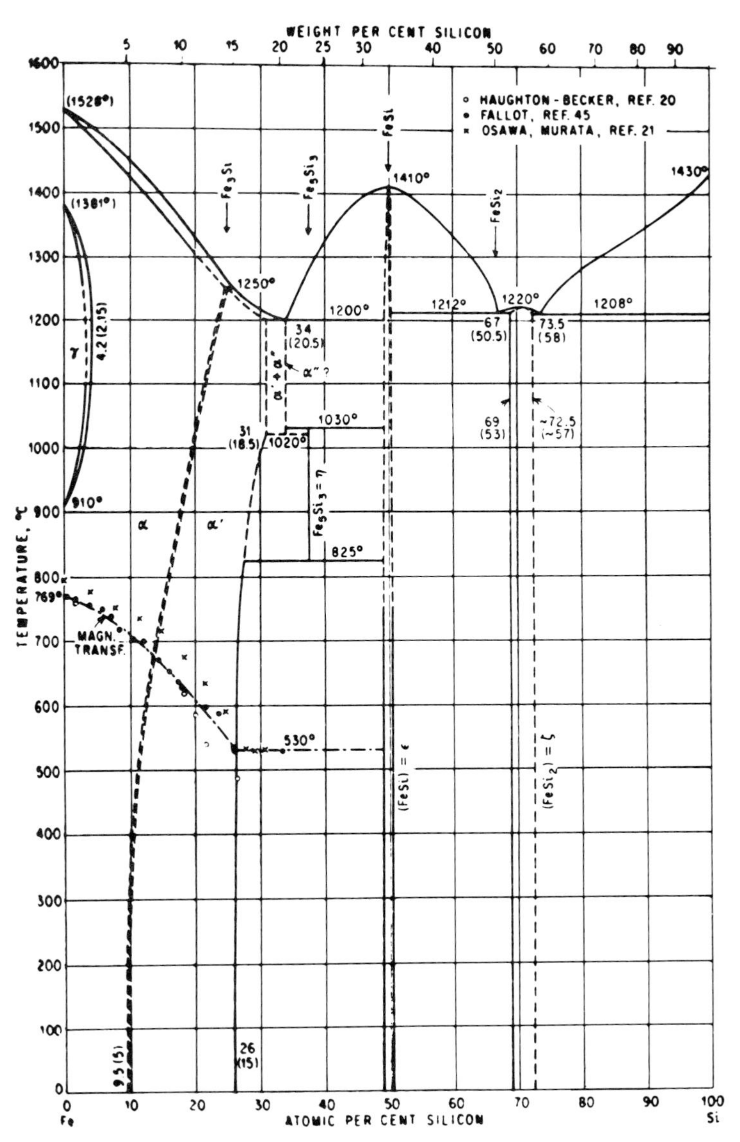

Figure 8-15 Fe–Si binary diagram.[36] (From *Constitution of Binary Alloys* by M. Hansen. Copyright © 1958 McGraw-Hill Book Company. Used with the permission of McGraw-Hill Book Company)

oxidation resistance and structural stability. The spheroidal graphite alloys have higher strength and ductility. Alloying may be made to enhance properties, of some interest being Si–Mo spheroidal graphite cast iron which has been considered as an alternative to Cr–Mo steel in a number of heat-resistant applications.

In the silicon cast irons, a problem is the impact transition temperature. At silicon contents higher than 5 per cent, the impact transition temperature for ductile to brittle behaviour is above room temperature. At less than 4 per cent silicon, the oxidation resistance is reduced, and above 6 per cent silicon, the decrease of toughness is pronounced. The Si content of the alloys is therefore kept within the limits 4–6 per cent silicon.

An exception is the alloy series of 13–18 per cent silicon and 0–6 per cent molybdenum, a gray cast iron which is used for its pronounced corrosion resistance. In these alloys the η and ε phases appear. Increased corrosion resistance is thus obtained at the expense of mechanical properties.

Spheroidal Graphite Silicon Cast Irons[37]

The heat-resistant high silicon spheroidal graphite cast irons have compositions in the range 4–5 per cent silicon and 0.5–1.5 per cent molybdenum. The Mo additions are made to improve the high temperature strength, and lead to the presence of carbides in the structure which are not removed by annealing.

The carbon content is decreased from 3.8 to 2.9 per cent as the silicon is increased from 4 to 5 per cent. This is to maintain the alloys as hypo-eutectic and avoid primary graphite formation and flotation. The manganese content is maintained below 0.5 per cent. Since the oxidation resistance is provided by the built-up layer of oxide, structural transformation of α to γ, on raising the temperature, must be avoided, and the range of use is limited by the phase boundary for the $\alpha \rightarrow \gamma$ transition. The temperature for this transition increases with Si and becomes 900 °C at 5 per cent silicon.

The alloys show marked resistance to thermal cycling at slow cycles of heating and cooling, a property which is related to high strength at elevated temperature. This is discussed in Chapter 11.

In the review of Fairhurst and Röhrig[37] reference was made to an alloy having 4–5 per cent silicon and 1.0 per cent molybdenum. This is used in exhaust manifolds, supercharger castings, gas turbine components, and glass moulds. It is also an inexpensive material for industrial furnace applications. An alloy with improved high temperature properties has 4 per cent silicon, 3–3.5 per cent manganese, and 1.5–2.0 per cent molybdenum. This has an annealed hardness of 300 Brinell at room temperature, allowing machining, and a hot hardness of 200 Brinell at 760 °C.

8-10 Manganese (Austenitic) Cast Iron

The thermodynamics of the solid phases in the system Fe–Mn–C were evaluated by Benz, Elliott, and Chipman.[38] Only limited applications of austenitic manganese cast iron alloys have been reported because of the problem of large carbide volume. Lambert, Poyet, and Dancoisne[39] published procedures for casting these alloys, followed by malleablising of the product. Subsequent heat treatment at 1050 °C followed by water quenching is similar to that for a Hadfield steel. They suggested that the latter is difficult to cast in thin sections but the manganese cast iron, in this respect, presented little problem.

8-11 Tin as an Alloying Element in Cast Iron

The element tin added to gray cast iron, up to a level of 0.20 per cent, is reported as increasing the tensile strength.[40] Higher tin contents lead to embrittlement—the hardness increases, while the tensile strength decreases.

Tin stabilizes the pearlite phase. Other aspects of the physical metallurgy of tin additions to cast iron are less clearly understood.

8-12 Copper in Cast Iron

Copper is a common alloying addition to gray, white, malleable, and spherulitic graphite cast irons.[41] It is an austenite stabilizer and acts to delay the start of the transformation to pearlite. It refines the interlamellar spacing in pearlite, and hardens ferrite by solution in the solid state. The overall effect in gray cast iron for copper present up to 0.35 per cent is an increase of the tensile strength, hardness, and corrosion resistance.

In white cast irons, for applications where the transformation of the γ phase to martensite is important, it has been reported that Cu can replace part of the Ni where this is used to delay the start of the pearlite transformation

As a component of Ni-resist austenitic cast irons, it can be present in concentrations between 5.50–7.50 per cent in alloys containing 13.5–17.5 per cent nickel, or up to 0.50 per cent in the higher Ni alloys.

References

1. Hansen, M.: *Constitution of Binary Alloys*, McGraw-Hill, New York, 1958.
2. Jackson, R. S.: *J. Iron Steel Inst. London*, **208**, 163, 1970.
3. Jackson, R. S.: *Br Foundryman*, **67**, 34, 1974.
4. Bungardt, K., E. Kunze, and E. Horn: *Archiv Eisenhuettenw.*, **29**, 193, 1958.
5. Lundberg, R., M. Waldenström, and B. Uhrenius: *Calphad*, **1(2)**, 159, 1977.
6. Sato, T., and T. Nishizawa: *Nippon Kinzoku Gakkaishi*, **20**, 340, 1956.
7. Kuo, L.: *J. Iron Steel Inst. London*, **173**, 363, 1953.
8. Beach, J., and D. H. Warrington: *J. Iron Steel Inst. London*, **204**, 460, 1966.
9. Bowers, J. E.: *J. Iron Steel Inst. London*, **183**, 268, 1956.
10. Fredriksson, H., and B. Remaeus: *Int. Foundry Congress*, Firenze, Congress Papers 44th, Paper 7, 1977.
11. *Metals Handbook*, Vol. 8, 8th ed., Am. Soc. Met., 1973.
12. Maratray, F.: *Trans. Am. Foundrymen's Soc.*, **79**, 121, 1971.
13. Colton, W. J.: *Br. Foundryman*, **56**, 237, 1963.
14. Dodd, J., and J. L. Parkes: *Int. Cast Metals Res. J.*, **5**, 47, 1980.
15. Fredriksson, H., and M. Hillert: *Br Foundryman*, **64**, 54, 1971.

16. Hughes, I. C. H.: *The Solidification of Metals*, Iron Steel Inst. Publ. 110, London, 1967.
17. *Nickel as an Alloy in Cast Iron*, Am Foundrymen's Soc., 1977.
18. Marsh, J.: *Alloys of Iron and Nickel*, McGraw-Hill, New York, 1938.
19. *Gray and Ductile Iron Castings Handbook*, Am. Foundrymen's Soc., Cleveland, 1971.
20. Cox, G. J.: *Foundry Trade J.*, **136**, 31, 1974.
21. Durman, R. W.: *Foundry Trade J.*, **134**, 645, 1973.
22. Harrison, G. L., and R. H. T. Dixon: *Br Foundryman*, **55**, 164, 1962.
23. Cox, G. J.: *Br. Foundryman*, **63**, 1, 1970.
24. *Engineering Properties and Applications of the Ni-Resists and Ductile Ni-Resist*, Int. Nickel Co., New York, 1976.
25. Gagnebin, A. B.: *The Fundamentals of Iron and Steel Castings*, Int. Nickel Co., 1957.
26. Flinn, R. A., M. Cohen, and J. Chipman: *Trans. Am. Soc. Met.*, **30**, 1255, 1942.
27. Flinn, R. A., and D. J. Reese: Am Foundrymen's Assoc. Preprint 41–4, 1941.
28. Gerlach, H. C.: *Metallurgia*, **172**, 215, 1965.
29. Schelleng, R. D.: *Trans. Am. Foundrymen's Soc.*, **77**, 223, 1969.
30. Dodd, J.: *Foundry Trade J.*, **147**, 963, 1979.
31. Yaker, J. A., L. E. Brynes, W. C. Leslie and E. V. Petitbon: *Trans. Am. Foundrymen's Soc.*, **84**, 305, 1976.
32. Walson, R. P.: *Trans. Am. Foundrymen's Soc.*, **85**, 51, 1977.
33. Morral, F.: *J. Iron Steel Inst. London*, **130**, 426, 1934.
34. Löhberg, K., and W Schmidt: *Arch Eisenhuettenw.*, **11**, 607, 1937.
35. Löhberg, K., and A. Ueberschaer: *Giessereiforschung* (in English), Vol. 21(4), p. 167, BCIRA, Alvechurch, 1969.
36. Hansen, M.: *Constitution of Binary Alloys*, McGraw-Hill, New York, 1958.
37. Fairhurst, W., and Röhrig, K.: *Foundry Trade J.*, March **146**, 657, 1979.
38. Benz, R., J. F. Elliott, and J. Chipman: *Metall. Trans.*, **4**, 1449, 1973.
39. Lambert, B., P. Poyet, and P. L. Dancoisne: *Br. Foundryman,* **68**, 179, 1975.
40. Rooney, T. C., C. C. Wang, P. C. Rosenthal, C. R. Loper, and R. W. Heine: *Trans. Am. Foundrymen's Soc.*, **79**, 189, 1971.
41. Pearce, J. G.: *Copper in Cast Iron*, Copper Development Assoc., London, Hutchinson, 1964.

Bibliography

Cardno, J. R., and D. R. Hayward-Shott: *An Introduction to the Cast Irons*, 2nd ed., Council Iron Foundry Assocs., 1974.

Cercle d'Études des Metaux, Colloque International, St Étienne, November, 1973.

Maratray, F., and R. Usseglio-Nanot: *Factors Affecting the Structure of Chromium and Chromium Molybdenum White Irons*, Climax Molybdenum Co., 1971.

Chapter 9
Phase Transformations and the Heat Treatment of Cast Iron

The relief of residual stress in cast iron is first discussed, with a description of both body and microstresses. The heat treatment of different alloys of cast iron to give pearlite, bainite, and martensite matrix structures is dealt with, referring to both TTT and CCT diagrams for the alloys.

Different topics in the heat treatment of cast iron are discussed. These relate to solute segregation in cast structure and the lack of uniformity of composition in austenitizing, the secondary precipitation of carbides in alloy irons, and destabilization of austenite, retained austenite and the M_S temperature.

Finally a review is given of surface hardening of cast iron by laser and electron beam methods, the control of structure, and variation in the depth of heat-treated material.

9-1 Internal Stresses in Cast Iron

In Orowan's 1948 paper[1] the main types of internal stress in castings were reviewed. These were classified in two groups:

(a) Body stresses. These arise from non-uniform external influences acting on the casting such as thermal gradients, non-uniform chemical composition, and mechanical stress.
(b) Textural stresses due to microscopic inhomogeneities. Orowan's classification in this respect was large. It included precipitates, grain boundaries, dislocations, and structural variations.

An internal stress was defined as one which existed in a body on which no external forces were acting. The typical internal stress resulting from solidification was the result of plastic deformation which occurred at some higher temperature. At a lower temperature, the plastically deformed region could no longer fit the rest of the body without the imposition of elastic strains. To these correspond the internal stresses between the two parts.

Figure 9-1 shows the typical stress pattern in a rectangular-shaped casting

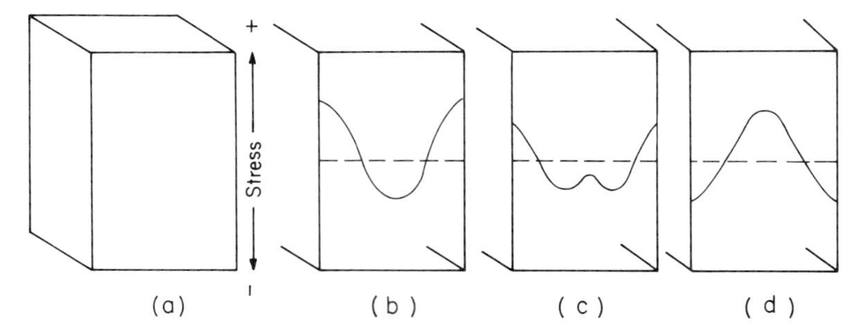

Figure 9-1 Thermal and residual stress in a casting. The casting (a) has just solidified. The outer region is in tension, the inner region is in compression (b). As the material cools, (c) and (d), the stresses become tensile in the inner region and compressive in the outer region.[2] (From *Mechanical Metallurgy* by G. E. Dieter. Copyright © 1961 McGraw-Hill Book Company. Used with the permission of McGraw-Hill Book Company

during solidification.[2] Since the inside of the casting is at a higher temperature than the outside, it deforms plastically under the contraction stresses of the outer layer. These are in tension, and the plastically deformed interior is in compression. As the inside cools, it contracts and the stress changes to tension (d). This must be balanced by the stresses of the exterior becoming compressive.

There are other possibilities for the existence of residual stresses after plastic deformation, e.g. in a microscopically small volume at the bottom of a notch or crack. This occurs in the vicinity of flakes or spheres of graphite.

Tesselated Stresses of Lászlò

A series of four papers on stresses round precipitates and inclusions and examining the graphite phase in cast iron were written by Lászlò,[3] were discussed by Nabarro.[4] Lászlò theorized that a mosaic type of stress pattern would exist in materials. Orowan suggested that a more appropriate term would be textural, as for example in grain boundaries and in phases like pearlite or graphite. An important contribution to this type of stress is conferred by differences in the coefficients of thermal expansion of constituents of the microstructures.

Lászlò's Calculations of Stresses Round Graphite

In cast iron, the stresses round graphite in flake or spherulitic form could be calculated from the differences in thermal expansion coefficient. For graphite, ferrite, and Fe_3C, the values of the coefficient are as follows:

Graphite	3×10^{-6} per °C
Ferrite	15×10^{-6} per °C
Fe_3C	12×10^{-6} per °C

Lászlò's calculations showed that the stresses round a spherulitic graphite crystal after cooling from the casting temperature would be very large, of the order of 1.378 GN m^{-2}. If the graphite formed lamellae, the stresses would be smaller by a factor of 10. However, Nabarro calculated that all stresses would be less than 0.1 GN m^{-2}.

Estimation of Residual Stress in Plastically Deformed Bodies

Residual stress can be calculated with some approximation in plastically deformed bodies, but there are some problems in doing this for castings. In a cast metal, the stresses are basically thermal and the temperature distribution for complex shapes is difficult to calculate. A further important problem in residual stress due to thermal stress is the role played by creep. This is discussed for the case of thermal cycling and thermal fatigue in Chapter 11.

Thermal stresses can be calculated for casting shapes of reasonably uncomplicated geometry,[5] from which residual stress might be computed. Some calculations of residual stress for different cast geometries was summarized by Angus.[6]

For a body subjected to purely mechanical forces and plastically strained, the calculation of residual stress was suggested by Nadai[7] as follows:

(a) Calculate the plastic stress distribution.
(b) Assume that plastic deformation ceases at the slightest reduction of the external forces and the body then behaves as if perfectly elastic while the forces are gradually removed.
(c) Calculate the stresses set up by the same external forces in a purely elastic body of the same material and same plastically deformed shape.
(d) The residual stress is obtained at any point on the body by subtracting the stress calculated in (c) from the stress calculated in (a).

A calculated residual stress pattern for a tube after being subjected to plastic deformation in internal pressure in the bore is shown in Figure 9-2.[7] This shows the compressive tangential residual stress at the bore, changing to

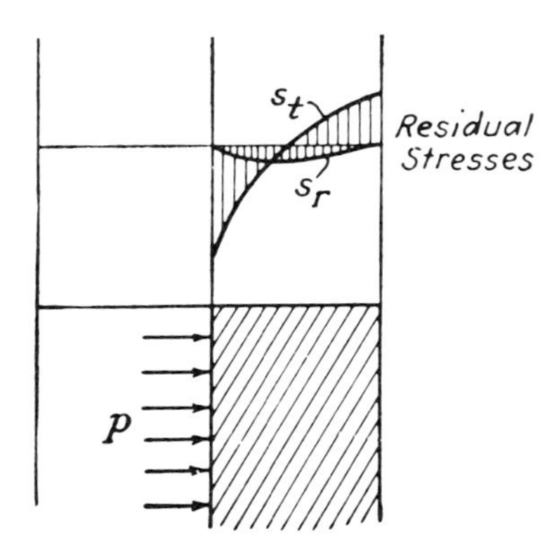

Figure 9-2 Nadai's calculation[7] for residual stress determination in a plastically deformed body. The stress–strain relationship assumed is for a perfectly plastic body. The stresses calculated for a thick-walled cylinder with an internal pressure p are shown. S_t is the tangential residual stress and S_r is the radial residual stress. (From *Plasticity* by A. Nadai. Copyright © 1931 McGraw-Hill Book Company. Used with the permission of McGraw-Hill Book Company)

a tensile residual stress at the outer surface. The radial stress is zero at the inner and outer tube surfaces and compressive at the centre line of the wall thickness.

Internal stress may be investigated by other methods including the photoelastic one.

Annealing of Cast Iron

Iron castings are annealed to reduce internal stresses due to the casting process. Although existing stresses will be reduced by treatment at an elevated temperature, stresses may again be generated as the casting is cooled to room temperature. A recommended graph for heat treatment temperatures for gray cast iron is given in Figure 9.3.[8] Thermal treatment at 450 °C is recommended for reducing the chances of cracking and distortion, while relief of all the internal stress is proposed at 600 °C.

These treatment temperatures may of course be accompanied by changes of microstructure and mechanical properties. The overall problem of residual stress is a very complicated one and depends on factors like casting design, mould properties, rate of cooling of the casting, and alloy composition. Heat treatment of alloy cast iron may change the residual stress pattern resulting

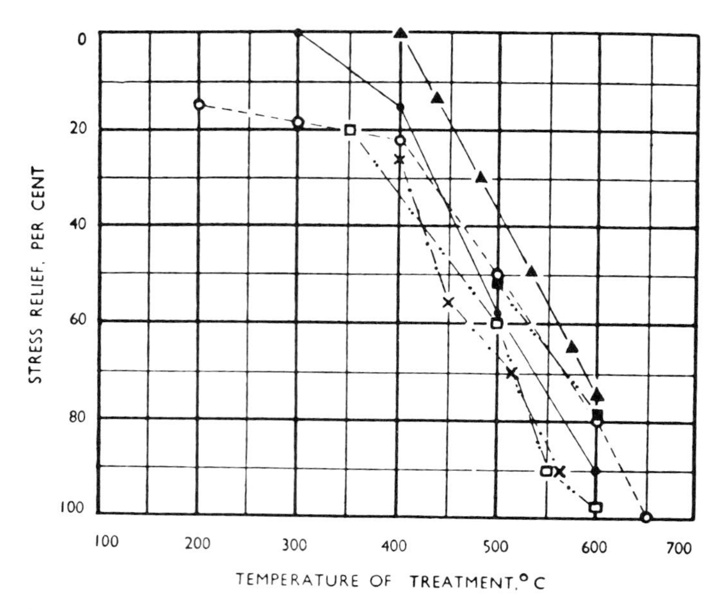

Figure 9-3 Data on effect of temperature of treatment on percentage stress relief of iron castings.[8] The different curves are for different sources of experimental data and times at temperature. (Reproduced by permission of The Metals Society)

from the casting process and introduce new residual stresses due both to the new heating and cooling cycle as well as to the stresses initiating in phase transformations.

Residual Stresses in Alloy Iron Castings

Some of the special problems which relate to high chromium molybdenum cast iron were described by Dodd and Parkes.[9] In addition to the thermally induced stresses of the cooling process after casting, transformation stresses may arise from the martensite transformation. This will occur due to the raising of the M_S temperature as slow cooling of the casting leads to precipitation of secondary carbides. This occurs in a temperature range of cooling down to 540 °C.

Cooling to room temperature in the mould may be made or the authors recommended removing the casting above the M_S temperature and cooling uniformly in a furnace or in insulation.

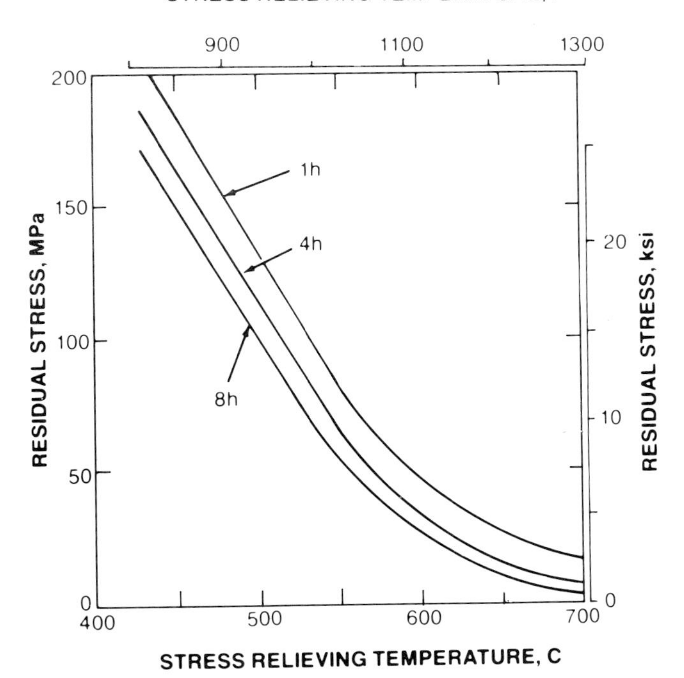

Figure 9-4 Curve of residual stress as a function of stress relieving temperature for an alloy steel.[9] The change of residual stress at any one temperature is shown for annealing times of 1, 4, and 8 hours (Reproduced by permission of American Foundrymen's Society)

Whichever way the residual stresses are generated, it becomes difficult to achieve their relief in these alloys other than by heating to approximately 600 °C. Because treatment at this temperature would transform the matrix structure to one of spheroidized carbides in α, the time and temperature of annealing must be carefully selected and some level of residual stress will remain.

Dodd and Parkes[9] showed a curve for residual stress in a steel casting at different times and stress relieving temperature (Figure 9-4). Longer times are required to satisfactorily reduce residual stress levels at lower temperatures. The curves shown are for a structural steel and it was suggested that much longer times or much higher temperatures must be used for effective stress relief in the high chromium white irons.

9-2 Phase Transformations in the Heat Treatment of Cast Iron

The heat treatment of cast iron is largely dependent on transformation of the γ phase. This phase in non-austenitic materials must be formed by heating above A_1 or A_3. The progress of the subsequent transformation on cooling is dependent on the content of carbon (which is provided by the carbide or graphite phases), the alloying elements in solution in the γ phase, and the cooling rates. In some treatments, the spheroidization of carbide phases is required.

In this section different aspects of these transformations are described.

Heating above A_1; Formation of Austenite and Dissolution of Cementite

For the phase transformations to pearlite, bainite, or martensite, heating must first be made above the A_1 critical temperature, when austenite will be nucleated. The original phases present in the cast iron structure at room temperature are normally ferrite, carbides, and graphite. The austenite phase will grow in volume, the α phase transforms to γ, and the carbon content of the γ phase will increase. In cast iron, the carbide phase can also eventually be transformed to graphite. The equilibrium carbon content of the γ phase should be established according to the phase diagram. In effect the rate of solution of carbon in the γ phase is dependent on the microstructure originally present (15) and on the kinetics of dissolution of the carbon-bearing phases.

Molinder[10] studied the dissolution of the carbide phase in a steel having the composition:

1.27 %C, 0.20 %Si, 0.30 %Mn, 0.19 %Cr, 0.03 %V

The initial structure was α and spheroidized carbide. Figure 9-5 shows the phase diagram for Fe–C. The course of the transformation, on heating to various temperatures above the critical, was divided into three periods:

Period 1. The initial structure remained untransformed. During this period, embryos of the γ phase were in formation.

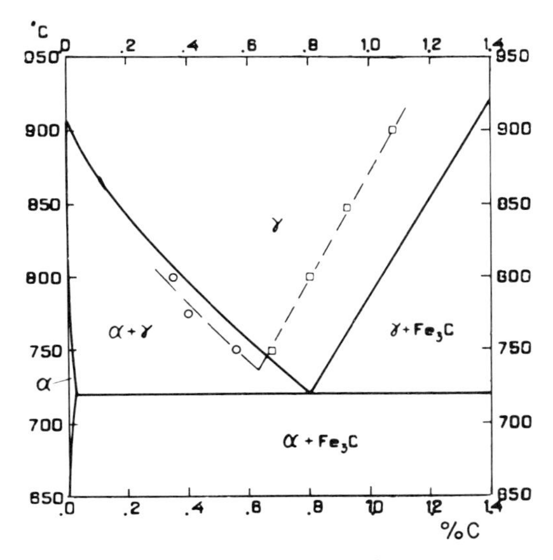

Figure 9-5 Graph by Molinder[10] showing progress of solution of carbide phases in γ above the critical temperature. Circles show composition of γ first formed on transformation of α to γ. Squares show final carbon concentration in γ. (Reproduced by permission of Pergamon Press Ltd.)

Period 2. The α phase transformed to γ with simultaneous rapid solution of Fe_3C. The carbon concentration of the γ phase first formed is shown in circles in Figure 9-5.

Period 3. When γ formation has been completed, Fe_3C dissolves further in the γ but at a slower rate, until the carbon concentration shown by the squares of Figure 9-5 is achieved. The figures are lower than for the equilibrium Fe–C diagram due to the composition of the alloy.

The estimate for the completion of period 2 at 900 °C was 0.2 s. Period 3 was slower and was related to the rate of dissolution of Fe_3C at the γ–Fe_3C phase boundary.

Problems of the rate of solution of elements other than carbon, e.g. Cr, Mo, in the γ phase must also be taken into account. These dissolution rates are comparatively slow and are responsible for inhomogeneity of microstructure.

Eutectoid Transformation: Pro-eutectoid Ferrite Formation; Pearlite Formation;[11] Problems Involved in Cast Iron

Figure 9-6(a) shows the eutectoid transformation range for a binary Fe–C alloy. This range of temperature and composition is shaded, as for the region of cooperative eutectic growth described in Section 3-2. Within this range,

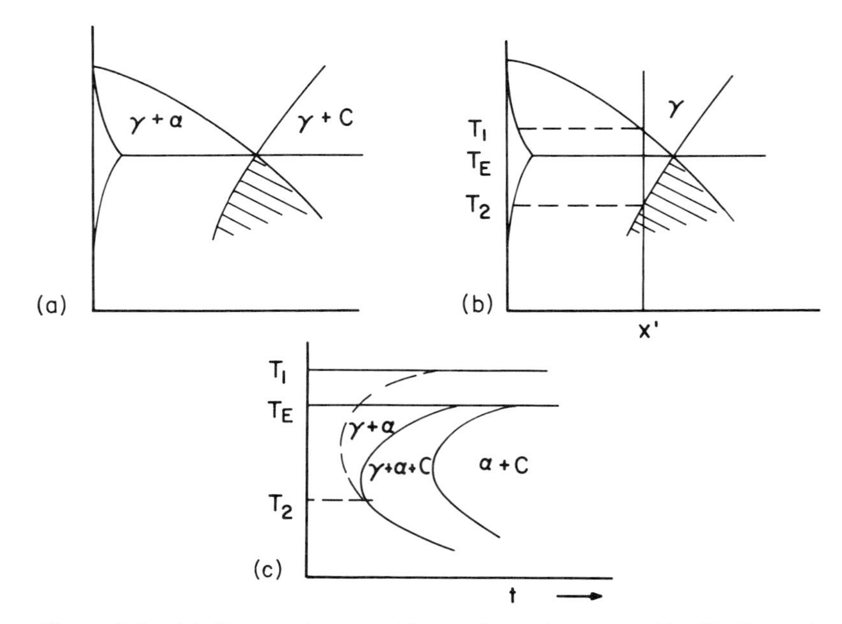

Figure 9-6 (a) Range of eutectoid transformation in an Fe–C alloy. The shaded region has a similar connotation with the coupled eutectic zone of Figure 3-14. Outside the shaded area a pro-eutectoid phase should separate before the eutectoid. (b) Between T_1 and T_E an Fe–C alloy of composition X' will precipitate α before the eutectoid. If alloy X' is quenched into the region $T_E - T_2$, pro-eutectoid ferrite will separate first but at T_2 the transformation should be directly to pearlite. (After Christian.[11] Reproduced by permission of Pergamon Press Ltd) (c) C curve for isothermal transformation of alloy X'

only the eutectoid phase should grow. To either side, the proeutectoid phases α or Fe_3C should form first.

Figure 9-6(b) shows a part of the binary Fe–C diagram with composition X' marked, commencing α precipitation from the γ phase at temperature T_1. Figure 9-6(c) shows the C curve for isothermal transformation of alloy X'. For this diagram, which is essentially for the Fe–C binary system, the formation of pro-eutectoid ferrite is possible down to temperature T_2.

The transformation behaviour of cast irons to ferrite and pearlite on cooling below the critical temperature is complex, and has been discussed by Cias.[12] It is influenced by the presence of the graphite phase, so that secondary graphite may precipitate and be followed by α. This leads to a more extended α precipitation than is indicated by the phase diagram. Influencing factors will be the alloying elements in solution, the cooling rate, and the inhomogeneity of composition of the γ phase. Cias has noted this in CCT diagrams of cast iron by not indicating transformation details of the α phase preceding the pearlite transformation.

The presence of Si in solid solution in the γ phase has a powerful influence on the formation of α in the microstructure. Thermodynamic reasoning for the growth of pro-eutectoid phases in the region of the ternary phase diagram where two-phase growth should occur was given in Chapter 4.

Pearlite and Bainite Transformations; Complexities of TTT Diagrams

The curves for the start and finish of isothermal transformations from the γ phase are basically of three forms, shown in Figure 9-7(a), (b) and (c).

From the eutectoid transformation temperature T_E, down to the straight line M_S indicating the temperature of the start of the martensitic transformation, two different types of transformation are possible:

(a) Eutectoid (pearlite), a discontinuous type of transformation involving the diffusion controlled growth of the α and Fe_3C phases in a cooperative manner.
(b) The bainite transformation, controlled by the growth of the α phase and involving an element of shear.

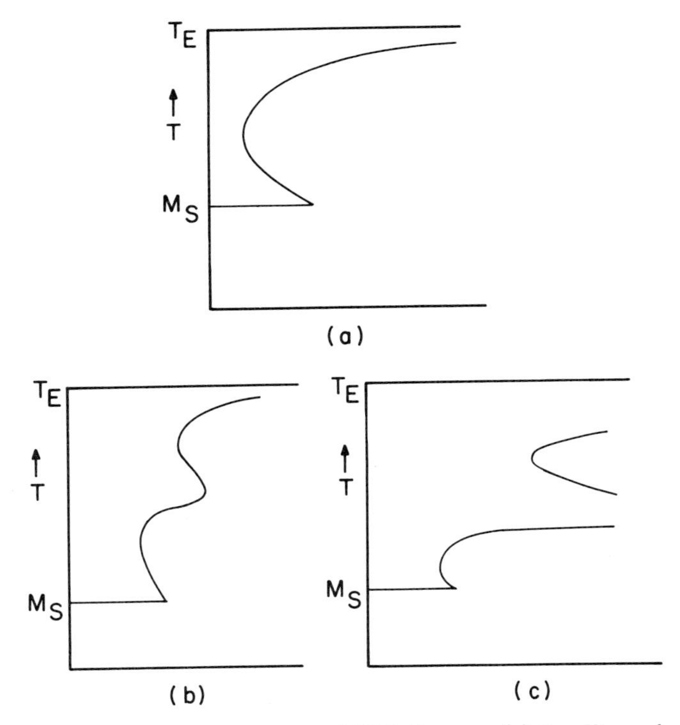

Figure 9-7 Different types of TTT diagram. (a) Pearlite and bainite transformations overlap, (b) the two transformations partially overlap, and (c) separation of the two transformations (After Christian[11] Reproduced by permission of Pergamon Press Ltd.)

The characteristics of the three types of diagram are as follows:

(a) The two transformations merge so that the start and finish of either transformation are not indicated by the diagram (Figure 9-7a).
(b) The two transformations partially overlap so that there is an intermediate temperature range in which the start of the pearlite transformation and the start of the bainite transformation are indistinguishable (Figure 9-7b).
(c) The two transformations are clearly separated by a temperature interval (Figure 9.7c).

The influence of the different alloying elements on the TTT curves is clear only in very limited cases. Otherwise the curves must be constructed individually or the influence of different combinations of elements on hardenability must be examined by special tests.

Kinetics of the Pearlite Transformation Cahn and Hagel[13] related the kinetic data for the pearlite transformation by an equation of the following form:

$$X = 1 - \exp(-bt)^{m}$$

where X is the fraction transformed in time t and m is a parameter related to the type of nucleation involved. For example, $m = 1$, 2, or 3 is related to nucleation sites on grain surfaces, edges, and corners respectively. Kinetic data can be plotted as $\log \ln[1/(1 - x)]$ versus $\log t$. The slope gives the parameter m and the intercept gives the parameter b.

From the relationship $X = 1 - \exp(-bt)^{m}$, the time interval between the beginning and end of an isothermal reaction on a TTT diagram can be computed. It is

$$\log t_{f} - \log t_{s} = \beta/m$$

The constant β depends on how much transformation is arbitrarily chosen for the beginning (X_{s}) and end (X_{f}) of the reaction and is determined from

$$\beta = \log \ln [1/(1 - X_{f})] - \log \ln [1/(1 - X_{s})]$$

The horizontal time interval between the beginning and end of the pearlite reaction is inversely proportional to m. On a logarithmic scale this interval is small at high temperatures, where m is large, and increases with decreasing T.

Cahn and Hagel[13] commented that an adequate theory for the influence of elements on the start of the pearlite transformation on the TTT diagram did not exist. Notwithstanding this, these diagrams can now be calculated. They emphasized that the elements have an important role on the growth rate of pearlite, slowing it in dependence on element type and quantity. This is related to composition with a thermodynamic requirement of alloy partition. An analysis of the behaviour of different alloy elements, Mn, Ni, Cr, and Mo, in the pearlite transformation was given by Puls and Kirkaldy.[14] An analysis due to Cahn was given in Chapter 4.

Bainite Transformation[11] Bainite is the microstructure resulting from the transformation of austenite in a temperature range below that for pearlite and above that for martensite. Immediately below A_1, austenite transforms to α and Fe_3C in a cellular transformation by the two phases forming a common interface.

When the growth rates of the α and Fe_3C phases are examined as a function of decreasing temperature, it is observed that the two-phase pearlite structure grows faster than the α phase above a temperatue termed T_B. At this temperature, the α phase can grow faster.

The nucleation rate of pro-eutectoid ferrite and carbide also go through a maximum as the undercooling increases, and at the knee of the TTT curve and just below, bainite begins to nucleate. It grows fast enough to transform part of the γ to bainite before transformation to pearlite.

Figure 9-8(a) shows the nucleation rate dependence on temperature as discussed by Hobstetter.[15] The nucleation rate is given by

$$I^* = K(T)\exp\left(-\frac{\Delta G^* + Q_D}{kT}\right)$$

where $K(T)$ = a pre-exponential term
ΔG^* = activation energy for nucleation
Q_D = activation energy for diffusion

At low temperatures Q_D is large and the value of I^* is small, being dependent on the diffusion term. At temperatures close to the transformation temperature T_e, the value of ΔG^* is large and the nucleation rate is again small, dominated by the activation energy for nucleation. Below T_e the nucleation rate increases as ΔG^* decreases. Hobstetter plots Figure 9-8(b) using the reciprocal of the nucleation rate as the abscissa to give a curve conforming to the C curve for the eutectoid transformation.

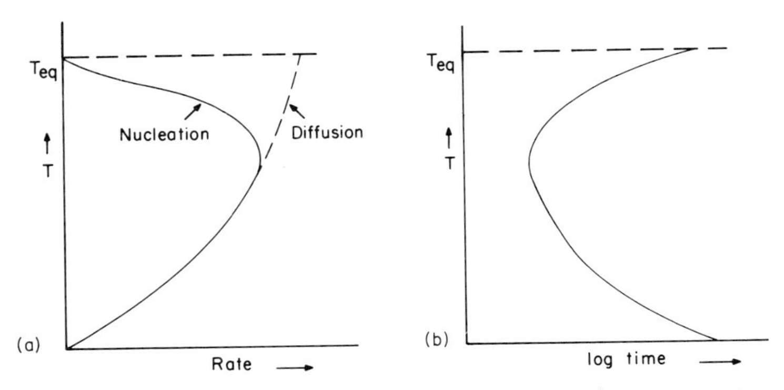

Figure 9-8 (a) Nucleation rate and diffusion related to temperature below T_e for pearlite transformation. (b) C curve derived from Figure 9-8(a).[15] (Reproduced by permission of The Metallurgical Society of AIME)

Crystallography and Metallography of Bainite[11] Bainite consists of a non-lamellar aggregate of ferrite and carbide. It appears as acicular, i.e. needle-like in character but the ferrite crystals are plate-like. At high temperatures, there is some tendency to form closely spaced parallel plates and the micro-structural appearance may then be described as feathery.

The bainite transformation range is divided into two parts on the basis of both structural and kinetic characteristics, known as upper and lower bainite respectively. In the description of J. W. Christian[11] it was suggested that it is not always possible to detect any sharp change in microstructural appearance, and the transition between upper and lower bainite occurs in a fairly narrow temperature range around 350 °C in Fe–C. This is apparently sensitive to carbon content. In lower bainite, the carbide is all contained within the elongated ferrite plates. It is probable that the freshly formed ferrite contains appreciable amounts of carbon—probably as much as in the original austenite—and the supersaturation is removed by carbide pecipitation. The low carbon content of the ferrite may, however, imply some partition of carbon between ferrite and austenite during the growth process.

In upper bainite, the cementite phase of the structure grows as plates which are larger in size than the plates of lower bainite. They appear to form directly from the austenite and are oriented parallel to the main growth direction of the structure. There is a marked carbon enrichment of austenite during the formation of upper bainite which grows first as nearly carbon-free ferrite. In the upper bainite range, ferrite plates may merge to form extended areas surrounded by dense clusters of cementite precipitated from the supersaturated austenite.

Similarities between Bainite and Martensite The bainite transformation, which is intermediate between the nucleation and growth mechanism of pearlite and the martensite transformation, has some characteristics of both. When the individual ferrite plates grow, they do so with an accompanying shape deformation.

Christian[11] summarized the bainite transformation as being characteristic of alloys having two components of widely differing mobilities. The more rapidly moving component would diffuse to establish the composition differences between the phases. The other component would ensure that structural changes are accomplished by a martensitic type transition.

Continuous Cooling Transformation Diagrams

Continuous cooling transformation diagrams[12] are available for cast iron, the alloyed gray irons, the alloyed white cast irons, and spheroidal graphite cast iron. A diagram for a Cr–Mo white cast iron[12] is shown in Figure 9-9(a).

Cias[12] has described the determination of these diagrams which also include, for each CCT diagram, a hardenability diagram (Figure 9-9b). The latter is presented in terms of half the cooling time, which for cast iron is that

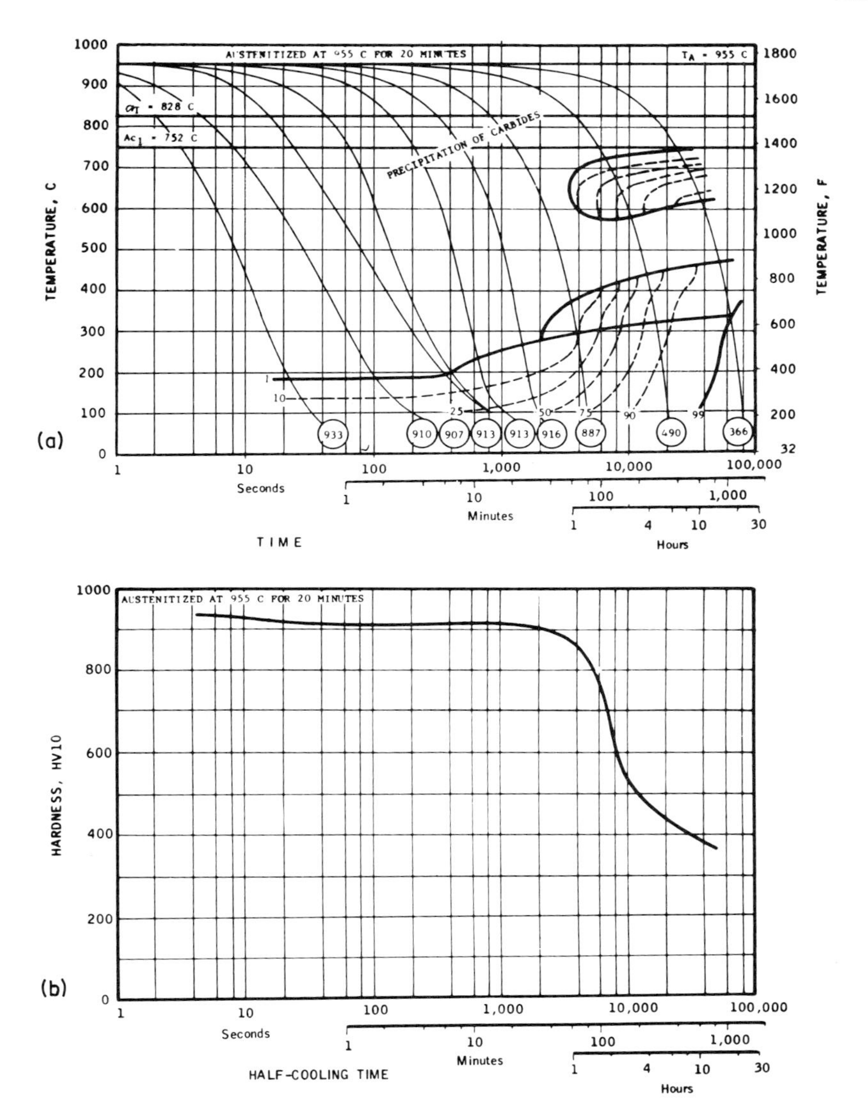

Figure 9-9 (a) CCT diagram for Cr–Mo white cast iron containing 2.93 per cent C, 0.59 per cent SI, 0.76 per cent Mn, 17.5 per cent Cr, 1.59 per cent Mo. Austenitized at 955 °C for 20 min. Pretreated for 2 hours at 955 °C and air cooled. (b) Hardenability diagram for the material of (a)[12] (Reproduced by permission of Climax Molybdenum Company)

required to cool from the temperature of austenitizing to a temperature midway to room temperature. CCT diagrams can now also be calculated.[16]

The CCT diagrams employ a wide range of cooling rates. Dilatometry is used to establish the transformation characteristics, supplemented by metallographic examination and hardness testing. Individual cooling curves are shown on the diagram of temperature versus the logarithm of time. For actual castings, cooling curves for different cross-sections might be difficult to determine and resort must be made to available curves for round or square sections in different cooling media.

The CCT diagrams show the Ac_1 and Ac_3 temperatures, the iso-percentage transformation lines, and the hardness of the material (in circles) at the end of each cooling curve. The designation α_T is the highest temperature at which the α phase exists.

The composition of the alloy in Figure 9-9(a), and (b) is:

2.93 %C, 0.59 %Si, 0.70 %Mn, 17.5 %Cr, 1.59 %Mo

The interval from the austenitizing temperature to that of the start of the pearlite transformation is characterized by the precipitation of secondary carbides. These are a component of the structure together with primary or eutectic carbides resulting from the liquid–solid transformation. The remaining structural feature is the austenite transformation product, which is pearlite, bainite, or martensite.

The precipitation of secondary carbides depletes austenite of carbon and alloying elements and changes the austenite transformation characteristics. Note that the M_S temperature is continuously raised. There is also a slight peak in hardness curve which corresponds either to enhanced volume of martensite or to the influence of secondary carbides.

Use of CCT Diagrams The cooling rate must be established for the section size of the casting being treated in the cooling medium. This enables the transformation to be established from the diagram. Cias[12] has given approximate cooling rates for the centres of round steel bars of various sizes cooled from 800 °C in water, oil, and still air. By prior agreement on the microstructures and hardnesses required for a material, it is possible to determine the cooling cycle which is necessary.

Variations of section size of the casting and the related variations of cooling rate will give different structures in different parts of the casting. A conservative estimate of structure and properties can be made using the cooling rate at the centre of a representative casting section. The hardness versus half-cooling time plot is itself a hardenability curve comparable with data given by Jominy testing.

CCT Diagram for Low Alloy Gray Cast Iron Figure 9-10[12] is a CCT diagram for a low alloy gray cast iron containing 0.30 per cent molybdenum and 0.60 per cent copper. After solidification, the structure consists of γ and graphite. Secondary graphite precipitation is indicated on the diagram in a temperature range close to Ac_1. Cooling into the nose of the pearlite transformation will give a mixed structure with bainite, as cooling subsequently progresses into the bainite region. Pearlite is obtained at the slowest cooling rates.

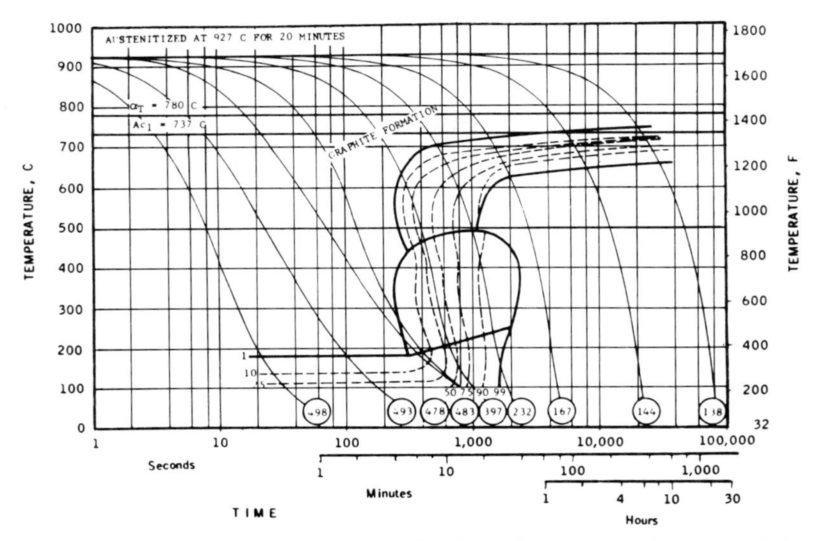

Figure 9-10 CCT diagram for low alloy Mo–Cu gray cast iron containing 3.17 per cent C, 2.0 per cent Si, 0.75 per cent Mn, 0.30 per cent Mo, 0.60 per cent Cu. Austenitized at 927 °C for 20 min[12] (Reproduced by permission of Climax Molybdenum Company)

Austempering of Ductile Iron

The type of cooling programme for an austempering heat treatment is shown in Figure 9-11[17] on a TTT diagram. It is also possible to make initial use of a CCT diagram for the type of alloy being austempered. This allows judgement

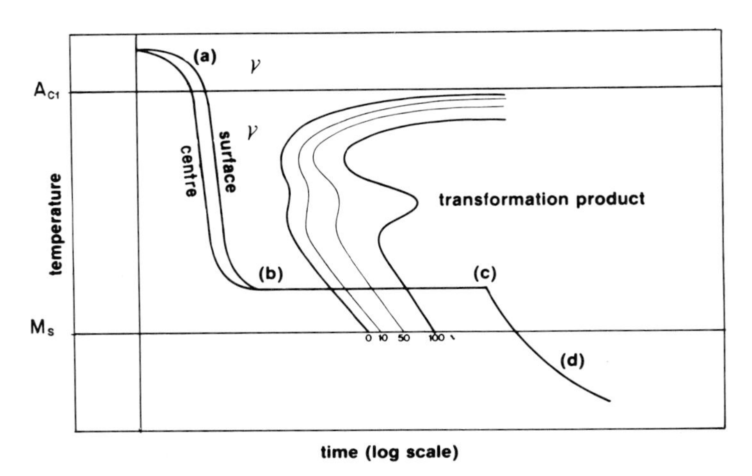

Figure 9-11 Austempering heat treatment shown on TTT diagram.[17] Reproduced by permission of Fuel and Metallurgical Journals Ltd.)

Table 9-1 Relationship between austenitizing, and austempering temperatures[17] Reproduced by permission from American Foundrymen's Society

Austenitizing temperature	%C	Austempering temperature
850 °C	0.73	420 °C
900 °C	0.93	400 °C
1000 °C	1.14	360 °C

of the cooling rate necessary to avoid transformation products preceding the bainite transformation.

The temperature of the transformation is selected to conform with the type of bainite, upper or lower, and the mechanical properties desired. The temperature of austenitizing is important since the carbon content of the austenite increases with this temperature. At the lower austenitizing temperature, the carbon content will be lower, giving greater ductility.

The austenitizing temperature has an important relationship with the austempering temperature. As the carbon content of the γ increases the temperature of the transformation for upper to lower bainite is depressed to lower tempeatures. Table 9-1 shows these effects. As the austenitizing temperature is increased and the carbon content increases, the selected austempering temperature for optimum properties is depressed.

In the transition from lower to upper bainite, maximum ductility and toughness is obtained with lower tensile and yield strength.

Small batches of iron castings of suitable alloy composition are austempered by quenching to the isothermal transformation temperature in hot oil. Transformation can be continued in the same bath. For larger quantities of castings, transfer is recommended to a hot air furnace for isothermal transformation. The times required for this may be two to three hours.

The M_S Temperature; Retained Austenite

Thermodynamically, the transformation of austenite to martensite should occur when $G_\gamma = G_M$ and $\Delta G = G_M - G_\gamma = 0$. G_γ is the molar free energy of austenite and G_M is the molar free energy of martensite.

Patel and Cohen[18] showed that M_S temperature occurs at a certain value of ΔG, e.g. for Fe–C alloys, $\Delta G = -290$ cal mol^{-1}. For Fe–Ni, the value is *f*200 cal mol^{-1} and for Fe–C–Ni, the value is -370 cal mol^{-1}. The reason for this departure from equilibrium is due to the nucleation requirement for martensite. The temperature must drop to allow ΔG^*, the critical free energy for nucleation of martensite, to achieve a specific value.

The temperature of the start of the martensite transformation M_S is influenced by different elements in solid solution in γ, principally carbon. In theory, any element which goes into solid solution in austenite should depress M_S. The problem of austenite retained in the structure after heat treatment is its instability. It may transform under mechanical stress or by further cooling.

Isothermal Transformations of Austenite in Cr and Cr–Mo White Cast Irons

Maratray and Usseglio-Nanot[19] studied the isothermal transformation of austenite in Cr and Cr–Mo white irons. These alloys which were in the composition range 2.0–4.3 per cent carbon, 11.0–26.0 per cent chromium, and 0–4.0 per cent molybdenum were substantially austenitic in thick sections as cast. Figure 9-12(a) shows the isothermal transformations observed. These are:

(i) $\gamma \rightarrow$ secondary carbides + γ (above Ac_3)
(ii) $\gamma \rightarrow$ secondary carbides + α + γ (below Ac_3)
(iii) $\gamma \rightarrow$ fine pearlite (in the temperature range 650–750 °C)

Figure 9-12(b) is for the transformation of γ after the precipitation of secondary carbides has been performed at a higher temperature. This isothermal transformation is to bainite.

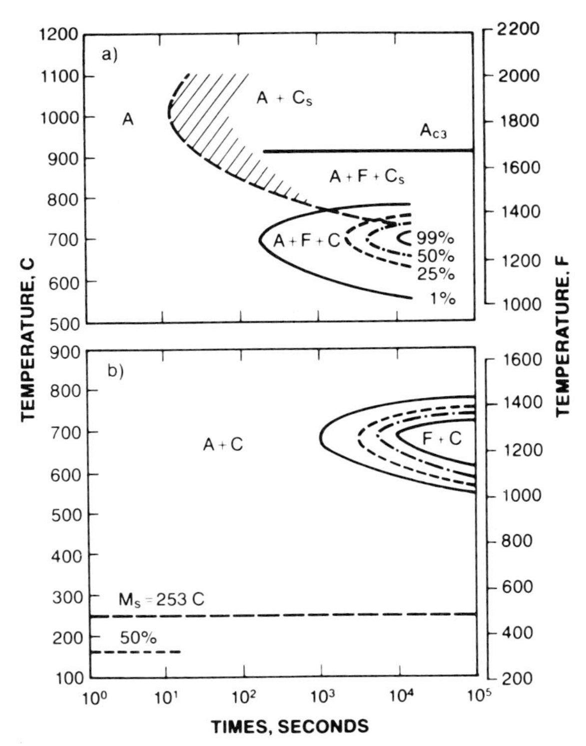

Figure 9-12 (a) Isothermal transformation diagrams for austenite in Cr–Mo white iron. Between 650 and 750 °C the transformation is to pearlite. (b) After treatment of γ to precipitate secondary carbides at high temperature, an isothermal transformation to bainite may be observed.[19] (Reproduced by permission of Climax Molybdenum Company)

On precipitation of secondary carbides at a high temperature, the composition of the austenite is changed and M_S is raised, resulting from a lower composition in carbon. Martensite can be obtained as a transformation product provided the rate of cooling is sufficient to avoid the bainite nose. This type of precipitation treatment is termed destabilization. Maratray and Usseglio-Nanot[19] derived the following expressions for white cast iron destabilized at 1000 °C for 20 minutes. For the maximum diameter of bar of Cr, C, and Mo white iron which would transform to martensite after destabilization:

$$\log_{10} D_{max}(\text{mm}) = 0.32 + 0.158 \frac{\text{Cr}}{\text{C}} + 0.385\ \text{Mo}$$

The time in seconds to the pearlite nose after the destabilization treatment above was:

$$\log \text{pearlite time (s)} = 2.61 - 0.51\ \text{C} + 0.05\ \text{Cr} + 0.37\ \text{Mo}$$

Ferritic and Pearlitic Heat Treatments for Spheroidal Graphite Cast Iron

The matrix structure of spheroidal graphite cast iron in the ordinary grades after solidification and cooling to room temperature will probably be a mixture of ferrite and pearlite. Ferrite will border the graphite. A completely ferritic structure can be obtained by heating at about 50 °C below A_1 for a period of time depending on the casting cross-section.

If as-cast carbides are part of the structure, annealing can be performed at 900 °C for a period of approximately 2 hours before slow cooling to 50 °C below A_1 and holding as above. If no alloying elements are present which stabilize pearlite, e.g. Cu or Mo, the casting can be obtained ferritically by slow cooling after solidification.

Pearlitic structures are best obtained by adding pearlite stabilizers, e.g. Cu or Mo. Heat treatment of the casting can be performed at 900 °C or 50 °C above A_3 and cooled at a rate obtained from a CCT diagram.

9-3 Surface Hardening

Surface hardening is an important aspect of the heat treatment of cast iron. In applications requiring resistance to wear and abrasion, exceptionally hard surfaces can be obtained on cast iron by heating into the austentic range followed by cooling. Control of this process leads to control of the transformation products and to variations in hardness and mechanical behaviour. Recent interest has been centred on glazing of cast iron surfaces.

In the present section, the different possibilities of electron beam and laser methods, are discussed with emphasis on the density of thermal energy available with the different techniques and control of the thermal cycle.

Electron Beam; Fusion Treatment

Hiller[20] has described the surface treatment of both gray and ductile irons by an electron beam. The sequence of physical events which takes place when an electron beam impinges on the cast iron surface is shown in Figure 9-13(a). These events are (i) heating, (ii) melting, (iii) penetration, and (iv) solidification. Figure 9-8(b) shows the variation of hardness in the three zones which result. These are the solidified metal, the heat affected zone, and the base metal.

In the surface melted zone which has a fine ledeburite structure, a hardness value of 920 VPN was reported for cast iron and a case depth of 1.0 mm. The heat affected zone extended for a further 0.5 mm. In spheroidal graphite cast iron, hardness figures of VPN 1000 were reported. The surface hardness of the ledeburite obviously depends on its microstructure and analysis and can be varied by the composition of the base iron and the rate of solidification of the liquid metal.

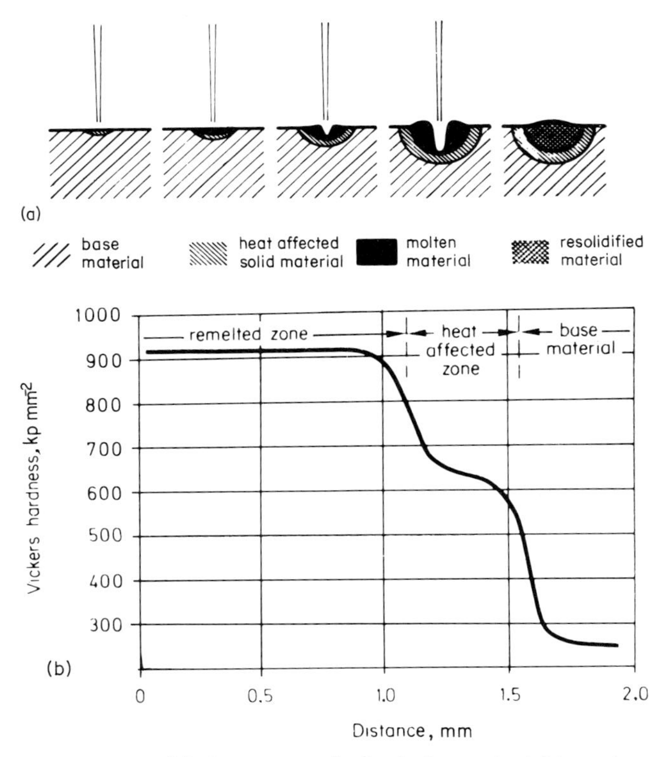

Figure 9-13 (a) Sequence of physical events taking place when an electron beam impinges on a cast iron surface. (b) Variation in hardness of cast iron from the surface resulting from the processes occurring in (a).[20] (Reproduced by permission of Georgi Publishing Co.)

Different techniques are available, e.g. spot melting or continuous seam melting. The problems are those normally associated with the freezing of liquid metals, i.e. cracking, etc.

Heating by the electron beam without melting, a process akin to induction hardening, has also been reported.[21]

Laser Beam Heat Treatment of Surfaces

The laser beam for surface heat treatment in metallurgy has been reviewed in a number of papers concerned both with steel and cast iron.[22] The energy of the beam is absorbed at the surface, regions of the solid remote from the surface being heated by thermal conduction.

The only lasers capable of heat treating at reasonable rates are CO_2 gas lasers. These present a problem in that metallic surfaces have a high reflectivity in the range of wavelengths emitted. Some of the references give figures for energy absorbed when coatings are employed on the metallic surfaces. Laser power presently employed is reported as between 1–2 kW for heat treatment purposes. Within this range, depth of hardening is approximately 0.75 mm using a surface coverage rate (A^2/t) of approximately 62 $cm^2\ min^{-1}$. Figure 9-14 shows a diagram for the laser hardening of cast iron.[23] The ordinate is the rate of surface coverage. This is by a beam which is transversely oscillated, the frequencies being indicated on the diagram. The abscissa is the case depth and it is noted that this is highest for surface melting. The region in which no melting occurs, recommended as desirable, is limited in depth.

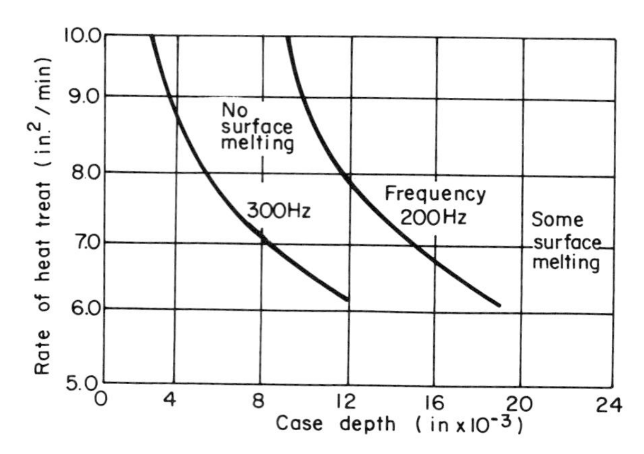

Figure 9-14 Diagram for laser hardening of cast iron. The ordinate is the rate of surface coverage. The abscissa is the case depth.[23] (Reproduced by permission of McGraw-Hill Book Co.)

Laser Beam Application to Surface Heat Treatment of Cast Iron

In the review of industrial applications of lasers by Ready,[22] reference was made to surface temperature and penetration depth, which can be varied by adjusting the beam power, by focusing, and by speed of traverse.

Figure 9-15 shows the relationship between hardness R_c and case depth for flake graphite cast iron treated with a CO_2 laser beam having the dimensions 1.7 cm width and 1.0 cm in the direction of motion.[24] The beam was moved over the surface at 50 mm s^{-1}.

Overlapping effects are noted in laser beam techniques, and these are similar to welding using multiple pass techniques. The problem is to achieve uniform hardening effects. Various scanning methods have been developed, e.g. traversing linearly using a defocused beam. Alternatively, use can be made of a more finely focused beam oscillated at high frequency in a direction transverse to the direction of traverse.

Harth Diagrams

Use has been made of Harth diagrams[25] as a means of controlling the hardening conditions for laser heat treatment. These plot curves of constant case

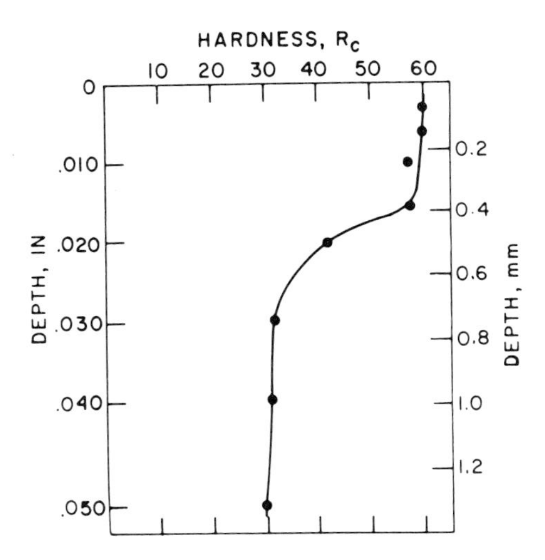

Figure 9-15 Relationship between depth and hardness for a cast iron treated with a CO_2 laser beam. Scan rate 375 cm^2 min^{-1}.[24] (From E. V. Locke and R. A. Hella *IEEE J. Quantum Electronics*, **QE.10**, 179, 1974. Reproduced by permission of IEEE. Copyright © 1974 IEEE)

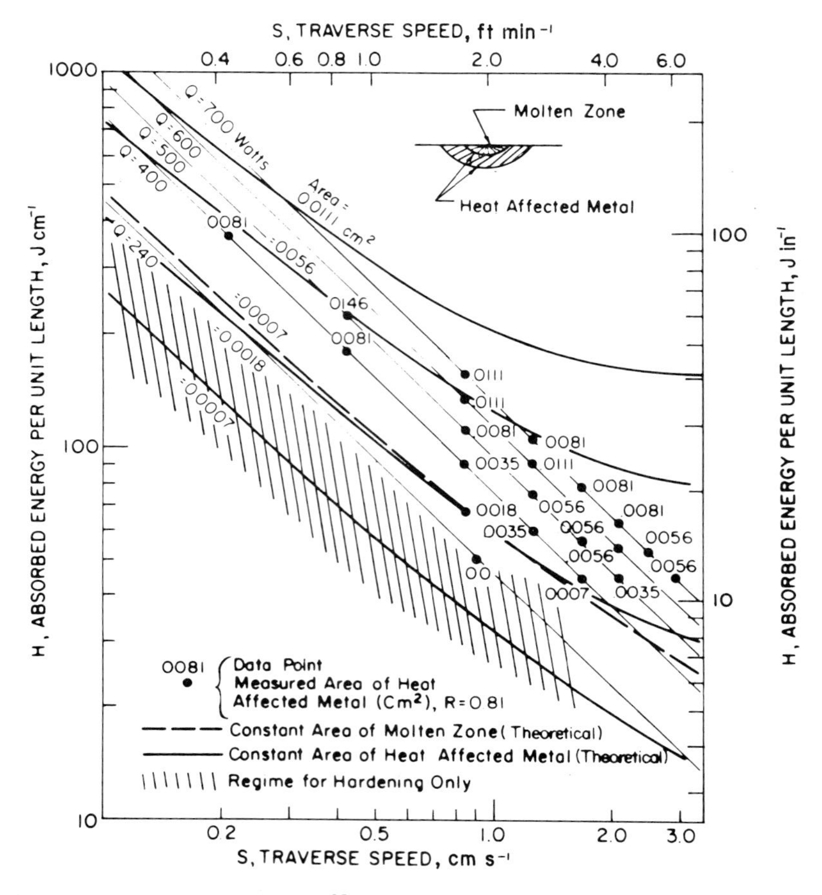

Figure 9-16 A Harth diagram[25] for laser hardening of a surface. (Reproduced by permission of The Metallurgical Society of AIME)

depth where the ordinate is absorbed energy density delivered by the laser and the abscissa is the feed rate of the work piece (Figure 9-16).

The diagrams were constructed following experimental work on the melting of surface layers and the extent of the heat-affected zone using a point source theory developed for welding by Rosenthal.[26]

These diagrams employ the following relationship:

$$H = \frac{R_a Q}{S}$$

where R_a = absorptivity
H = absorbed energy per unit length
Q = energy of the laser beam
S = traverse speed

These parameters determine the size of the hardened region in laser heat treatment. Any location within the diagram defines a unique set of laser heat treating conditions. In Figure (9-16), the dashed line is a calculation for the cross-section of molten metal. Above the dashed line, the heat-affected zone includes a molten area. The hatched region is the approximate area where hardening occurs without significant melting. This area is approximately 0.0007 cm^2.

In laser hardening, it is generally desirable to produce as deep a hardened case as possible and to avoid surface melting. It may be difficult to satisfy both requirements since significant amounts of heat-affected material are produced only when melting occurs. This suggests that laser heat treatment with small diameter beams results in shallow hardened layers. Harth *et al.*[25] suggested that the depth of hardening could be increased without melting by using a beam of larger diameter, i.e. by using a more powerful laser but keeping the energy and power density within the desired range.

Other Laser Techniques

Other laser techniques were reviewed by Ready.[22] Surface alloying employs powder which is melted into the surface by the laser beam. The treatment rate

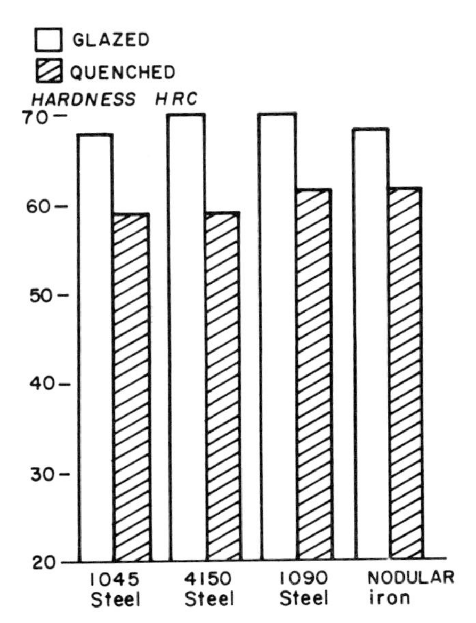

Figure 9-17 Surface heat treatment of different materials by lasing and glazing[27] (From *Metal Progress*, December 1981, p. 27, Fig. 6. Reproduced by permission of The American Society for Metals)

is somewhat slower than for transformation hardening because surface melting is desired. Elements introduced included boron, chromium, and silicon.

Gray[27] reviewed substitution and conservation technology for critical materials and gave the data of Figure 9-17 for surface hardening techniques with different steels and spheroidal graphite cast iron. The glazed structures were prepared using an electron beam technique, and the hardness was compared with conventional quenching. A hardness of R_c 67 was obtained with the spheroidal graphite iron while the ordinary quenched hardness is R_c 52.

References

1. Orowan, E.: *Symposium on Internal Stress in Metals and Alloys*, The Inst. of Metals, London, 1948.
2. Dieter, G. E.: *Mechanical Metallurgy*, 1st ed., McGraw-Hill, New York, 1961.
3. Lászlò, F.: *J. Iron Steel Inst. London*, **147**, 173, 1943.
4. Nabarro, F. R. N.: *Symposium on Internal Stress in Metals and Alloys*, The Inst. of Metals, London, 1948.
5. Grill, A., J. K. Brimacombe, and F. Weinberg. *Ironmaking Steelmaking*, **3**, 31, 1976.
6. Angus, H. T.: *Cast Iron, Physical and Engineering Properties*, 2nd ed., Butterworths, 1976.
7. Nadai, A.: *Plasticity*, McGraw-Hill, New York, 1931.
8. Inst. Br. Foundrymen.: Sub-Committee T.S. 17, *Symposium on Internal Stress in Metals and Alloys*, The Inst. of Metals, London, 1948.
9. Dodd, J., and J. L. Parkes: *Int. Cast Metals Res. J.*, **5**, 47, 1980.
10. Molinder, G.: *Acta Metall.*, **4**, 565, 1956.
11. Christian, J.: *The Theory of Transformations in Metals and Alloys*, Pergamon, 1965.
12. Cias, W. W.: *Austenite Transformation Kinetics of Ferrous Alloys*, 1978.
13. Cahn, J. W., and W. C. Hagel: *Decomposition of Austenite by Diffusional Processes* (Eds. V. F. Zackay and H. I. Aaronson), Met. Soc. AIME, Interscience Publ., 1962.
14. Puls, M. P., and J. S. Kirkaldy: *Metall. Trans.*, **3**, 2777, 1972.
15. Hobstetter, J. N.: *Decomposition of Austenite by Diffusion Processes* (Eds. V. F. Zackay and H. I. Aaronson), Met. Soc. AIME, Interscience, 1962.
16. T. Ericsson (Ed.): *Computers in Materials Technology*, Pergamon Press, 1981.
17. Dodd, J.: *Modern Castings*, **68**, 60, 1978.
18. Patel, J. R., and M. Cohen: *Metall.*, **1**, 531, 1953.
19. Maratray, F., and R. Usseglio-Nanot: *Factors Affecting the Structure of Chromium and Chromium—Molybdenum White Irons*, M. 458E, Climax Molybdenum Co., 1971.
20. Hiller, W. *The Metallurgy of Cast Iron* (Eds. B. Lux, F. Mollard and I. Minkoff), Georgi Publ. Co., Switzerland, 1975.
21. Discussion of Ref. 20 in *The Metallurgy of Cast Iron*, Georgi Publ. Co., Switzerland, 1975.
22. Ready, J. F.: *Industrial Application of Lasers*, Academic Press, New York, 1978.
23. Engel, S. L.: *Am. Machinist*, **1976**, 107, May 1976.
24. Locke, E. V., and R. A. Hella: *IEEE J. Quantum Electronics*, **QE. 10**, 179, 1974.
25. Harth, G., W. C. Leslie, V. G. Gregson, and B. A. Sanders: *J. Met.*, **28**, 5, April 1976.
26. Rosenthal, D.: *Weld. J. Suppl.*, **20**, 2205, 1941.
27. Gray, Allen G.: *Metal Progress*, **1981**, 18, December 1981.

Bibliography

Denton, A. A.: *Determination of Residual Stress, Met. Reviews*, Vol. II, p. 1, Inst. Metals, London, 1966.

Jack, D. H., and J. Nutting: *Internat. Met. Reviews*, Vol. 19, p. 90, Met. Soc. and Am. Soc. Met., 1974.

Shewmon, P. G.: *Transformations in Metals*, McGraw-Hill, 1969.

Chapter 10 *Graphitizing Reactions in the Production of Malleable Cast Iron*

The subject matter of this chapter is related to the solid state reactions occurring during the malleablizing process of white cast iron. Of practical interest are the rates of graphitization of the white structure and the form of the graphite. These rate processes are dependent on a number of factors related to composition and structure and involve dissolution of the carbide phase, nucleation, and growth of the graphite phase.

The first stage of graphitization occurs in the region of the phase diagram above A_1 where the γ phase coexists with either carbide or graphite. The first-stage reaction is one of nucleation and growth and is generally written

$$Fe_3C \rightarrow \gamma + \text{graphite}$$

The different models for this reaction are presented. In the second stage of graphitization, secondary carbides are transformed to graphite below A_1.

The influence of alloying elements on these two transformations is discussed, and the factors which determine the morphology of graphite are presented.

10-1 Division of Malleable Iron Types

Three types of malleable iron are generally manufactured:

(a) Blackheart malleable iron with a fully ferritic matrix.
(b) Whiteheart malleable iron in which graphitization occurs during progressive decarburization of the white iron structure, a carbon gradient existing from the interior towards the surface. The matrix may range from pearlitic at the centre to ferritic at the edge, or it may be completely pearlitic.
(c) Pearlitic malleable iron. In this type of material, a pearlitic matrix is obtained in a material produced by the blackheart process, in which only

the first-stage graphitization process is undertaken. The chemical composition is adjusted so that pearlite either forms during furnace cooling, air cooling, or air quenching.

Other heat treatment procedures are also possible at the end of first-stage graphitization, e.g. oil quenching and tempering.

10-2 Kinetics of Graphitization

The kinetics of graphitization, i.e. the rate at which the carbide phase is transformed to graphite, are important industrially. The kinetics are studied experimentally in order to determine factors which control the process and how these are influenced by composition. The factors are distribution and dissolution of the carbide phase, nucleation of graphite, and its rate of growth. The dependence of graphite morphology on these factors is an important consideration.

The overall transformation can be followed by different techniques such as length changes of reacted specimens and metallographic point counting techniques on specimens reacted for a specific length of time. The curves obtained for the overall progress of the transformation are sigmoidal in form. The curves for growth rates of individual nodules have been interpreted in different ways.

First-stage Graphitization Kinetics—Burke and Owen

Burke and Owen[1] studied the kinetics of first-stage graphitization in Fe–C–Si alloys of composition varying between 2.0–3.5 per cent carbon and 0.68–1.37 per cent silicon. The experimental method employed was the measurement of length increases of isothermally heated specimens at different temperatures. The progress of the reaction was taken as the increase in length of the specimens expressed as a fraction of the total increase in length for the complete reaction. This fraction was plotted against the log of time. Sigmoidal form curves were obtained.

A detailed presentation of different methods employed in analysing isothermal transformation kinetics is found in the text of Christian.[2] Burke and Owen used a modified Zener equation of the following form to analyse the data:

$$y = 1 - \exp\left(-\frac{t}{K}\right)^n \tag{10-1}$$

or, differentiating:

$$\frac{dy}{dt} = \frac{n}{K^n}(1 - y)t^{n-1} \tag{10-2}$$

where y = fraction transformed at time t
K = temperature-dependent rate constant
n = constant.

In the original analysis by Zener[3] the transformation rate for a precipitation reaction was described by a power law as follows:

$$\frac{dy}{dt} = Bt^m$$

where B is a temperature-dependent rate constant and m is a constant which depends on the shape of the precipitate. To account for impingement, i.e. mutual interference between particles in claiming available solute, the decrease in rate is assumed proportional to the fraction transformed, $(1 - y)$. The equation then becomes:

$$\frac{dy}{dt} = B(1 - y)t^m$$

The quantity m is determined by the geometry of the precipitate and by the kinetics of the nucleation process and is assumed independent of temperature. The constant B is a temperature-dependent term of the following exponential form:

$$B = C_1 \exp(-Q/RT)$$

where Q = activation energy for the reaction
T = absolute temperature
C_1 = constant
R = gas constant

Burke and Owen, using Equation (10-1), could obtain K directly from the transformation curves as follows:

$$\ln \frac{1}{1 - y} = \left(\frac{t}{K}\right)^n \tag{10-3}$$

From Equation (10-3), when

$$y = \frac{e - 1}{e} \simeq 0.63$$

then $K = t$. The value of K at a particular temperature is given by the time at which $y = 0.63$. Writing (10-1) in the form:

$$\log \log \frac{1}{1 - y} = n \log t - n \log K - \log 2.3$$

the value of n could be obtained from the straight line obtained by plotting log log $1/(1 - y)$ against $\log t$. Burke and Owen obtained an average value of $n = 4.07$ with no definite trend for varying temperature and composition.

The value of Q_K was obtained from experimental determinations of K by plotting $\log K$ against $1/T$ and measuring the slope. A second possibility was to plot $1/T$ against $\log t$ for a given fraction transformed, say 50 per cent. The activation energy Q_K for the first-stage graphitization, obtained by averaging all the determinations, was 27.2 kJ mol^{-1}.

Growth Model

The growth model used by Burke and Owen[1] is shown in Figure 10-1(a). The graphite phase grows in conditions of carbon supersaturation determined by the difference in composition of austenite at the cementite interface C_α and that at the graphite interface C_e. This is shown in Figure 10-1(b) for the temperature of graphitization T; C_p is the composition of graphite.

Assuming a linear concentration gradient and coefficient of diffusion D independent of composition, the rate of growth of a spherical particle is:

$$\frac{\delta r}{\delta t} = \frac{D}{L}\left(\frac{C_\alpha - C_e}{C_p - C_e}\right) = G, \text{ a constant}$$

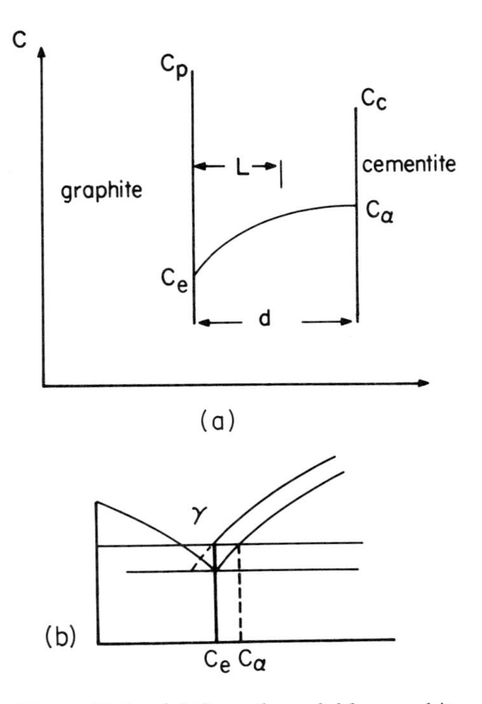

Figure 10-1 (a) Growth model for graphitisation used by Burke and Owen.[1] (Reproduced by permission of The Metals Society. (b) Conditions of carbon supersaturation determined by difference in composition of austenite at cementite interface C_α and at graphite interface C_e

If N is the nucleation rate per unit volume of material, the expression obtained for the fraction transformed is:

$$y = 1 - \exp\left[- c' \frac{\pi}{3} \frac{D^3}{L^3} \left(\frac{C_\alpha - C_e}{C_p - C_e} \right)^3 N t^4 \right]$$

where c' is a constant. When compared with

$$y = 1 - \exp\left(- \frac{t}{K} \right)^n$$

the experimentally obtained value of $n = 4$ is seen to conform to the spherical growth model.

Burke and Owen pointed out that this might be partly fortuitous since the growing phase was not spheroidal. Furthermore, a constant growth rate dependence with time was assumed which was not in accordance with the Zener model of a diffusion controlled, parabolic rate of growth process. The values for activation energy of the different controlling factors were calculated to be the following:

Overall graphitization reaction:	$Q_K = 27.2$ kJ mol^{-1}
Nucleation only:	$Q_N = 31.6$ kJ mol^{-1}
Growth only:	$Q_G = 25.2$ kJ mol^{-1}

Three separate diffusion processes were considered for growth in the Fe–C–Si alloys examined:

(a) Carbon diffusion through austenite to the graphite.
(b) Silicon diffusion from the area of the growing graphite phase.
(c) Iron diffusion from the area of the growing graphite phase.

The three different activation energies for diffusion are respectively 11.2, 22.4, and 29.6 kJ mol^{-1}. Therefore it was suggested that for Q_G experimentally determined, either silicon or the self-diffusion of iron controlled the graphite growth rate, but this could not be unequivocally established.

For the nucleation process, graphical plots were made of the number of nodules against time for varying silicon contents. These curves showed a maximum, suggesting that the smallest nodules redissolved with time. The influence of silicon was to increase the nucleation rate and increase the number of nuclei, moving the maximum to shorter times and more nodules.

Further Observations of the First-stage Graphitization Process

Owen and Wilcock[4] made a further analysis of the reactions involved in first-stage graphitization. The constant diffusion distance initially assumed appeared justified by the observation of nucleation on the interface between γ and eutectic carbide and the dissolution behaviour of the Fe_3C. A number of observations were made.

The movement of the Fe_3C–γ and graphite–γ interfaces was complicated. In particular the Fe_3C–γ interface was not symmetrically spaced round the graphite nodule. An activation energy of 31.2 kJ mol^{-1} was evaluated for the rate of solution of Fe_3C in γ.

Over most of the reaction, it was noted that the dissolution of Fe_3C kept pace with the growth of graphite and the growth curves again suggested that the graphite growth rate was approximately constant.

For the nucleation event, the γ–Fe_3C interfacial energy and its variation with composition was pointed out as important. Once established direct transfer of carbon atoms across the interface might occur. With time the carbide phase dissolves, but it was noted that carbides remote from the graphite phase commenced dissolution. This suggested that dissolution of the carbide phase might be a somewhat slower process than the growth of graphite. It was noted that there was a sufficient excess of Fe_3C during the early stages of the reaction to maintain the supply of carbon, and it was not until the volume of Fe_3C became small that the Fe_3C dissolution mechanism could start to exercise a control over the reaction. Progress of the graphitization process was further complicated by the re-solution of small graphite nodules so that the controlling factors in the transformation could be continuously varying.

10-3 Other Graphitization Models

Birchenall and Mead[5] examined different analyses presented for the first stage of graphitization while confining their own analysis to the growth of a nodule only. They used data from the research of Burke and Owen[1] described above and suggested that the growth rate was better fitted by a parabola if an initial induction period was considered.

In the model adopted, a spherical graphite nodule grows in a matrix consisting of a mixture of γ and Fe_3C, the particles of cementite being uniformly distributed. There is a spherical shell round the graphite from which the Fe_3C has been depleted. The model is shown in Figure 10-2.

The inner shell diameter is R_1 and the carbon concentration is C_1 while the outer diameter is R_2 and the carbon concentration is C_2. The concentration of carbon at the boundary of the γ envelope is $C_\propto$. The graphite is denoted by C_0. The rate of increase of carbon in the sphere is:

$$(C_0 - C_1)4\pi R_1^2 \frac{\mathrm{d}R_1}{\mathrm{d}t} \tag{10-4}$$

The rate of transfer of carbon to the sphere is:

$$4\pi R_1^2 D \left(\frac{\delta c}{\delta r}\right)_{r=R_1} \tag{10-5}$$

These two quantities must be equal.

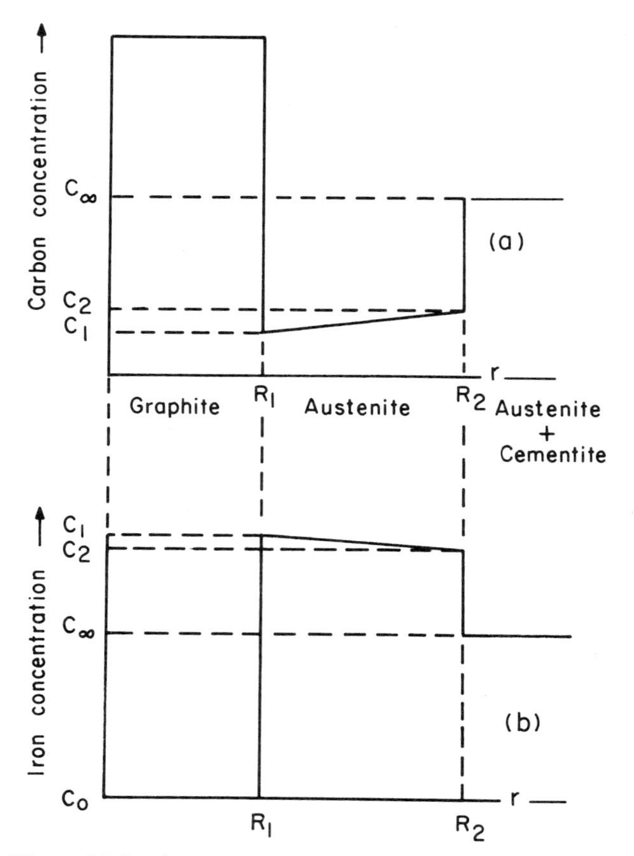

Figure 10-2 Growth model of Birchenall and Mead.[5] (Reproduced by permission of The Metallurgical Society of AIME)

Similarly, the growth of the ring depleted of cementite will depend on the rate at which carbon is removed by diffusion at R_2:

$$(C_\infty - C_2)4\pi R_2^2 \frac{dR_2}{dt} = 4\pi R_2^2 D \left(\frac{\delta c}{\delta r}\right)_{r=R_2} \tag{10-6}$$

The flux of material through a shell of radius r between R_1 and R_2 is:

$$4\pi r^2 D \left(\frac{\delta c}{\delta r}\right)_{R_1<r<R_2} = 4\pi D \frac{C_2 - C_1}{1/R_1 - 1/R_2} \tag{10-7}$$

The dependence of the concentration $C(r)$ between R_1 and R_2 on r is given by:

$$C(r) = C_2 - (C_2 - C_1)\left(\frac{R_1}{r}\right)\left(\frac{R_2 - r}{R_2 - R_1}\right) \tag{10-8}$$

If volume changes are neglected, the amount of any material inside the sphere of radius R_2 at any time must be the same as that initially present, since no concentration changes have occurred beyond R_2. Thus:

$$C_\infty \tfrac{4}{3}\pi R_2^3 = C_0 \tfrac{4}{3}\pi R_1^3 + 4\pi \int_{R_1}^{R_2} r^2 C(r)\, \mathrm{d}r \tag{10-9}$$

Inserting Equation (10-8) in (10-9) and integrating yields:

$$(C_\infty - C_2)R_2^3 = (C_0 - C_1)R_1^3 + \tfrac{1}{2}(C_1 - C_2)R_2 R_1(R_2 + R_1) \tag{10-10}$$

Neglecting the last term, which leads to a negligible error:

$$\frac{R_1}{R_2} = \left[\frac{C_\infty - C_2}{C_0 - C_1}\right]^{1/3} \tag{10-11}$$

Combining (10-4), (10-5), (10-7), and (10-11) and integrating yields:

$$R_1^2 = \left[\frac{2(C_2 - C_1)D}{(C_0 - C_1)\{1 - [(C_\infty - C_2)/(C_0 - C_1)]^{1/3}\}}\right] \times t \tag{10-12}$$

Combining (10-6), (10-7), and (10-11) yields:

$$R_2^2 = \left[\frac{2(C_2 - C_1)D}{(C_\infty - C_2)[(C_0 - C_1)/(C_\infty - C_2) - 1]^{1/3}}\right] \times t \tag{10-13}$$

if R_1 and $R_2 = 0$ at zero time. The values in brackets are parabolic rate constants and have fixed values at given temperature.

Calculation was made of the rate constants by evaluating C_0, C_1, C_2, and C_∞ from different research sources. Diffusion coefficients for carbon and iron were used in the calculations.

The results showed that, with the model employed, the rate constants with carbon diffusion agreed with experimentally measured rates, within a factor of about 5 to 10. This was considered as a good agreement. The rate constants based on iron diffusion were very much lower than the experimental values. Birchenall and Mead therefore suggested that carbon diffusion through an austenite shell around the graphite particle controlled the rates in experiments for which the data were used. These experiments were by Brown and Hawkes[6] and Burke and Owen.[1]

10-4 The Influence of Alloying Elements on Graphitization Kinetics in Blackheart Malleable Iron

Difficulties are involved in analysing graphitization reactions because of the complexity of the process, but if the data for individual steps of the transformation are examined some knowledge of controlling mechanisms may emerge. Sandoz[7] classified some known influences as follows:

(a) Elements which accelerate the first stage: Cu, Ni.
(b) Elements which retard the first stage: Mn, Cr, V.

The data presented in Section 10-3 seemed to indicate that the graphite growth process in first-stage graphitization is controlled by carbon diffusion, with a limited influence of the alloying elements on the diffusion process. The influence of the alloying elements which retard the first stage can thus be related to their stabilization of the carbide phase. They do this by dissolving in this phase. Some of the theoretical aspects will be discussed in Section 10-5.

In this section different experimental evaluations of the influence of composition on graphitization will be described.

Effects of Manganese and Sulphur on Graphitization in Blackheart Malleable Iron

The data of Rehder[8] showing the influence of the Mn to S ratio at varying S concentrations are shown graphically in Figure 10-3. The analysis of iron for these experiments was made from two compositions:

A: 2.45 %C, 1.20 %Si, 0.06 %Mn, 0.098 %S, 0.06 %P

B: 2.45 %C, 1.21 %Si, 0.07 %Mn, 0.205 %S, 0.06 %P.

The variations of the Mn to S ratio were made by melting the alloys A and B with varying quantities of ferromanganese. The results show that when Mn

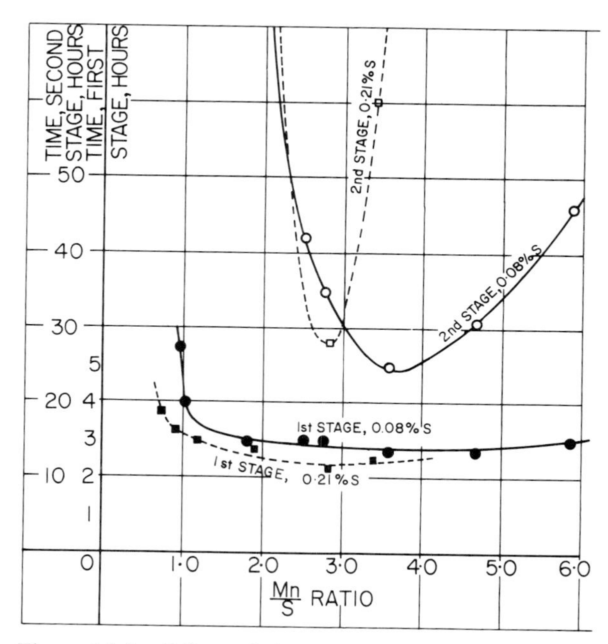

Figure 10-3 Effect of the Mn to S ratio on the first and second stages of graphitization in low and high sulphur content white cast irons.[8] (Reproduced by permission of American Foundrymen's Society)

and S are present together, the first-stage graphitization is little effected by analysis other than at low Mn content. The higher S irons complete the first stage more rapidly than the irons of lower S content.

The second-stage process is shown to be very dependent on the Mn to S ratio and on the S content. The second-stage annealing curves for high S contents have a sharp minimum, the curves for higher S having a steeper curvature.

Rehder's data have been discussed in relation to the infleunce on graphitization of the excess Mn or S present required to form MnS. When present individually, either Mn or S influence graphitization kinetics, probably by stabilizing Fe_3C. The element Mn appears to have a pronounced effect on second-stage graphitization, as noted from Figure 10-3. The element S is noted to retard both first- and second-stage graphitization.

10-5 Stabilization of Carbides

Elements in solution stabilize the carbide phase if they dissolve in it. One effect is to decrease the free energy of the phase, as shown in Figure 10-4(a).

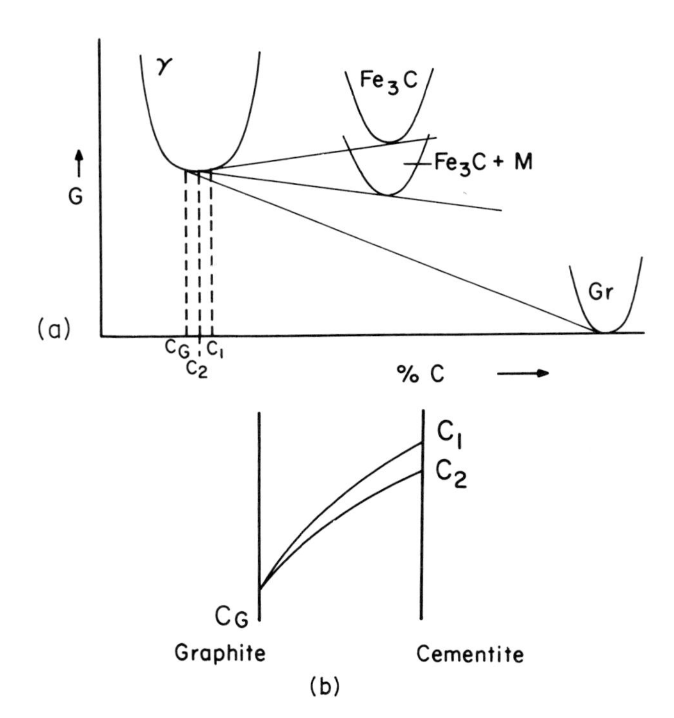

Figure 10-4 (a) Lowering of free energy of a carbide phase by a soluble element. (b) Change in supersaturation due to change of carbon in γ in equilibrium with Fe_3C and alloy carbide

New solubility relationships are created between the Fe_3C and graphite phases in equilibrium, as shown in Figure 10-4(a), and a smaller equilibrium difference exists in the γ phase between the interfaces with Fe_3C and graphite (Figure 10-4).

In addition to the new solubility relationships established, the movement of the carbide interface now involves the diffusion of the solute element, e.g. Cr, which may control the dissolution process. The kinetics of the graphitization process may therefore be changed.

This should be compared with elements like Ni and Cu which enhance the graphitization process. They dissolve in the austenite and lower its free energy, changing the stability of the carbide phase and graphite in relation to austenite.

Graphitization Controlled by the Rate of Dissolution of Cementite

Burke[9] examined possible controlling mechanisms for graphite nodule growth. He developed equations for the various controlling mechanisms examining the following:

(a) The solution of cementite.
(b) The transport of C atoms through austenite.
(c) Attachment of C atoms to the graphite interface.
(d) Diffusion of Fe and alloying elements away from the growing nodule.

The problem was determination of the slowest process. To ascertain the controlling mechanism, growth must be calculated for each of the above processes and comparison made with the experimental data. Only diffusion control equations by Birchenall and Mead,[5] described above, had been developed to examine any of these processes individually.

Burke intentionally avoided the usual method of estimating the controlling mechanism, i.e. to derive the activation energies for the growth rate and compare the different values determined experimentally.

Considering Birchenall and Mead's Equations (10-12) and (10-13) above, Burke suggested that the treatment given applying these equations to his own data with Owen[1] was too approximate. It also involved some misquotation of data. Therefore agreement between the theoretical and experimental values using this calculation, suggesting diffusion control by carbon, were of dubious significance.

Growth Equations by Burke[9] for Graphitization Control by Solution of Cementite

Burke used the model for graphite growth of Figure 10-5. A spherical graphite nodule of radius R_1 grows in a γ envelope of radius R_2. The envelope is bounded by a $\gamma + Fe_3C$ phase area of composition C_a. The carbon concentration at the graphite–γ boundary is C_1 and at the γ–$\gamma + Fe_3C$ boundary it is C_2. The density of graphite is ρ.

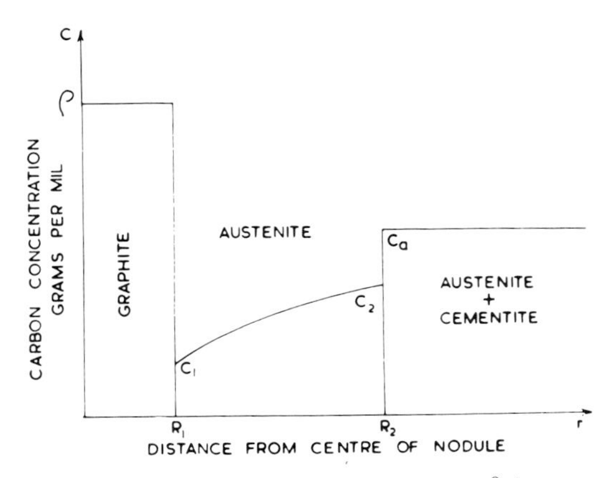

Figure 10-5 Growth model of Burke.[9] (Reproduced by permission of Pergamon Press Ltd.)

Initially, the relationship between R_1 and R_2 is found. A carbon balance due to the movement of R_1 and R_2 is given as follows. The carbon transferred to graphite is $4\pi R_1^2(\rho - C_1)\mathrm{d}R_1$. The carbon transferred by Fe_3C dissolution is $4\pi R_2^2(C_a - C_2)\mathrm{d}R_2$. The balance is given by:

$$R_1^2(\rho - C_1)\,\mathrm{d}R_1 = R_2^2(C_a - C_2)\,\mathrm{d}R_2 \tag{10-14}$$

On integrating, this gives:

$$\frac{R_2}{R_1} = \left(\frac{\rho - C_1}{C_a - C_2}\right)^{1/3} = \alpha \tag{10-15}$$

Equation (10-15) is independent of the mechanism controlling growth. The rate of growth $\mathrm{d}R_1/\mathrm{d}t$ is related to the flux J as follows:

$$4\pi R_1^2 \frac{\mathrm{d}R_1}{\mathrm{d}t}(\rho - C_1) = J \tag{10-16}$$

The isothermal rate of solution of cementite is taken to be S gm of carbon per unit area of interface per unit time. The carbon flux J is equal to S times the area of cementite. Taking this area to be equal to the outer surface of the γ shell:

$$J = 4\pi R_2^2 S \tag{10-17}$$

Using Equation (10-15):

$$J = 4\pi R_1^2\alpha^2 S \tag{10-18}$$

and Equation (10-16) becomes:

$$\frac{\mathrm{d}R_1}{\mathrm{d}t} = \frac{\alpha^2 S}{\rho - C_1} = \beta \tag{10-19}$$

This gives a constant radial rate of growth of a nodule controlled by Fe_3C dissolution. Using an impingement factor,

$$\frac{dR_1}{dt} = \beta(1 - y) \tag{10-20}$$

Relating Equation (10-20) to the Johnson–Mehl equation[10] with $n = 4$:

$$\frac{dR_1}{dt} = \beta \exp[-(t/K)^4] \tag{10-21}$$

To integrate Equation (10-21), the exponential is expanded in a series. The quantity K is equal to the time when $y = 0.63$ (see Equation 10-3). Up to this fraction of graphitization, (t/K) is less than unity and the series is convergent. Therefore:

$$\frac{dR_1}{dt} = \beta\left[1 - \left(\frac{t}{K}\right)^4 + \left(\frac{t}{K}\right)^8 \cdots\right] \tag{10-22}$$

Integrating:

$$R_1 = \beta\left(t - \frac{t^5}{5K^4} + \frac{t^9}{9K^8}\right) \tag{10-23}$$

This equation for the growth rate of the graphite nodule predicts a linear relationship, except when the impingement terms become important, i.e. long times or small K.

Burke suggested that the experimental data available could be analysed in terms of a two-part reaction. In the first part, the rate of growth is governed by diffusion. The graphite nodule grows in saturated γ. As the available carbon from the saturated γ in the vicinity of the graphite nodule is used up, solution of cementite will commence; if it is the slowest process, it will assume control. Growth rate curves from Burke and Owen,[1] analysed according to their reaction sequence, are shown in Figure 10-6.

It was pointed out that the parabolic diffusion controlled model for the growth rate curve could still be applicable. However, other factors of the growth kinetics were to be examined if proof of the prevailing mechanism was to be established.

This was done by re-examining Equation (10-1):

$$y = 1 - \exp\left(-\frac{t}{K}\right)^n \tag{10-24}$$

The value of n obtained from transformation curves is 4. A model of the growth process must account for this value. Examining the nucleation part of the process, this can be written:

$$N = k_N t^a$$

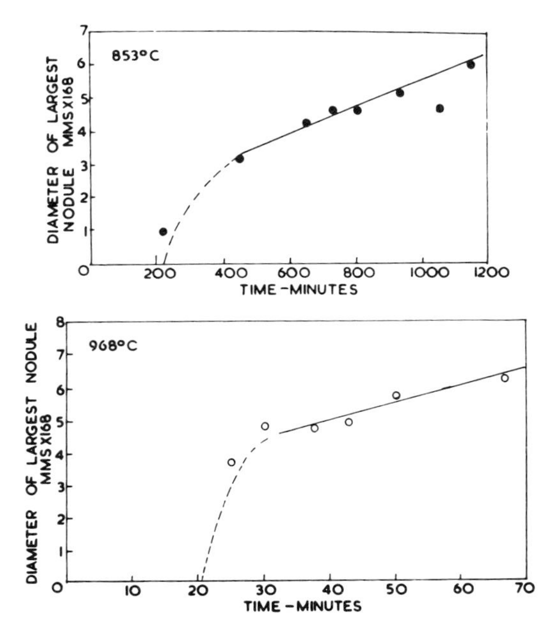

Figure 10-6 Growth rate curve from Burke and Owen[1] analysed by Burke.[9] The data were from high purity Fe–C–Si alloys and were treated as two-stage curves. (Reproduced by permission of Pergamon Press Ltd.)

where N = the number of nodules per unit volume at time t measured from the end of the incubation period
k_N = a temperature-dependent constant
a = a constant equal to 1, up to 50–60 per cent. of the reaction, and $a = 0$ thereafter (i.e. the nucleation rate becomes zero)

Examining the growth equation this can be written $R_1 = k_R t^b$, where R_1 is the rate of growth of a single particle.

Following a techniques of Johnson and Mehl,[10] for the first part of the reaction corresponding to $a = 1$, $n = 3b + 1$. For $a = 0$, $n = 3b$. For diffusion control, the growth rate is parabolic and $b = \frac{1}{2}$.

In this case $n = 2.5$, which is contrary to the experimental finding and therefore this suggested that growth was not diffusion controlled. Equation (10-19) gives a linear rate of growth so that $b = 1$ and the value of n is 4 as obtained experimentally. From this Burke suggested that the model proposed with solution of cementite as a rate controlling process was correct.

10-6 Dissolution of Cementite; Influence of Alloying Elements

The dissolution of cementite was studied by Hillert, Nilsson, and Törndahl[11] for the case of steel but some of the analyses presented might be applied to carbide phase dissolution during malleablizing or in the austenitizing heat treatment of alloy cast iron. Different models were presented in which the α or γ phases play a role in dissolution. Reactions will be discussed in which the γ phase surrounds a carbide particle and the carbide phase dissolves by diffusion in the γ phase. Control of the dissolution may be either by diffusion of carbon or diffusion of alloying elements.

Alloying Elements and Diffusion

An alloying element in the Fe–C–M system may affect the diffusion of carbon if it affects the activity of carbon. To handle this in cases where the rate of transformation depends on a difference in carbon content between a point in an advancing phase interface and some other point in the system, the carbon activity at the point of interest was evaluated. The composition difference was then read on the binary Fe–C side of the ternary diagram at points corresponding with the carbon activities. In Hillert's notation, the quantity U_M is used instead of mole fraction of M where $U_M = X_M/(1 - X_C)$; X_C is the mole fraction of carbon.

Low Alloy Contents and Cementite Dissolution

At low alloy contents, dissolution may be mainly controlled by carbon diffusion. Hillert presented the case of cementite dissolving in a γ envelope which had the α phase at its boundary. The model is illustrated in Figure 10-7(a) and a schematic ternary phase diagram is shown in Figure 10-7(b).

The alloy is initially in a state of α, composition a_0 and cementite, composition C_0. The γ phase grows into the cementite and has a composition at the $\gamma/(\gamma$ + cementite) phase boundary equal to g_3. The γ phase growing into the α will have a composition at the $\gamma/(\gamma$ + cementite) boundary equal to g_4. For the case of a plain carbon material, the phase compositions in equilibrium would be g_1 and g_2 and the rate of carbon diffusion through the γ shell would be dependent on $g_1 - g_2$.

For an alloyed material, the difference must be taken for the carbon activity between g_3 and g_4. Drawing the isoactivity line in the γ phase by dashed lines, the calculation of composition difference was made for the points g_5 and g_6 where the isoactivity lines cut the Fe–C diagram.

Using this model, calculations were made for carbide dissolution and γ shell growth in a steel containing 0.5 per cent manganese. The calculations indicated rates governed by carbon diffusion alone.

Cementite Dissolution Controlled by Alloy Elements

A dissolution reaction involving alloy elements was analysed for three stages. In the first, carbon is transferred from cementite to the γ phase and the latter inherits the alloy content of the cementite. This stage is controlled by carbon diffusion alone.

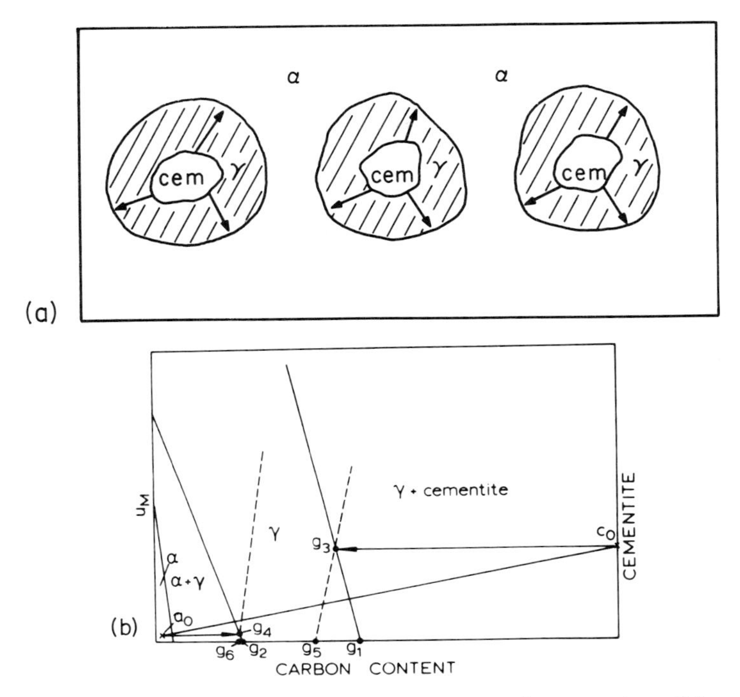

Figure 10-7 (a) Model for dissolution of cementite. (b) Schematic phase diagram showing initial α composition a_0 and cementite composition C_0. Dashed lines are isoactivity lines for carbon at the γ–α and γ–cementite interfaces.[11] (Reproduced by permission of The Metals Society)

The reaction then proceeds by two possible stages controlled by diffusion of the alloy element. One stage involves diffusion of the alloy element into the carbide phase and the other stage involves diffusion of the alloy element into austenite.

The ternary diagram showing the γ and (γ + cementite) regions at the temperature of the reaction is given in Figure 10-8(a).

The γ phase has the initial composition g_0 and the carbide phase has composition C_0. By the same construction as before, the point g_3 is found. This is the composition of the γ phase which inherits the alloy content of the cementite. The carbon content of γ can then increase up to g_4 on the dashed isoactivity line through g_3 without involving long-range transport of the alloy element.

In the second stage of the reaction, a further increase of carbon in the γ phase can take place from g_4 to g_8. This requires that the alloy content at the γ–cementite interface changes from g_3 to g_7 by diffusion. The requirements of local equilibrium determine the concentration versus distance profiles of Figure 10-8(b). The full lines show the initial concentration profile with C_0 and g_0

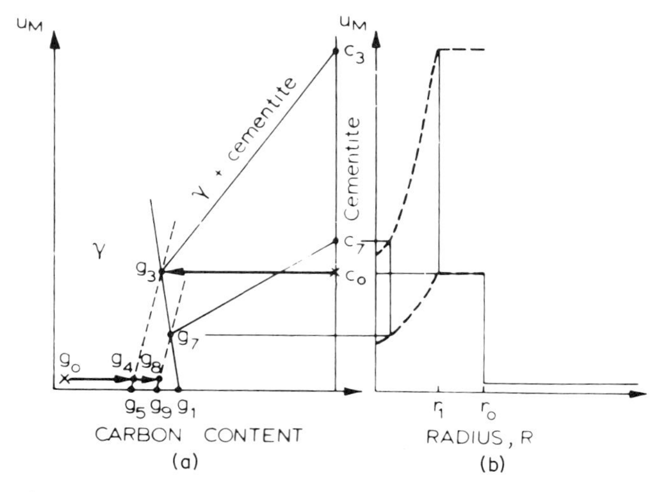

Figure 10-8 (a) Schematic phase diagram showing γ and $\gamma+$ cementite regions. γ has the initial composition g_o and the carbide phase is C_o. (b) Concentration versus distance profiles for dissolution reaction involving alloy elements. Full lines show initial concentration profiles with C_o and g_o in equilibrium. Growth occurs from r_o to r_1 when C_3 is in equilibrium with g_3. At r_1 cementite can continue to dissolve at the rate determined by lowering of alloy content at the interface along the dashed line.[11] (Reproduced by permission of The Metals Society)

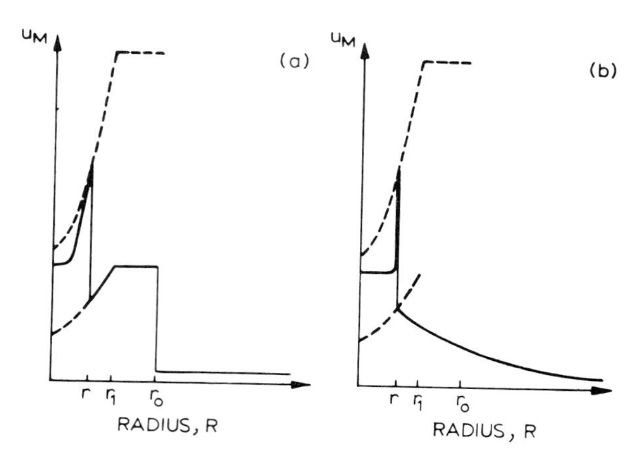

Fig. 10-9 Profile for diffusion of alloy element into (a) the cementite and (b) the γ phase.[11] (Reproduced by permission of The Metals Society)

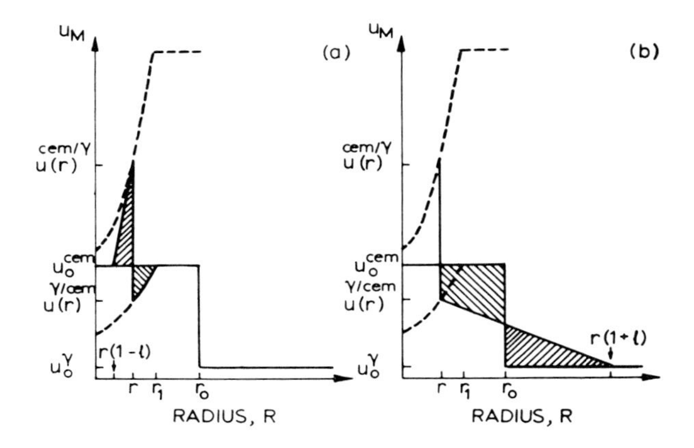

Figure 10-10 Approximate straight lines for diffusion into (a) the cementite and (b) the γ phase.[11] The slopes of the profiles of (a) and (b) can be evaluated by equilibrating the alloy contents represented by the shaded areas. (Reproduced by permission of The Metals Society)

existing at r_0. Growth occurs rapidly from r_0 to r_1 and γ phase at g_3 is in equilibrium with cementite of composition C_3. Note that $g_3 - C_3$ is a tie-line in the two-phase field γ + cementite.

At r_1, the first rapid reaction stops and the particle can continue to dissolve only as fast as the alloy content at the interface can be lowered along the dashed lines of Figure 10-8(b). This can occur by diffusion in two possible ways, which are into the cementite or into the γ matrix.

Figure 10-9(a) and (b) shows the diffusion profiles for the two possibilities, while Figure 10(a) and (b) shows approximate straight lines for the diffusion profiles into the cementite and into the austenite. The slopes of the profiles of Figure 10-10(a) and (b) can be evaluated by equilibrating the alloy contents represented by the shaded areas.

10-7 Shape of the Graphite Phase in Malleable Iron

Theoretically, the shape of the graphite phase in the malleablizing process should be determined by the same factors that determine graphite shape in growth from the liquid, i.e. by instability effects. From the limited observations made studying extracted crystals in different stages of growth, this seems to be the case. The investigation of Bolotov[12] was described in Chapter 2. Here the electron microscope was employed to study extracted graphite crystal surfaces and the published figures show how the (0001) surfaces of initially formed hexagonal platelet crystals of graphite become unstable. They move over the crystal edges, somewhat in the manner observed for the case of solute effects (e.g. boron) in the growth of graphite from the melt (see Figure 7-3).

The main factors affecting the shape of solid state grown graphite have been suggested to be the S to Mn ratio in the alloy and the hydrogen content of the annealing atmosphere. In the research of Hultgren and Östberg[13] the study was made of the formation of flake, compact graphite, and spherulitic forms during annealing. The technique employed for this study was two-dimensional optical microscopy, which imposed a handicap in observing what is basically a three-dimensional phenomenon. The experiments of Hultgren and Östberg were performed by annealing irons mainly of a high S to Mn ratio at 900–1150 °C. It was determined that any of the three types of graphite could be produced by varying the S to Mn ratio, together with the hydrogen content of the annealing atmosphere and the annealing temperature. When annealing was performed at 1100 °C in nitrogen, the graphite assumed the configuration of dispersed flake aggregates or nests. This form changed as the S to Mn ratio increased or as the hydrogen content of the annealing atmosphere increased. This pointed to an apparent influence of the sulphur, either alone or in combination with hydrogen, on the graphite morphology.

Hultgren and Östberg suggested that sulphur and/or hydrogen had an effect on the growth kinetics of graphite, high sulphur tending to favour C axis growth. From the research reported on graphite growth in Chapter 3, the influence of sulphur, and also hydrogen, is on the graphite interface.

In the research of Hultgren and Östberg, a first requirement was to establish a curved interface between graphite and the matrix. In their observations, graphite did this by nucleating and growing as a flake, then curving back on itself. They associated this with the physical character of the nucleating particle. Spherulites were often found to contain core particles consisting of iron sulphide or manganese sulphide. In their view, these particles provided oriented nucleation as the graphite grew along the surface of the particle. This created a radial texture which had to be preserved in growth by the influence of the elements sulphur and hydrogen in the iron. In Bolotov's description,[12] a flake grew and became unstable so that the (0001) planes continued growth round the crystal. If the results of Hultgren and Östberg are reconsidered, the role of the surface-active elements sulphur and hydrogen is in allowing graphite growth to become unstable at small supersaturations. As has been noted, this is different from the conditions in spherulitic growth from the liquid. In the solid state graphitization transformation, sulphur is a requirement for spherulite formation, while in the liquid, a prior requirement in the normally practised treatments is that it be removed. In the liquid, the supersaturation required for instability is obtained by a reactive element addition. This remove sulphur. In the solid state graphitization transformation, it is experimentally noted that the presence of sulphur promotes spherulitic growth. It can do this by unstabilizing surfaces at small existing supersaturations.

References

1. Burke, J., and W. S. Owen: *J. Iron Steel Inst. London*, **176**, 147, 1954.
2. Christian, J.: *The Theory of Transformations in Metals and Alloys*, Pergamon, 1965.
3. Zener, C.: *J. Appl. Phys*., **20**, 950, 1949.
4. Owen, W. S., and J. Wilcock: *J. Iron Steel Inst*., **182**, 38, 1956.
5. Birchenall, C. E., and H. W. Mead: *J. Met*., **8**, 1004, 1956.
6. Brown,. B. F., and M. F. Hawkes: *Trans. Am. Foundrymen's Soc*., **59**, 181, 1951.
7. Sandoz, G.: *Recent Research on Cast Iron* (Ed. H. Merchant), Gordon and Breach, New York, 1968.
8. Rehder, J. E.: *Trans. Am. Foundrymen's Soc*., **56**, 138, 1948.
9. Burke, J.: *Acta Metall*., **7**, 268, 1959.
10. Johnson, W. A., and R. F. Mehl: *Trans. AIMM*, **135**, 416, 1939.
11. Hillert, M., K. Nilsson, and L. E. Törndahl: *J. Iron Steel Inst. London*, **209**, 49, 1971.
12. Bolotov, I. Ye: *Phys. Met. Metallogr*. (Engl. Transl.), Vol. 20(2), p. 86, Pergamon, 1967.
13. Hultgren, A., and G. Östberg: *J. Iron Steel Inst. London*, **176**, 351, 1954.

Bibliography

Gilbert, G. N. J.: *Engineering Data on Malleable Cast Irons—SI Units* BCIRA, 1974.

Malleable Iron Castings, Malleable Founders' Soc., Cleveland, 1960.

Chapter 11
Strength and Fracture of Cast Iron

This chapter reviews research on the failure of cast iron under uniaxial or biaxial stress. The representation of fracture by Orowan is discussed. Flake graphite cast iron was suggested as a material whose behaviour under stress would be that of a brittle type solid with Griffith cracks. Subsequent examination pointed to differences between theory and the observed relationships between compressive and tensile strength.

Research on the strength of cast iron at ordinary temperature was later concerned with analysis of plastic behaviour around cracks developing in the structure. The prediction of properties was attempted. Investigation was directed to understanding the different contributions of phases in the microstructure to strength processes.

Some of this research is related to fracture toughness testing and some to ductile–brittle fracture in ductile iron as a function of temperature. This has enabled assessment of various mechanisms of fracture. Both the fracture of flakes and the ductile extension of spherulites is described. In spheroidal graphite cast iron, interest lies in the dual role of the spherulites both in fracture and as possible crack arresters. The contribution of ferrite and pearlite to strength is described.

11-1 Orowan's Model of a Brittle Solid

Orowan[1] provided a geometrical representation of the biaxial fracture criterion of Griffith.[2] This is shown in Figure 11-1. The condition for fracture of Griffith used the solution for the stress distribution around an elliptical hole in a plate given by Inglis.[3] If P and Q are the principal stresses, the fracture condition can be related to the tensile strength K for uniaxial stressing. For $P > Q$ and tensile stresses taken as positive, the following fracture condition obtains:

$$\text{Fracture occurs when } P = K \text{ if } 3P + Q > 0 \qquad (11\text{-}1)$$

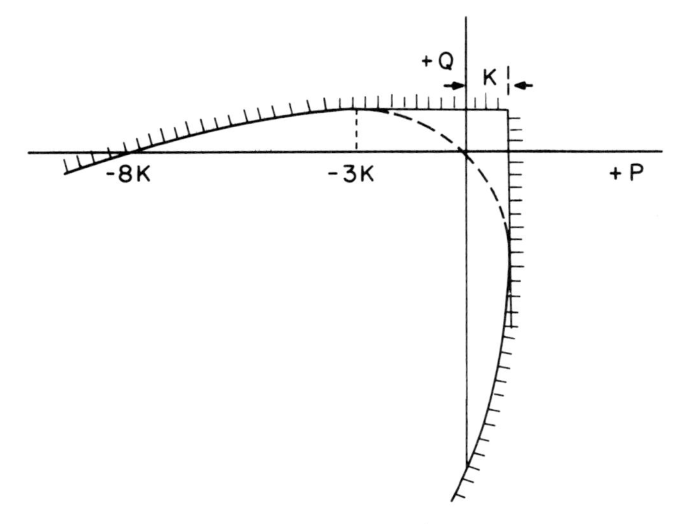

Figure 11-1 Orowan's representation[1] of the biaxial fracture criterion of Griffith.[2] P and Q are the principal stresses. K is the tensile strength for uniaxial loading. Fracture will occur at a state of stress crossing the curve towards the shaded areas. For uniaxial compression, the strength is $8K$ (Reproduced by permission of the Institute of Physics and the Physical Society)

If $3P + Q < 0$, fracture occurs when

$$(P - Q)^2 + 8K(P + Q) = 0 \tag{11-2}$$

Orowan's graphical representation shows P and Q plotted as rectangular coordinates. Equation (11-2) is a parabola and Equation (11-1) is the vertical tangent to the parabola.

For $3Q + P > 0$ and $Q > P$ the fracture conditions are completed by $Q = K$, i.e. by the horizontal tangent to the same parabola, and the complete fracture curve is that drawn with full lines. If the state of stress, represented by a point on the diagram, crosses the curve towards the shaded area, fracture will occur. For uniaxial compression ($P = Q$, $Q < 0$), the strength predicted by Orowan was $8K$, i.e. eight times the strength in tension.

Orowan thought that this was approximately fulfilled in cast iron. This was on the basis that cast iron is a brittle solid. However, the testing work described in the following paragraphs shows that while Orowan's brittle fracture envelope is obtained for cast iron in form, the actual value of the compressive strength is closer to $3K$.

11-2 Stress–Deformation Models of the Mechanical Behaviour of Cast Iron

Cast iron with flake graphite behaves in an apparently brittle manner because of the mode of deformation resulting from the flakes. Different models have been proposed of this behaviour conforming to the experimental observations. On the experimental side, a vast amount of data has been collected from mechanical tests and analysed statistically. This has enabled mechanical behaviour of cast iron to be reasonably predicted from the analysis or from the microstructure.[4,5,6] Collected data are available according to cast iron type, i.e. flake or spherulitic, matrix microstructure, quantity of graphite.

The strength of cast iron under multiaxial stress has been reasonably well evaluated and some of the early papers presented on this subject are given here.

Fisher's Model

Fisher[7] proposed a model for the fracture stress of cast iron based on the stress for yielding. Zener[8] had calculated the stresses in slip bands in plastically deformed material at grain boundaries or at the boundaries of second phases. The stresses concentrated at the extremity of a slip band could exceed the fracture stress and Fisher suggested that the stress for first yielding at the tip of a graphite flake would be the fracture stress for cast iron.

To evaluate yielding, the distortion energy criterion was used with the values of the stresses concentrated near the tips of the graphite flakes. The distortion energy theory is stated as follows:

$$(S_1 - S_2)^2 + (S_2 - S_3)^2 + (S_3 - S_1)^2 = 2S_0 \qquad (11\text{-}3)$$

where S_0 is the yield stress in simple tension and S_1, S_2, and S_3 are the three principal stresses. The square root of one half of the left-hand side of Equation (11-3) is known as the effective stress.

Local plastic flow will occur in the matrix of cast iron when the conditions of Equation (11-3) are satisfied by the stresses in the neighbourhood of suitably oriented flakes. Fisher approximated the shape of flakes in cast iron to ellipsoids of revolution (Figure 11-2a and b) and applied the calculations of Neuber[9] to evaluate the stress concentration factors. The axial direction for application of the stress is the axis of rotation of the ellipsoid. The maximum stress concentration occurs at the tip periphery and is calculated to be approximately 3 for the geometry analysed, the radial and tangential stresses for this case being approximately zero.

Figure 11-2(b) shows the stresses at the ellipsoid periphery on applying a stress S at right angles to the axial direction. Fisher's model was based on tensile stresses only being concentrated.

The following cases were examined

Case 1: Cast Iron Subjected to a Biaxial Tension The most severe stress state is that for which the largest tensile stress acts in the axial direction relative to the graphite flake. From Figure 11-2(a) and (b), there is a point on the

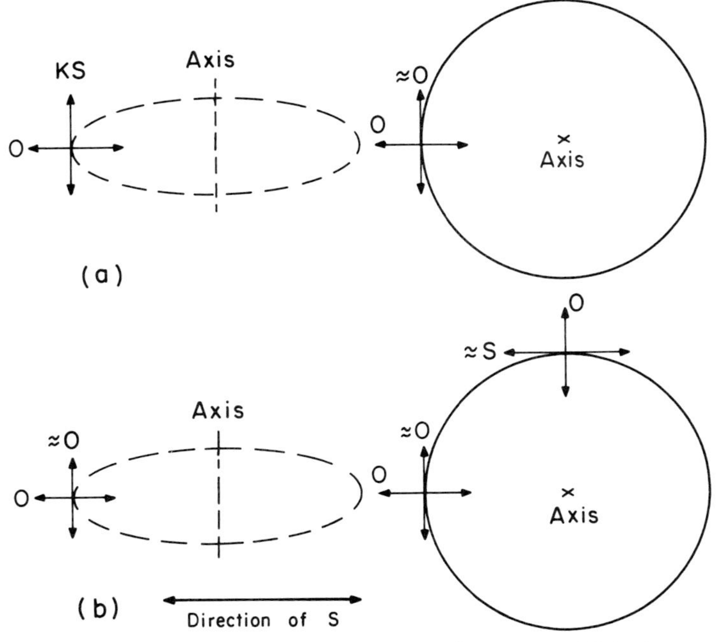

Figure 11-2 (a) Approximation by Fisher[7] of shape of flake in cast iron by ellipsoid of revolution. Stress is applied in direction of axis of rotation of ellipsoid. For this case, maximum stress concentration K at tip periphery is approximately equal to 3. Radial and tangential stresses are zero. (b) Stresses at ellipsoid periphery on applying a stress S at right angles to the axial direction. (Copyright, American Society for Testing and Materials, 1916 Race Street, Philadelphia, PA. 19103. Reprinted, with permission)

periphery of the plate where S_2 and S_3 are zero. Equation (11-3) then reduces to:

$$KS_1 = S_0 \tag{11-4}$$

where S_1 is the maximum tensile stress for the initiation of plastic flow in cast iron subjected to biaxial tension and the distortion energy criterion for biaxial tension here coincides with the maximum stress criterion.

Case 2: Cast Iron Subjected to a Tensile Stress S_1 and a Compressive Stress S_3; The Intermediate Stress $S_2 = 0$ The most severe stress state is that for which the tensile stress acts in the axial direction relative to the graphite plate, as in case 1 above. From Figure 11-2(b), there is a point at the periphery of the plate where the tangential stress is S_3 and the radial stress is zero. Equation (11-3) then becomes

$$(KS_1)^2 - (KS_1)(S_3) + (S_3)^2 = S_0^2 \tag{11-5}$$

Case 3: Biaxial Compression In the absence of stress concentrations, Equation (11-3) is unchanged:

$$(S_1 - S_2)^2 + (S_2 - S_3)^2 + (S_3 - S_1)^2 = 2S_0^2$$

Using Equations (11-4), (11-5) and (11-3), the yield curve of Figure 11-3 was plotted with a yield stress of S_0 = 620 N mm^{-2} (90,000 psi). This was compared with experimental values of the fracture of cast iron under biaxial stress determined by Grassi and Cornet.[10]

Fisher suggested that the agreement between his theory and experiment was striking. This suggested that the distortion energy theory could be used to predict the failure of cast iron on the basis of a flow criterion. The fracture of cast iron would be initiated when the matrix underwent plastic flow in the neighbourhood of the graphite flakes. Tensile stress produced stress concentrations, while compressive stresses remained unchanged.

Coffin's Model

Coffin[11] made an extensive theoretical and experimental evaluation of frac-

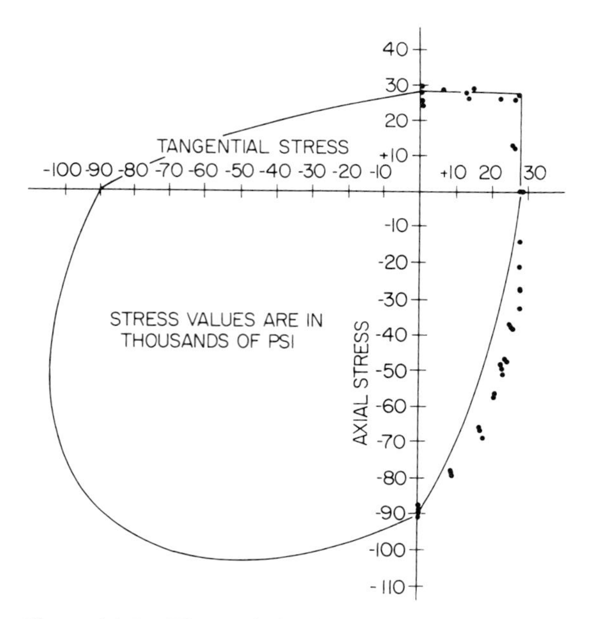

Figure 11-3 Theoretical yield curve plotted by Fisher[7] analysing failure of gray cast iron tubes subjected to internal pressure and end load (solid curve). Experimental points are fracture stress values of Grassi and Cornet.[10] (Copyright, American Society for Testing and Materials, 1916 Race Street, Philadelphia, PA. 19103. Reprinted, with permission)

ture of a cast iron under combined stresses, using thin-walled tubes. He wanted to show what laws were followed when a brittle substance flows and what laws could be obtained for fracture. A cast iron was employed having the following analysis:

3.08 %TC, 2.04 %Si, 0.5 %Mn, 0.112 %S, 0.33 %P

Coffin criticized the proposal of Fisher that plastic flow, no matter how small, initiated fracture. The test results of Coffin showed that large-scale plastic flow could occur in cast iron prior to fracture for all states of biaxial stress. This suggested that the simple flow criterion of Fisher was incorrect.

Comparing curves of effective stress versus effective plastic strain, particularly wide divergences were obtained between simple tension and simple compression with no uniqueness in relationships. The plastic effective strain at fracture increased considerably as the degree of compression increased. Coffin analysed his data, using the concept of a graphite flake acting as a crack, as did Fisher, but assuming that residual tensile stresses acted at the edge of the plate. These acted in a direction normal to the plate and were the result of differences in the coefficients of thermal expansion between the iron and the graphite. The graphite is compressed in cooling and residual tensile stresses remain at the flake edges. According to Coffin's estimate, these would have a value of approximately 200 N mm^{-2}.

Correlation between Brittle Fracture Theory and Cast Iron Fracture by Clough and Shank

Clough and Shank[12] extended the experimental programmes for examining the mechanical behaviour of cast iron and the development of theory, as commenced by Fisher,[7] Coffin,[11] and others.[13,14]

Clough and Shank examined the fracture surfaces and showed that a brittle cleavage fracture may exist, but the major portion of the surface is of a different character. One way in which ductile fracture originates was shown to be by the separation of graphite flakes. Highly deformed regions could also exist in the matrix between flakes appropriately oriented in the structure. Cracks became interconnected and the overall structure became spongy and porous.

The problem of analysing fracture in a brittle mode for a plate-like phase existing in a ductile matrix is further discussed in the following sections.

The fracture envelope obtained from the biaxial test programme of Clough and Shank is given in Figure 11-4(a) and compared with test results of Coffin (Figure 11-4b) showing data also for the compression–compression region.

It is noted that the experimental envelope forms a straight line between the point of fracture under equal tension–tension stresses and the point of fracture under numerically equal tension–compression stresses. The envelope in the tension–compression region forms a curve which is tangent, or nearly so, to the straight line section from $\sigma_t = \sigma_z$ to $\sigma_t = -\sigma_z$. The ratio of fracture stress in pure compression to the fracture stress in pure tension is approximately

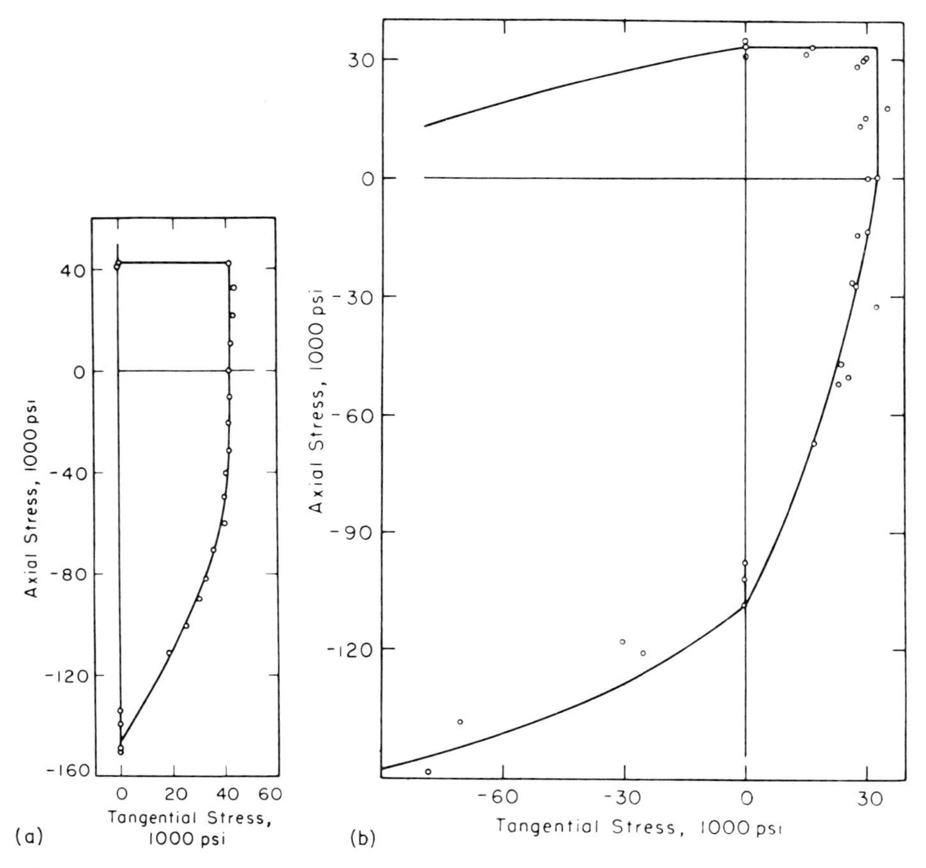

Figure 11-4 (a) Fracture envelope obtained by Clough and Shank[12] for perlitic gray cast iron. (b) Fracture data and theory of Coffin[11] for gray cast iron. (Reproduced by permission of American Society for Metals)

3.5. Knowing the experimental values of the strength in pure tension and in compression, it should be possible to calculate the fracture envelope in the regions of different stress combinations. In the compression–compression region, a useful but crude assumption would be to consider that compressive stresses are transmitted unaltered by the graphite flake. The distortion energy theory of Fisher could then be applied in this region

Clough and Shank: Comparison of Experiment with Brittle Failure Theory

Clough and Shank [12] remarked on the striking comparison between the fracture envelope for gray iron, experimentally determined, and the envelope for a purely brittle Griffith material as plotted by Orowan. In a brittle material, the ratio of compressive strength to tensile fracture strength is theoretically

8.0 and the envelope forms a straight line for $\sigma_z = \sigma_t$ to $\sigma_z = -3\sigma_t$. The remainder of the envelope in the tension–compression region of Orowan's diagram is parabolic and exactly tangent to $\sigma_z = -3\sigma_t$.

The nature of failure by Griffith cracks in a brittle material and failure by the action of flakes in cast iron was discussed. For Griffith cracks, fracture under compressive loading must still occur under the influence of tensile stresses set up at the cracks. In cast iron, in the tensile compressive region. The compressive stresses may set up tensile stresses at properly oriented flakes and this would cause failure at smaller values of the applied tensile stress. In pure uniaxial compression, a high stress is required to set up the necessary tensile stresses at the crack.

Density decrease as measured in cast iron under load are significant and are due to the opening up of voids at the graphite–metal interface. Significant density decreases are also marked in purely compressive samples.

Critique of Failure Theories

Clough and Shank criticized some of the proposed failure theories of cast iron. They thought that such theory could not be based purely on a yielding criterion since this occurred soon after initial loading and long before fracture.

With respect to applying Griffith theory[2] the similarity between the fracture envelope for a fully brittle Griffith material and the fracture envelope obtained for cast iron was probably deceiving. The two cases were different. Griffith had analysed a fully brittle elastic material. Cast iron required a mathematical solution for a plastic material with a random distribution of saucer-like voids. A criterion for ductile rupture of a plastic matrix was necessary.

Despite this, Clough and Shank pointed out that if the fracture stresses for gray cast iron were known for simple tension and simple compression, a useful fracture envelope for design purposes could be constructed for the tension–tension and tension–compression regions.

For the compression–compression region the compression stresses could be considered as transmitted unaltered by the graphite flakes. This was a very crude assumption. Nevertheless, the distortion energy theory could be applied as Fisher had done and this seemed to fit the data presented by Coffin for the compression–compression region.

More recently developed theories of failure examining the surfaces of cast iron in fracture and extension are given in Section 11-7.

11-3 Stress–Strain Behaviour of Cast Iron

The stress–strain behaviour of gray cast iron is that for a material containing a uniform dispersion of discontinuities which act as cracks. Plastic deformation occurs at the tips of the cracks at small values of the stress and a non-linear

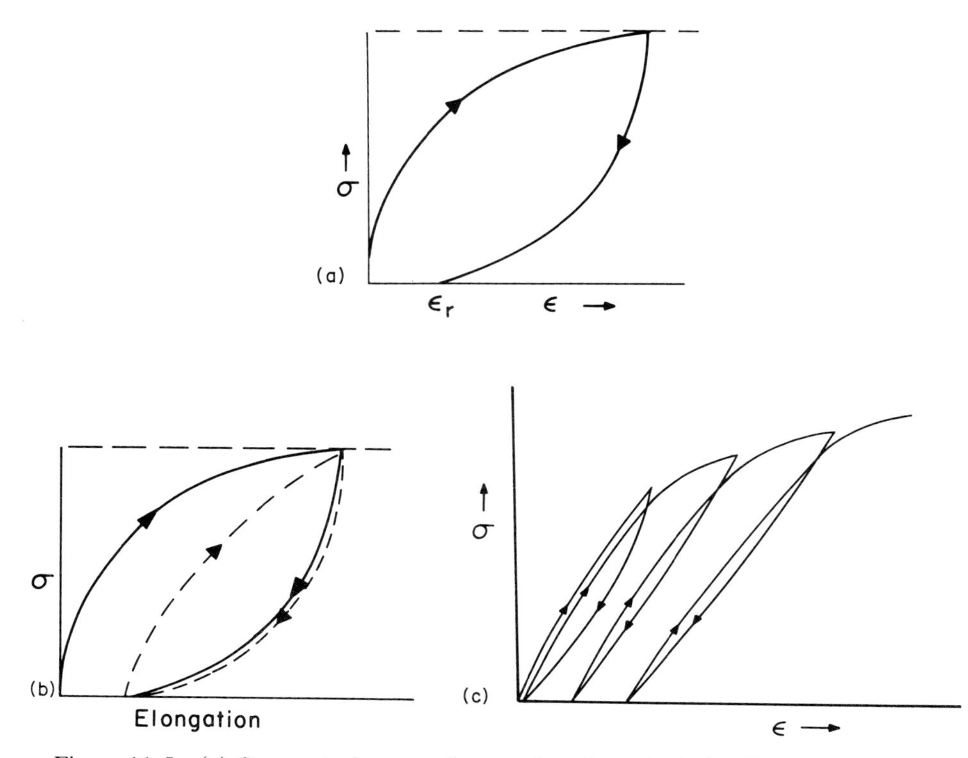

Figure 11-5 (a) Stress–strain curve for cast iron in a single loading and unloading cycle. (b) Second loading–unloading cycle with same value of maximum stress. (c) Curve for reversals of stress with gradually increasing maxima. (Reproduced by permission of Iron Castings Society)

stress–strain relationship is obtained. Permanent deformation is noted after unloading. Part of the strain recorded is related to the extension of voids.[12]

Figure 11-5(a) shows a typical stress–strain curve for cast iron in a single loading and unloading cycle. There is a departure from linearity in loading at the source, and the unloading curve shows permanent deformation ε_r. Figure 11-5(b) shows a second loading–unloading cycle with the same value of the maximum stress. This second unloading curve returns to the origin at the pre-loading value of the strain, such behaviour being termed stabilization or accommodation.

For reversals of stress with gradually increasing maxima, the curve shown in Figure 11-5(c) is obtained. The slope of the curve at the beginning of each cycle decreases and the hysteresis shown by each cycle is progressively reduced. The energy corresponding to the hysteresis of the stress–strain cycle is a measure of the damping capacity of the material in cycles of stress reversal.

11-4 Stress at the Periphery of a Graphite Flake

The departure from elastic behaviour, and the hysteresis of the σ–ε curve, can be related to the plastic deformation of the metal matrix and of the graphite flakes. From the analysis of Inglis,[3] if σ_L is the applied stress and a crack of elliptical geometry exists having a half length c, the maximum stress at the periphery of the crack is given by

$$\sigma \simeq \sigma_L \times 2(c/\rho)^{1/2} \tag{11-6}$$

where ρ is the radius of curvature of the crack and $\rho \ll c$.

The difficulty in making reliable estimates of the stress concentration for cast iron is in the evaluation of ρ. A single graphite flake is flat rather than elliptical and the flake edge is blunt. In the eutectic cell, the morphology is complex (see Section 1-4). A maximum estimate of the diameter of a cell would be 1 mm and the flake thickness might be 20 μm. This would give a value for the stress concentration factor approximately 10. Then if the yield strength of the matrix is 140 N mm^{-2} the first local deformation would occur at a loading stress of only 14 N mm^{-2}.

The plastically deformed zone is estimated to be approximately 10 times ρ. Hence, the matrix at the edge of the graphite flake is deformed over a distance of 100 μm. Since the distance between flakes is approximately 10 to 15 times the flake thickness, most of the matrix may undergo plastic deformation.

11-5 Modulus for Cast Iron

The non-linear character of the stress–strain curve for cast iron offers several choices for assigning a modulus relating stress and strain.[4,15] Some constructions are shown in Figure 11.6.

In Figure 11-6, the constructions are:

(a) Tangent to the origin of the curve at O.

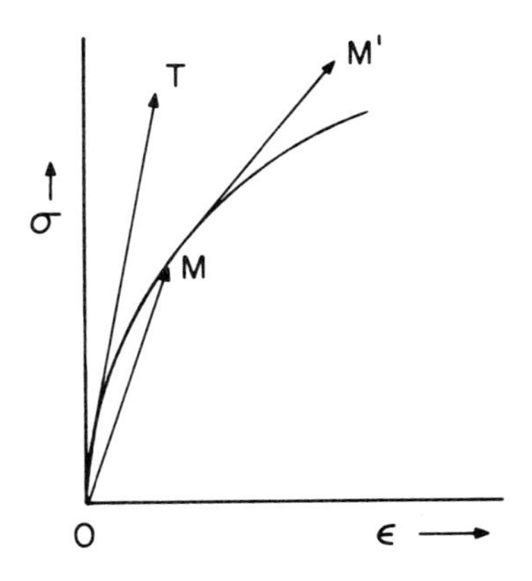

Figure 11-6 Constructions for assigning a stress–strain modulus to cast iron. OT is a tangent modulus at the origin. OM is a secant modulus at M. MM is a tangent modulus at M

(b) Secant modulus OM at M. The location of M is generally chosen to be 25 per cent of the maximum tensile stress but any other value of M can be selected.
(c) Tangent modulus MM′ at M.

Static and Dynamic Modulus

Dynamic procedures can be employed to measure Young's modulus. The dynamic elastic modulus (DEM) relates stress and strain in a standard linear solid when the stress and the strain are periodic. It may be defined as the ratio of the stress to that part of the strain which is in phase with the stress. It is lower than the elastic modulus measured under static conditions by a factor due to anelasticity.

The percentage deviation between the measured dynamic modulus and the static modulus varies as a function of the structure of the iron. This is shown in Figure 11-7.[15] The maximum deviation is approximately 10 per cent for a coarse structured iron.

As discussed in Chapter 12, advantages are to be obtained in determining the dynamic modulus since there is correlation between this figure and the tensile strength.

Influence of Graphite Form and Quantity on Modulus of Cast Iron

The form and quantity of the graphite in the structure will influence the modulus as follows:

(a) The smaller the graphite length, the higher the stress required to depart from a linear stress–strain relationship and the higher the effective modulus.
(b) The greater volume of graphite, the greater the volume of matrix deformed and the more rapid the departure from linearity.

Figure 11-8 shows the forms of the curves obtained for flake iron of different compositions and different cooling rates.[15] As the cooling rate decreases to the right, the greater the average length of the graphite and the lower the modulus. As the amount of graphite increases, the curves are displaced to lower values of the modulus.

Calculation of the Elastic Moduli of Gray and Nodular Cast Iron

The elastic moduli of cast iron, either gray or nodular, are dependent both on volume of the graphite and on morphology. Plenard[15] employed theoretical relationships for two-phase materials to enable the calculation of Young's modulus for spheroidal graphite cast iron. The calculation for flake graphite iron is more problematic since the flake length must be taken into account.

Speich, Schwoeble, and Kapadia[16] have examined the evaluation of the

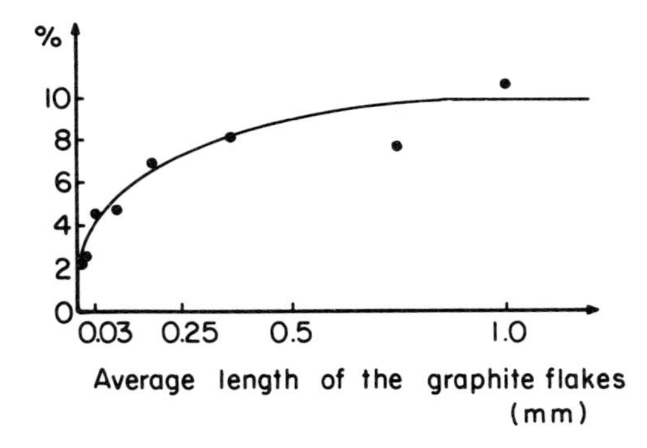

Figure 11-7 Percentage deviation between measured dynamic modulus and the static modulus as a function of average length of the graphite flakes.[15] The measured static modulus is normally higher than the dynamic. The maximum deviation shown in approximately 10 per cent. (Reproduced by permission of Gordon and Breach Science Publishers Inc.)

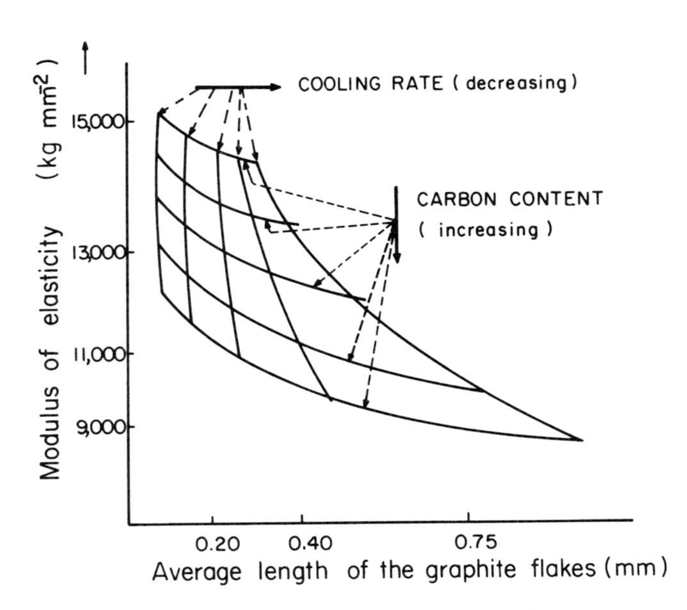

Figure 11-8 Influence of graphite length (cooling rate) and quantity (carbon content) on modulus of cast iron[15] (Reproduced by permission of Gordon and Breach Science Publishers Inc.)

elastic moduli of cast iron from theoretical relationships for two-phase materials. They have shown that, making some elementary assumptions about elastic constants and the dispersion of the phases, calculations may be made using theory for two-phase materials which are in adequate agreement with experiment. The temperature dependence of the elastic moduli can also be calculated and this agrees well with measurements. Figure 11-9(a) shows the theoretical and experimental comparison of Young's modulus for nodular and gray cast iron at 25 °C and Figure 11-9(b) shows the temperature dependence of elastic moduli.

The values of the elastic moduli were determined by dynamic methods, using either rod or plate specimens. For the former, a 100 kHz pulse-echo technique was used, while for plate specimens, a 1 MHz pulse-echo overlap technique was employed. For measurements involving the temperature dependence of the moduli, a tungsten wire was spot-welded to the rod specimen, both to transmit and also receive elastic pulses from the specimens. For these measurements, the following relationships were employed:

$$E = \rho V_l^2$$

$$G = \rho V_s^2$$

where V_l = velocity of longitudinal waves
V_s = velocity of shear waves
ρ = density

For gray cast iron where plate specimens were employed, G was evaluated from $G = \rho V_s^2$ and E was evaluated from the following:

$$E = \frac{G(3\lambda + 2G)}{G + \lambda}$$

where $\lambda + 2G = \rho V^2$, λ being the Lamè constant $= \nu E/(1 + \nu)(1 - 2\nu)$. For calculating the moduli for spheroidal graphite irons from the theory of two-phase materials, equations of Hashin[17] were applied. These relate to spherical particles. The equations for the bulk and shear moduli were approximated by the following:

$$K = K_{Fe}\left[1 - \frac{3(1 - \nu_{Fe})(1 - K_C/K_{Fe})}{2(1 - 2\nu_{Fe}) + (1 + \nu_{Fe})K_C/K_{Fe}} f_C\right] \cdots \qquad (11\text{-}7)$$

$$G = G_{Fe}\left[1 - \frac{15(1 - \nu_{Fe})(1 - G_C/G_{Fe})}{(7 - 5\nu_{Fe}) + 2(4 - 5\nu_{Fe})G_C/G_{Fe}} f_C\right] \cdots \qquad (11\text{-}8)$$

where K = bulk modulus
G = shear modulus
Subscript Fe refers to the matrix
Subscript C refers to graphite
ν = Poisson's ratio
f_C = volume fraction of graphite

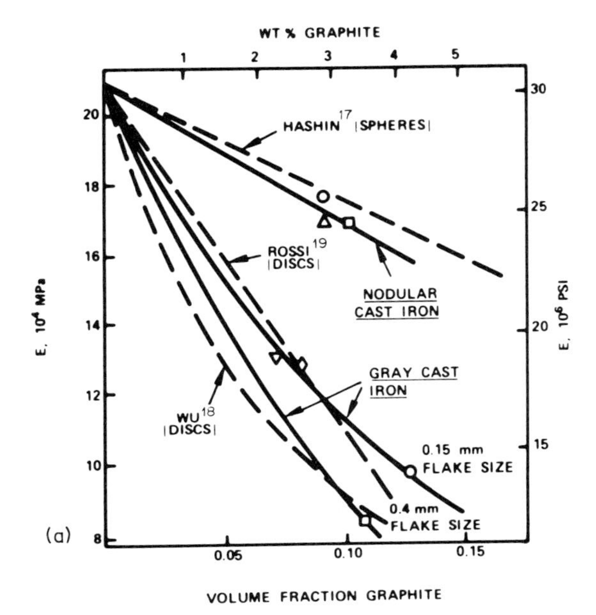

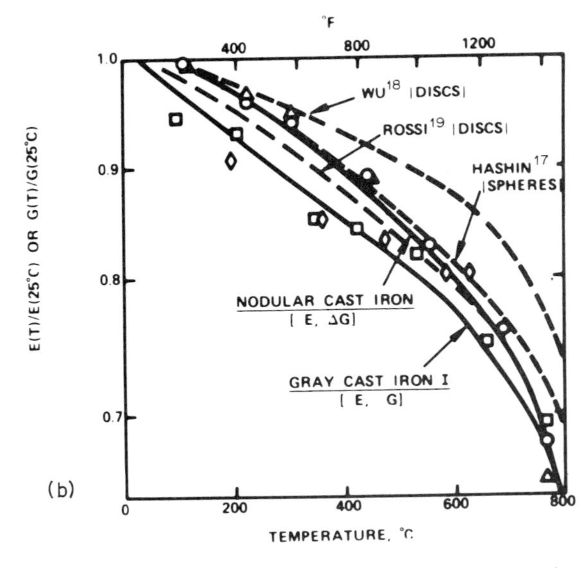

Figure 11-9 (a) Theoretical and experimental comparison of Young's modulus for nodular and gray cast iron at 25 °C. (b) Temperature dependence of elastic moduli: theoretical and experimental comparison.[16] (From G. R. Speich, A. J. Schwoeble and B. M. Kapadia, *J. Appl. Mech.*, **47**, p. 821, 1980, by permission of American Society of Mechanical Engineers)

To calculate E for spheroidal graphite iron, Equations (11-7) and (11-8) were used with the following relationship between the isotropic elastic constants:

$$E = \frac{9KG}{3K + G}$$

For flake graphite cast iron, solutions of the problem for needle or disc-shaped particles due to Wu[18] and Rossi[19] were taken. The following simple form of the equations of Wu[18] was used:

$$E = E_{Fe}\left[\frac{2 - (1 - E_C/E_{Fe})f_C}{2 - (1 - E_{Fe}/E_C)f_C}\right]$$

This was based on the assumption that the value of ν for the grahite phase, the matrix, and the composite material are equal to 0.2.

An equation used by Rossi[19] was also evaluated. This relationship is given as:

$$E = E_{Fe}(1 - Mf_C)$$

where M is a stress concentration factor which is determined by the axial ratio of the disc-shaped particles and by the ratio $E_{graphite}/E_{matrix}$. These results are shown in Figure 11-9(a).

The Influence of Particle Size on Elastic Moduli

The theoretical solutions of the equations for the elastic moduli used for flake graphite cast iron are based on distribution of plates. These do not take into account particle size. The graph of Plenard[15] (Figure 11-8) typifies the observations of the influences of increasing graphite flake size in decreasing the value of E.

In the discussion by Speich, Schwoeble, and Kapadia,[16] the influence of particle size was considered on the stress fields round individual particles. They proposed that as the flake size increased, the stress fields round individual particles began to overlap, resulting in lower elastic moduli. For small particle size, this is not experienced.

It is of interest to note that the theoretical analyses are based on a dispersion of isolated particles in a matrix. While this is true for ductile iron, it is not true for gray iron, where the flakes are interconnected, resulting from branching mechanisms. This must lead to some disparity between the theoretical models employed and the experimental values.

Temperature Dependence of the Elastic Moduli of Cast Iron

The temperature dependence of the elastic moduli of cast iron (Figure 11-9b) is derived from the temperature dependence of the elastic moduli of iron and of graphite. The variation of f_C with temperature is neglected, as are also the values of ν_{Fe}, K_C, and G_C.

11-6 Tensile Failure of Cast Iron

Different modes of failure of cast iron are encountered. In flake gray iron, the fracture takes place without significant extension by prior fracture of the graphite flake, followed by local plastic deformation of the matrix. Whether in tension or compression, considerable tensile stresses are concentrated at the graphite flakes. The propagation of the crack is dependent on microstructural details such as the spacing between flakes, the phases present (ferrite, pearlite, or other transformation products of austenite), solute elements in solid solution, segregated elements, and precipitates, as well as on temperature.

In the case of spheroidal graphite cast iron, the failure behaviour is not very different from that of steel of similar matrix structure but some modification occurs due to the distribution of spherical graphite particles.

Some of the fracture behaviour, and in particular the dependence on microstructure (i.e. graphite form, volume, ferrite grain size, pearlite content, etc.), can be related to tests based on fracture mechanics.

Low temperature behaviour of spheroidal graphite cast iron in the ductile–brittle transition range has been described, as well as the high temperature behaviour of cast iron alloys in conditions of creep.

Fracture of Gray Cast Iron under Static Conditions

The stress–strain curves for gray cast iron under normal testing conditions are demonstrative of failure with minimum elongation. This is because of the distribution of discontinuities in the form of graphite flakes and the confined, and hence intense, plastic deformation leading to pseudo-brittle failure.

The ordinary measurement made is of the tensile stress on the specimen measured at failure; this is referred to as an ultimate stress. The figure is presently used in static conditions for design, using an appropriate safety factor. For multiaxial loading, the curves or equations referred to in Section 11-2 are employed.

The possibility of analysing properties using fracture mechanics and correlating the different aspects of microstructure are in the process of development.

Different ways exist of showing the dependence of the tensile properties of cast iron on the analysis. The classification may be related to some aspect of composition such as the carbon equivalent but the freezing rate must be introduced to define the graphite distribution.

Figure 11-10 shows one of the methods of presenting the data in the diagram of Collaud,[20] taken from the text of Angus.[4] Taking composition along the abscissa, the strength is related to the graphite volume and distribution as determined by the section size.

Difference between Cast Iron and Griffith Type Brittle Materials

It is seen that the mechanical properties of cast iron differ from those of a brittle material related by crack size in the equation of Griffith.[2] For the

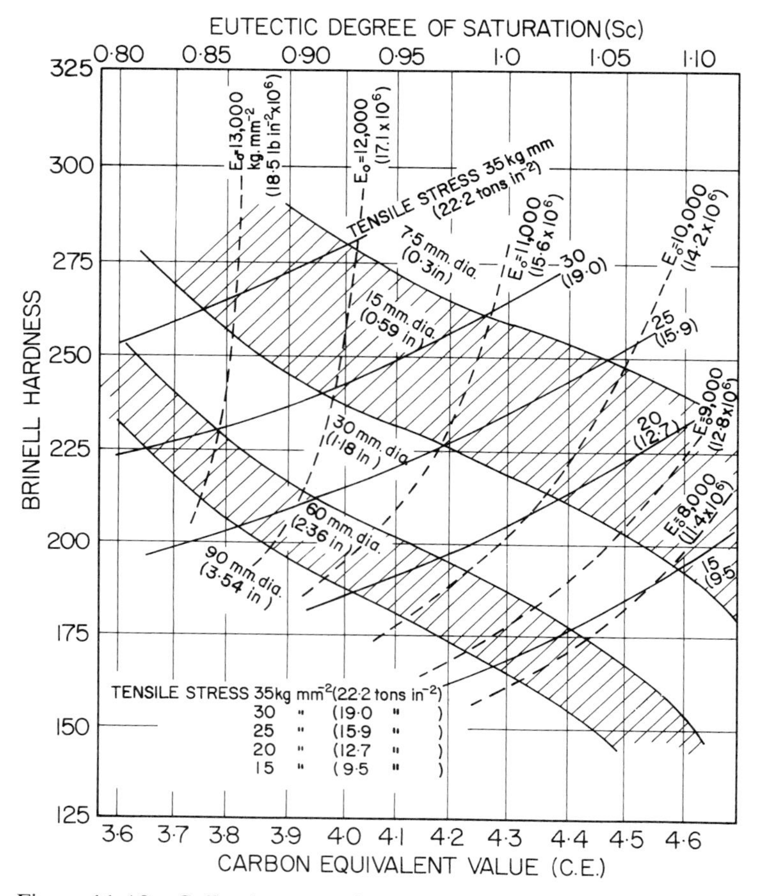

Figure 11-10 Collaud type of diagram showing dependence of mechanical properties on the carbon equivalent and diameter of test specimen (From Angus.[4] Reproduced by permission of Butterworths, London)

latter, the strength is given by $\sigma = kc^{-1/2}$, where k is a material property involving surface energy and Young's modulus, and c is half the crack length. For cast iron, the volume of the graphite phase is introduced as a factor $(1 - V_g)$, where V_g is the volume of the graphite.[21] Therefore:

$$\sigma/\sigma_U = k(1 - V_g)c^{-1/2}$$

For an ordinary material without graphite ($V_g = 0$), σ becomes the strength of the matrix σ_U where c is now half the length of a matrix fault. In all other cases, the specimen strength is related to both graphite volume (V_g) and flake size (c). Large graphite volumes decrease the space between flakes. The plastic work for fracture is reduced and the strength is reduced.

Siefer and Orths[22] published an extensive review of the problems involved in assessing strength in Fe–C alloys having a graphite phase. In their analysis they related the physical properties of malleable iron and spheroidal graphite cast iron to a combination of tensile strength and elongation.

First Stage of Failure: Fracture of a Graphite Flake

The defect structure of a graphite flake was described in Section 1-2. This structure is the same whether the graphite grows as a primary phase or as a phase of the graphite–γ eutectic. The principal defects are rotation boundaries, edge and screw dislocations, twin boundaries, and low angle tilt boundaries.

The fracture of graphite may occur along the rotation boundaries by separation into sections or by slip on (0001) when pile-up of dislocations at a twin or tilt boundary initiates cleavage. The structure of a twin boundary was shown in Figure 2-16 and twin boundaries in a crystal of graphite were shown in Figure 2-15(a) and (b). The habit plane is $(11\bar{2}1)$ and these make a hexagonal trace with the (0001) graphite crystal surfaces. Cracks are noted to be located at twin boundaries.[23] These cracks are the result of stresses due to dislocation pile-ups in the graphite crystal at the twin boundary. The stresses required for cracking in graphite have been noted to be in the range of approximately 40 N mm^{-2}.[23] The hexagonal patterns formed by the cracks along internal graphite boundaries, noted by Glover and Pollard[24,25,26] in scanning electron micrograph studies of cast iron failure, are due to this fracture mechanism. The parting by cleavage of the graphite flake is the first stage of cast iron tensile failure.

Behaviour of Graphite Spherulites during Deformation

Different observations have been made of the deformation of graphite spheruiites under tensile load; see, for example, the illustrations in the paper of Rickards.[27] In low temperature thermal treatment of spheroidal graphite cast iron for the decomposition of the carbide phase in pearlite, a graphite layer crystallizes on the spherulite. This rim of deposited graphite may crack when deformed.

The main body of the spherulite, however, exhibits appreciable ductility and shape change, which is in contradistinction to the limited ductility of a flake. This is related to the difference in defect structure between flake and spherulite. In polarized light, the spherulite shows the pattern of a Maltese Cross while the flake shows the presence of twins. The spherulite has low angle tilt boundaries, and dislocation movement, as well as boundary movement, appears to be reasonably unrestricted. Freise[23] noted that tilt boundaries in flakes were mobile while the twin boundaries were immobile. The tilt structure may allow large shape changes of the spherulite.

Considerable plastic flow of ferrite takes place in the neighbourhood of the spherulite, with eventual ferrite–graphite separation along the interface. In gray cast iron the cavities which open up between graphite and ferrite join and the intervening ferrite ruptures as part of a process of ductile fracture of the iron.

Fractographic Observations of Cast Iron: Fracture of the Ferrite Matrix in Flake Graphite Irons

Tensile fracture of the ferrite matrix in flake graphite irons takes place either transgranularly by cleavage or in an intergranular manner.[25] If it is the first mode, characteristic river patterns are observed. In an intergranular mode, some ductile tearing may be observed.

From observations of Chang, Heine, and Worzala[28] on the impact toughness of cast irons, cracks open up in regions of intense plastic flow confined to the regions of flake extremities or between flakes. The different stages of fracture were observed as hole formation (56 per cent of the fracture load), followed by yielding and then development of real cracks.

Fracture Toughness and Failure Observations

Fracture toughness has become a growing field of research in cast iron and has been applied to most of the types of iron in use. The data are useful in allowing comparison of such factors as matrix structure, graphite or carbide volume, and graphite shape, on the resistance to crack propagation.

A definition of the terms involved and some methods of measuring important quantities are given in Chapter 12.

Comprehensive observations of Glover and Pollard[25] were made on a series of cast irons with differing analyses, heat-treated to give varying amounts of ferrite. Tensile tests were performed, as well as four-point slow-bend tests, to determine K_{1c} values. Fractographs were taken from the slow-bend and impact test specimens next to the tip of the notch.

Figure 11-11 shows the effect of increasing ferrite content on the fracture toughness of flake graphite cast irons. Ferritization of these cast irons was performed by two different treatments:

(a) Above A_1, giving ferrite growth in the presence of γ.
(b) Below A_1, giving pearlite decomposition in the presence of α.

The curve shows that for a pearlitic cast iron there is a small increase in fracture toughness as heat treatment is performed to give an initial formation of ferrite round the flakes. Thereafter the fracture toughness slowly decreases in all materials tested. Observations obtained from the speciments removed from the test rig before complete separation revealed that the graphite flakes were cleaved ahead of the main crack front. In the low ferrite cast irons, ferrite was always associated with the graphite flakes. Propagation of the crack after failure of the graphite was into the ferrite. This fracture of the ferrite was transgranular and showed brittle fracture river patterns.

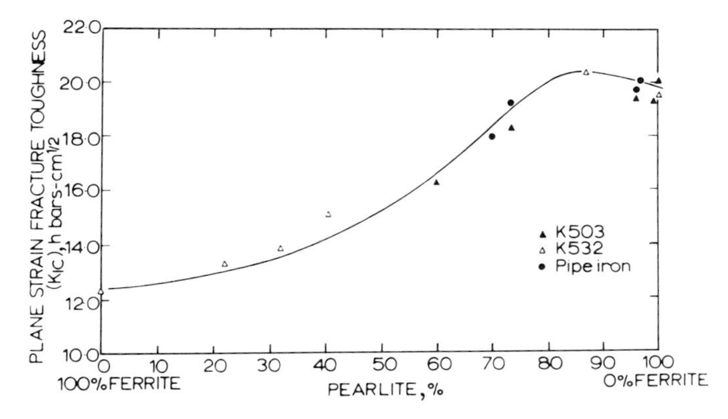

Figure 11-11 Effect of increasing ferrite content on plane strain fracture toughness of flake graphite cast iron.[25] (Reproduced by permission of The Metals Society)

In mixed ferrite–pearlite regions, cleavage failure in the ferrite propagated into the pearlite. When pearlite was associated directly with the graphite, the crack propagated across the pearlite lamellae. Some plastic deformation of the pearlite was observed, including formation of dimples. If separation of a graphite–pearlite interface occurred, the crack was contained entirely within the graphite.

Figure 11-11 shows a gradual decrease of fracture toughness as the ferrite content increases. The ferrite grain size has been increased by heat treatment and the fracture mode changed to intergranular. The fracture could propagate either along the ferrite grain boundaries bordering on the prior γ phase (in the heat treatment related to the γ range) or to the subcritically formed α boundaries. It was observed that failure did not follow the graphite eutectic cell boundaries. Some association of fracture path with segregation was suggested, e.g. with Fe_3P and Fe_3S type segregated phases at prior γ grain boundaries.

In their summary, Glover and Pollard suggested that fracture of a cast iron matrix depended on the orientation of the pearlite with respect to the crack, the percentage of ferrite, the nature of the heat treatment, and the association of the ferrite with either the graphite or pearlite.

Deformation and Fracture of Pearlite

Puttick[29] studied the plastic deformation and fracture of pearlite in steel using optical and electron microscopy. Two-stage replica techniques were employed for electron microscopy and photogrammetry of stereoscopic pairs.

The deformation mode observed for the pearlite indicated fine slip occurring in the ferrite lamellae parallel to the lamellar direction. In unfavourable oriented colonies, slip could also occur transverse to the lamellae. In this case, the α slip is accommodated by plastic deformation of Fe_3C and α–Fe_3C interfacial slip. The deformation of Fe_3C had a considerable component lying in the plane of the lamellae.

The fracture surface of coarse pearlite had the characteristics of brittle cleavage. There was a small patch of fibrous appearance at the origin of fracture, while most of the area had a bright texture simulating the pearlite microstructure.

Puttick suggested that the sequence of plastic flow and fracture of pearlite might be as follows. Favourable oriented colonies yield by slip parallel to the lamellae. Unfavourable oriented pearlite colonies deform by slip initiating in ferrite followed by slip at growth faults, producing large shears. These might be accompanied by deformation of Fe_3C and α–Fe_3C interfaces. Cracks initiate at Fe_3C or α–Fe_3C interfaces and grow to critical dimension.

Glover and Pollard,[24] in their observation of fracture in pearlite in cast iron specimens, noticed both crack propagation transverse to the cementite lamellae or within a single ferrite phase. In the latter case typical river patterns were observed. Crack propagation in pearlite was thought to be difficult. When initiated, cracks in pearlite might propagate only short distances. Thus as-cast materials with a pearlitic matrix give a high fracture toughness related to the crack-arresting properties of the pearlitic areas. As the amount of ferrite in the structure increased, it was the fracture of this latter phase which became the controlling mechanism.

11-7 Ductile–Brittle Failure of Spheroidal Graphite Cast Iron

The ductile–brittle failure of spheroidal graphite cast iron parallels in behaviour that observed in steel, with some modification due to the presence of graphite spherulites. In steel, a sudden decrease in ductility is observed at some low temperature.

Mogford, Brown, and Hull[30] examined the behaviour of spheroidal graphite cast iron. The previous research of Pellini[31] was reviewed, in which, as the number of graphite particles in ferritic nodular cast iron increased, the impact transition temperature and the energy absorbed in ductile impact fracture both decreased. Mogford, Brown, and Hull confirmed these data. They showed the following:

(a) At small values of interparticle spacing between graphite nodules, the flow stress of iron–graphite alloys decreased as the interparticle spacing decreased (Figure 11-12). This was attributed to the stress concentration at graphite nodules.
(b) The temperature dependence of the yield stress of ferritic nodular cast iron was the same as for an Fe–3 per cent Si alloy (Figure 11-12b).
(c) The transition temperature for ductile–brittle fracture of spheroidal graphite cast iron decreased as the interparticle spacing between graphite nodules decreased. The increased resistance to brittle fracture was attributed to the arrest of microcracks by graphite spherulites which delays the attainment of general cleavage. Figure 11-12c).

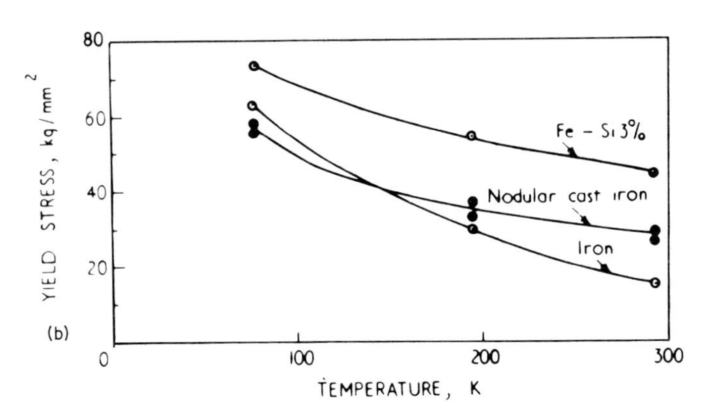

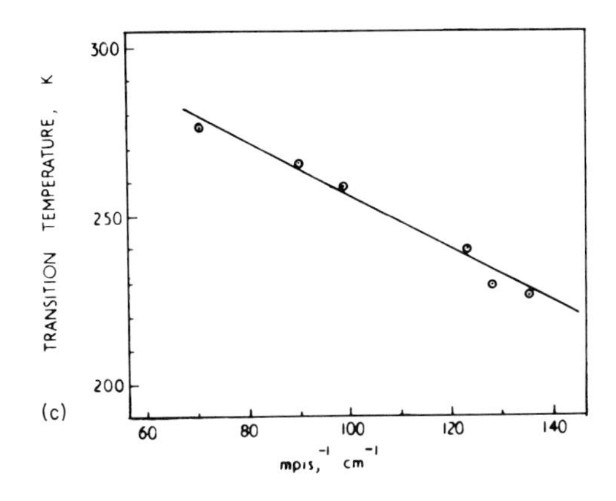

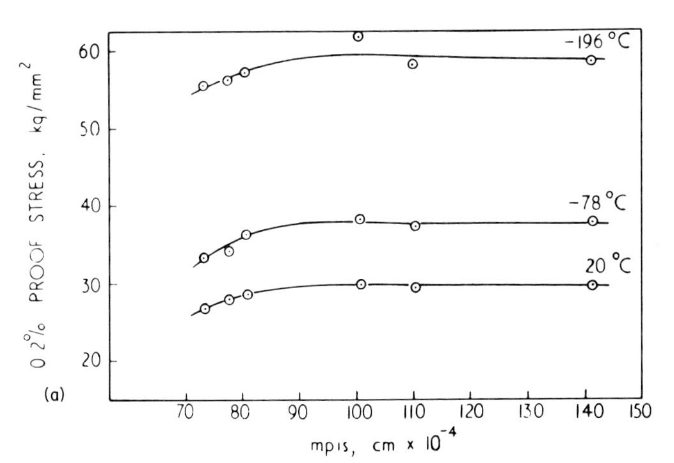

Figure 11-12 (a) Decrease of 0.2 per cent proof stress in spheroidal graphite cast iron at different temperatures as interparticle spacing decreases (mips = mean interparticle spacing). (b) Temperature dependence of yield stress of a ferritic spheroidal graphite cast iron is the same form as for an Fe–3 per cent Si alloy. (c) The transition temperature for ductile–brittle fracture of spheroidal graphite cast iron decreases as the interparticle spacing between graphite nodules decreases.[30] (Reproduced by permission of The Metals Society

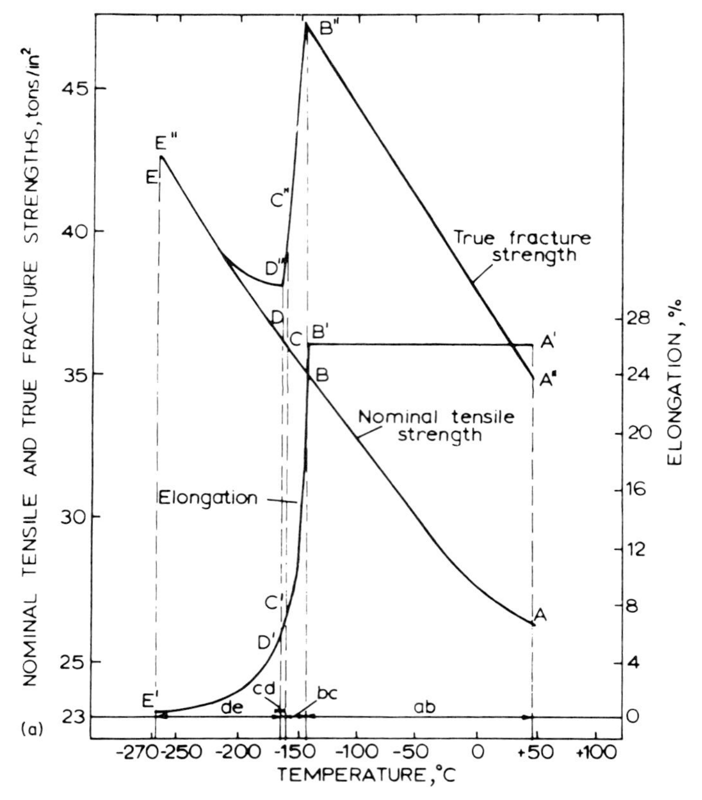

Figure 11-13 Tests performed by Rickards[27] on two analyses of ductile iron showing mechanical behaviour at low temperatures. (Reproduced by permission of The Metals Society)

(d) Cavitation was observed at the graphite spherulites after ductile fracture. This was related to decohesion at the weak interface between primary and secondary graphite and was responsible for the low energy required for ductile fracture of impact specimens.

Rickards[27] tested a series of ductile irons and published data which included the results of mechanical tests, shown in Figures 11-13(a) and (b). Details of the two irons for which the data of Figure 11-13 are published are given in Table 11-1.

The curves for nominal tensile strength and the observations of fracture as a function of temperature demonstrated the following. The materials were completely ductile above A. Below A, fracture commenced in a ductile manner but became partly cleavage. At C, the fracture was mostly cleavage and the total elongation had decreased to about 5 per cent. Below C, arrested cleavage microcracks were found and towards D brittle cracks propagated. In the interval DE, brittle fracture was initiated after critical amounts of plastic

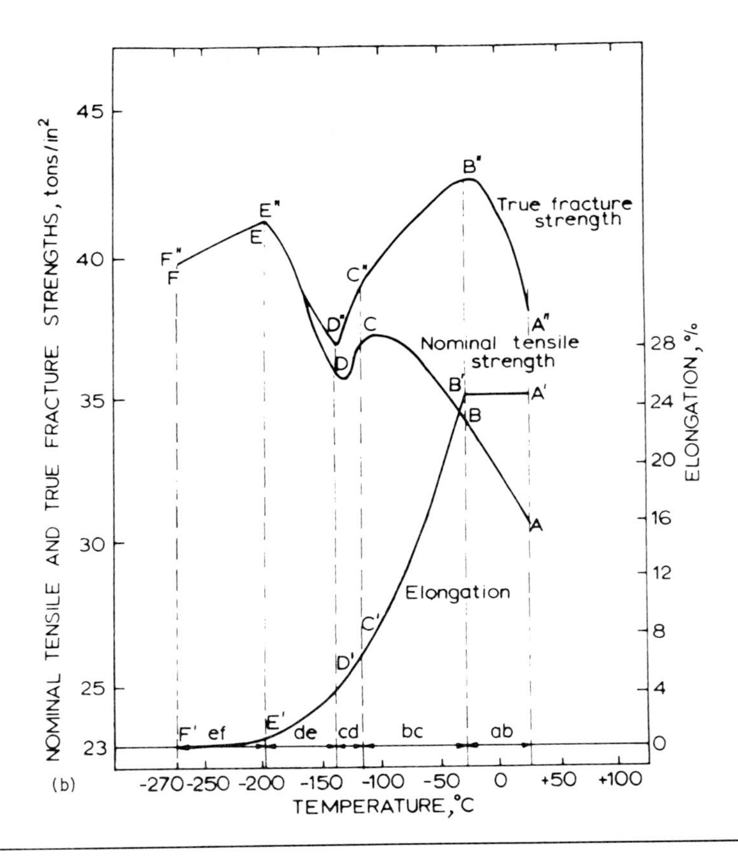

strain. Below DE, twinning was associated with brittle cleavage. Fracture in this range in these irons is primarily by cleavage through the ferrite grains. Relatively few nodules were observed in the fracture path. At normal temperatures, it was suggested, these irons can accommodate stress concentration at nodules by deformation. At low temperatures, fracture initiation is associated with grain boudaries and the stresses are relieved by cleavage.

Table 11-1 Analyses of irons tested by Rickards[27] (Reproduced by permission of The Metals Society)

Element	Iron 3, %	Iron 5, %
TC	3.46–3.69	3.45–3.65
Si	2.05–2.10	2.74–2.81
Mn	0.36–0.38	0.34–0.39
S	0.014–0.018	0.021–0.026
P	0.016–0.022	0.084–0.094
Ni	0.79–0.80	0.80–0.82
Mg	0.063–0.077	0.059–0.073

11-8 Thermal Fatigue

Thermal fatigue of cast iron results from temperature cycling and the associated cycling of thermal stress. It is dependent on a number of factors which include the modulus of elasticity of the alloy, thermal conductivity of the two-phase structure, and the behaviour under stress at high temperature, in particular the time dependence of strain in the material. It has been studied extensively in such applications as cast iron alloys for diesel engines

Thermal stress in a body can be expressed by the following relationship:[32]

$$\sigma_{th} = [\alpha E \Delta T/(1 - \mu)]F$$

where
E = modulus of elasticity
α = coefficient of thermal expansion
ΔT = temperature gradient in the body
μ = Poisson's ratio
F = shape factor, which varies as the body is constrained or unconstrained

Stress relaxation can occur by either plastic deformation or creep at high temperature, and at the end of each cycle of heating and cooling, there is a residual tensile stress. Different approaches have been made to evaluate the important parameters and classify cast iron in its thermal fatigue resistance.

Bertodo[33] analysed the then-existing knowledge of the high temperature behaviour of gray cast iron in diesel engine combustion chamber components. He made a systematic examination of the mechanical and physical properties of 166 plain and alloy cast irons and recommended that the important factor in choice of composition was as follows:

$$\frac{f_v f_r}{E H_B}$$

where
f_v = UTS
f_r = relaxation stress for the conditions of operation
E = Young's modulus
H_B = Brinell hardness

Bertodo recommended an ideal alloy cast iron which had the alloying additions 1.3 per cent (Cu plus Ni) and 0.4 per cent molybdenum.

Different experimental arrangements have been proposed for studying thermal fatigue. In their research, Coffin and Wesley[34] used a completely clamped specimen which was heated by passing current through it and cooled by a radial gas stream. Figure 11-14 shows their graphical representation of the cycles of stress and strain for a ductile material clamped at its ends and cycled between temperatures T_1 and T_2. For a specimen clamped into position at temperature T_2, point 2 is the origin of the σ–ε curve. As cooling occurs, path A is followed to 1. The deformation is at first elastic and then becomes increasingly inelastic. Little relaxation of stress occurs at T_1 and the reverse temperature cycle leads to the σ–ε curve, again denoted by A. The stress at

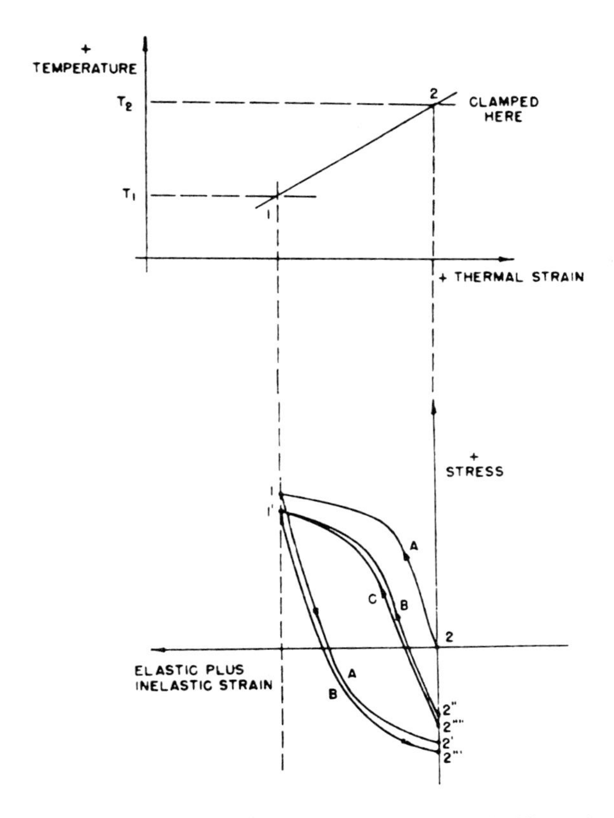

Figure 11-14 Graphical representation by Coffin and Wesley[34] of cycles of stress and strain in a ductile material clamped at its ends and cycled between temperatures T_1 and T_2. For specimens clamped into position at T_2, point 2 is the origin of the σ–ε curve (from *ASME Trans*, **76**, p. 923, 1954, by permission of American Society of Mechanical Engineers)

first follows an elastic path and then an inelastic path to $2'$. At T_2 the stress relaxes to $2''$. Path B is now followed on cooling until $1''$. Again no relaxation occurs at T_1 and on reheating to T_2 path B is followed to $2'''$. Relaxation then occurs to $2''''$ and on cooling path C is followed. A steady-state stress–strain loop is then obtained. This loop depends on a number of factors which were listed as follows:

(a) Temperatures T_1 and T_2.
(b) Time at T_2, and to a lesser extent at T_1.
(c) Rate of heating and cooling.
(d) Temperature effect on elastic properties.
(e) Inelastic properties (including creep) and their temperature dependence.

Gundlach[35] used a rigidly clamped test specimen heated by induction and cooled by conduction of heat to the water cooled grips. The thermal stresses developed were monitored by a load cell installed in one of the grips holding the specimen. Gundlach examined a series of pearlitic gray cast irons alloyed with various combinations of Cr, Mo, V, Sn, Ni, and Cu. In order to determine the values of the elastic moduli at temperature a pulsed-echo approach was employed (see Chapter 12), while further studies were made of stress relaxation in compressive loading at 500 °C.

Gundlach obtained an expression for log cycles to failure in the thermal fatigue test, arriving at the following by regression analysis:

$$\log N = 0.41 + 0903(\text{UTS in ksi}) + 1.89(\%\text{V}) + 1.79(\%\text{Mo}) + 0.11(\%\text{Cr}) - 0.14(\%\text{Ni} + \%\text{Cu})$$

He suggested that the resistance to creep appeared to be the most important property for resistance to thermal fatigue, and the influence of molybdenum or vanadium in the above expression, producing the biggest improvements, was related to this.

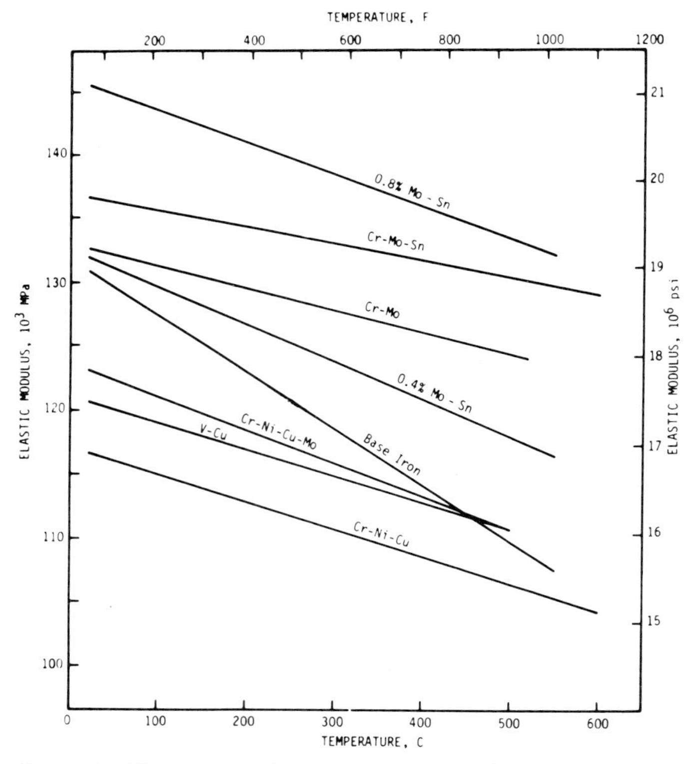

Figure 11-15 Curves of elastic modulus of different cast irons as function of temperature obtained by Gundlach[35]. (Reproduced by permission of American Foundrymen's Society)

The curves obtained for elastic modulus as a function of temperature are shown in Figure 11-15. Alloying reduced the rate at which the modulus decreased with temperature, but a number of irons show modulus increases with alloying. An increase in strength and resistance to stress relaxation were suggested to have more than compensated for this increase in elastic modulus.

Ranking in Thermal Fatigue Tests

Recent results of different thermal fatigue tests, which classify the resistance of different types of cast iron, were discussed by Röhrig.[32] Figure 11-16(a) shows a table of results for thermal fatigue tests on iron with different graphite types and different amounts of ferrite in the matrix. These results

(a)

Class of Iron	Matrix Constituents,† %	Composition, %					Cycles to First Crack	
		C	Si	Cr	Mo	Sn	Minor	Major
Ductile	39 P, 61 F			--	--	--	700	800
	53 P, 47 F	3.7	2.17	--	0.25	--	700	800
	81 P, 19 F			0.26	0.25	--	360	370
Compacted Graphite	60 P, 40 F			0.20	--	--	430	460
	60 P, 40 F	3.7	1.8	0.20	0.44	--	430	510
	95 P, 5 F			0.20	--	0.047	340	380
	95 P, 5 F			0.20	0.57	0.047	350	380
Gray	90 P, 10 F	3.84	1.65	0.15	--	--	170	220

†P = Pearlite, F = Ferrite

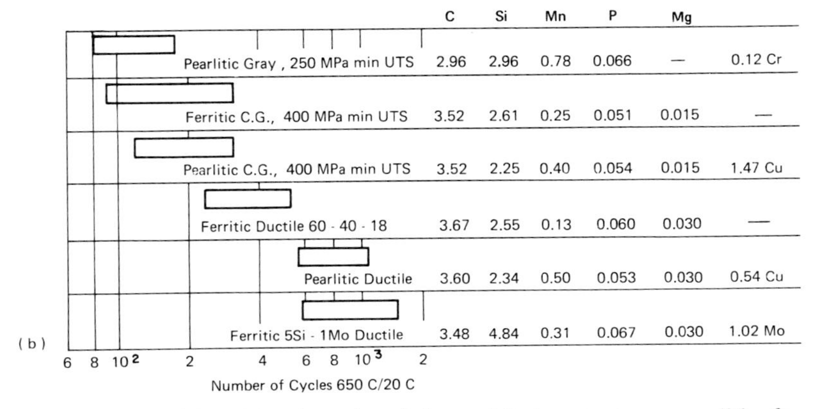

Figure 11-16 (a) Table of results of thermal fatigue tests on pearlitic–ferritic ductile and compacted graphite irons. Specimens cycled between 690 and 240 °C.[32] (b) Number of cycles to produce cracking in various cast irons cycled between 650 and 20 °C using test samples by Buderus AG (quoted by Röhrig[32]). (Reproduced by permission of American Foundrymen's Society)

indicate that the spheroidal graphite cast irons have highest thermal fatigue resistance and that for both spheroidal graphite and compacted graphite cast irons, this resistance increases as the ferrite volume increases. With gray cast iron, the opposite is generally the case and increasing pearlite volume tends to increase thermal fatigue resistance.

The manner of testing plays some role in the ranking of the different cast irons. Röhrig[32] reproduced the results of a different mode of testing by Buderus AG in Wetzlar, Germany (Figure 11-16b). These results show that a pearlitic compact graphite iron has better thermal fatigue resistance than the ferritic type. However, the resistance of a ferritic 5 per cent silicon–1 per cent molybdenum spheroidal graphite cast iron was superior.

References

1. Orowan, E.: *Rep. on Prog. Phys.*, **12**, 185, 1949.
2. Griffith, A. A.: *Phil. Trans. R. Soc., Ser. A.*, **221**, 163, 1920.
3. Inglis, C. E.: *Trans. Inst. Naval Arch. London*, **55**, 219, 1913.
4. Angus, H. T.: *Cast Iron. Physical and Engineering Properties*, 2nd ed., Butterworths, 1976.
5. Collaud, A.: Réflexions sur les Propriétés Mécaniques des Fontes Grises. Saint-Paul, Fribourg/Suisse, 1975.
6. Gilbert, G. N. J.: *Engineering Data on Grey Cast Irons. SI units*, BCIRA, Alverchurch. Birmingham, 1977.
7. Fisher, J. C.: *Am. Soc. Test. Mater. Bulletin*, **1952**, 74, April 1952.
8. Zener, C.: *Fracturing of Metals*, Am. Soc. Met., 1948.
9. Neuber, H.: *Kerbspannungslehre*, J. Springer, Berlin, 1937.
10. Grassi, R. C., and I. Cornet.: *Mechanical Eng.*, **70**, 918, 1948.
11. Coffin, L. F.: *J. Appl. Mechanics*, **17(3)**, 233, September 1950.
12. Clough, W. R., and M. E. Shank: *Trans. Am. Soc. Met.*, **49**, 241, 1957.
13. Flinn, R. A., and R. J. Ely.: Am Soc. Test Mater. Special Techniques Pub. 97, 1950.
14. Iitaka, I., and I. Yamagishi.: *Sc. Pap. Inst. Phys. Chem. Res. Tokyo*, **34**, 1025, 1937.
15. Plenard, E.: *Recent Research on Cast Iron* (Ed. H. Merchant), Gordon and Breach, New York, 1968.
16. Speich, G. R., A. J. Schwoeble, and B. M. Kapadia: *J. Appl. Mech.*, **47**, 821, 1980.
17. Hashin, Z.: *ASME J. Appl. Mech.*, **29**, 43, 1962.
18. Wu, T. T.: *Int. J. Powder Metallurgy*, **1**, 8, 1965.
19. Rossi, R. C.: *J. Am. Ceram. Soc.*, **51**, 433, 1968.
20. Collaud, A.: *Schweiz. Arch.*, **21**, 65, 1955.
21. Minkoff, I.: Unpublished research.
22. Siefer, W., and K. Orths: *Giessereiforsch* (in English), **21**, 140, 1969.
23. Freise, E. J.: Ph.D. Dissertation, Cambridge University, March 1962.
24. Glover, A. S., and G. Pollard: *Fracture Toughness of High Strength Materials*, Iron Steel Inst. Publ. 120 1970).
25. Glover, A. S., and G. Pollard: *J. Iron Steel Inst. London*, **209**, 138, 1971.
26. Glover, A. S., and G. Pollard: *Proc. Second Int. Conf. Fract.* (Ed. P. L. Pratt), Chapman and Hall, 1969.
27. Rickards, P. J.: *J. Iron Steel Inst. London*, **209**, 190, 1971.
28. Chang, Y-W., R. W. Heine, and F. J. Worzala: *Trans. Am. Foundrymen's Soc.*, **185**, 161, 1957.

29. Puttick, K. J.: *J. Iron Steel Inst. London*, **185**, 161, 1957.
30. Mogford, I. L., I. L. Brown, and D. Hull: *J. Iron Steel Inst. London*, **205**, 729, 1967.
31. Pellini, W. S.: *Trans. Am. Soc. Met.*, **46**, 418, 1954.
32. Roehrig, K.: *Trans. AFS,* **86**, 25, 1978.
33. Bertodo, R.: *J. Strain Analysis*, **5(2)**, 98, 1970.
34. Coffin, L. F., and R. P. Wesley *ASME Trans.*, **76**, 923, 1954.
35. Gundlach, R. B.: *Trans AFS*, **87**, 551, 1979.

Bibliography

Dieter, G. E.: *Mechanical Metallurgy*, 2nd ed., McGraw-Hill, 1976.
Kelly, A.: *Strong Solids*, Clarendon Press, Oxford, 1966.
Knott, J. F.: *Fundamentals of Fracture Mechanics*, Butterworths, 1973.
Mogford, I. L.: *Deformation and Fracture of Two-Phase Materials, Met Reviews*, Vol. 12, p. 49, Metal and Metallurgy Trust, London, 1967.
Roll, F.: *Handbuch der Giesserei Technik*, Vol. 1/2, Springer-Verlag, 1959.
Tetelman, A. S., and A. J. McEvily: *Fracture of Structural Materials*, Wiley, 1967.

Chapter 12
Testing and Inspection

A summary is given of some testing procedures for cast iron, reviewing initially mechanical tests for properties and behaviour, followed by ultrasonic testing. Emphasis is placed on some of the more recent methods for evaluating properties, including tensile strength from dynamic elastic modulus and fracture toughness testing. The scope of testing and inspection is large. The intention of this chapter is to give a brief introduction to some of the procedures.

12-1 Mechanical Testing

Brinell Hardness Test

The Brinell test for cast iron uses a constant load of 3,000 kg applied on a 10 mm ball. Different times of loading recommended are 10–15 s for gray cast iron and 3–5 s in the medium and high hardness range for ductile ferritic and malleable cast iron. Various reviews have been published, such as the application of the standard Brinell test in controlling the manufacture of as-cast spheroidal graphite and heat-treated pearlitic malleable iron.[1]

Because of the two-phase nature of cast iron alloys, it is necessary to calibrate the hardness test for any particular set of alloys being tested and obtain relationships between tensile properties and hardness. Maranda and Kearnes,[1] for example, gave curves for pearlitic malleable iron, oil quenched and drawn, as-cast pearlitic nodular iron, and as-cast ferritic nodular iron. Common pitfalls and sources of error for this type of testing procedure were discussed.

Wedge Penetration Test

A review of experience gained with application of the wedge penetration test in flake graphite cast iron was presented by Standke.[2] Figure 12-1 shows schematically the experimental arrangement. Two wedges with parallel cutting edges press on opposite sides of the sample, which is a disc or round bar, exerting tensile forces on the material. The wedge angle is 90°. For disc-

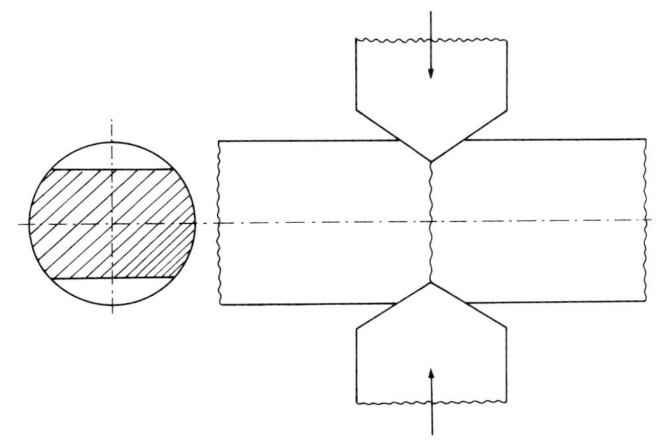

Figure 12-1 Wedge penetration test.[2] (Reproduced by permission of Giesserei-Verlag)

shaped specimens, the diameter recommended is 20 mm and the thickness is 6 mm. The top radius of the wedges is less than 0.2 mm.

This method has been described as rapid, inexpensive, and reliable. The tensile strength can be calculated with a high degree of reliability from the wedge penetration strength, using regression equations especially calculated for the particular production conditions. Linear relationships exist between tensile strength σ_B and wedge strength σ_K. Correlation coefficients range from over 80–97 per cent with standard deviations from 12 to 20 N mm^{-2}. The testing can be performed either on machined or on rough as-cast bars. In the latter case, the scatter is greatest.

For disc samples, the formula giving best general agreement was that of Ebner:[3]

$$\sigma_B = -77 + 1.9\sigma_K \pm 15.4 \qquad \text{N mm}^{-2}$$

For rough as-cast bars, the formula was:

$$\sigma_B = -41.4 + 1.92\sigma_K \pm 19.0 \qquad \text{N mm}^{-2}$$

For ductile materials, the relationship between tensile strength and wedge penetration strength remained linear up to an elongation of about 3 per cent. At high elongations, the wedge penetration strength is displaced, at the same tensile strength, to higher values. It was suggested as inadvisable to attempt using the test for ductile materials.

12-2 Fracture Toughness Testing

Evaluation of fracture toughness data is intended both to provide data for design and also to analyse the individual aspects of microstructure as they influence fracture. Some current research aspects of fracture toughness testing were discussed in relation to fracture in Chapter 11. Published results with details of testing methods are described in the following.

Fracture and Fracture Toughness

The data obtained from fracture mechanics can provide critical flaw dimensions and related strength, as well as allowing comparison of such factors as matrix structure, graphite volume, and shape, on the resistance to crack propagation. Some of the parameters evaluated are as follows.

K_c: Critical Stress Intensity Factor The critical stress intensity factor K_c is a rearrangement of the Griffith fracture criterion as follows:

$$\sigma = \sqrt{\left(\frac{E\gamma}{\pi c}\right)} \qquad \text{(Griffith fracture criterion)}$$

or

$$\sigma\sqrt{(\pi c)} = K_c = \sqrt{(E\gamma)}$$

where E = Young's modulus
γ = surface free energy
c = half crack length for internal flaw

K_{1c} is the critical stress intensity factor for a tensile mode of crack propagation. It can be measured by introducing a cut in a specimen and measuring the applied stress necessary to propagate the crack. Different matrix structures measured and placed in decreasing order of fracture toughness are as follows:

As-cast ferrite
Annealed ferrite
As-cast pearlite
Normalized steel
Austempered steel

COD: Crack Opening Displacement For a crack in a tensile loaded plate of sufficient thickness that plane strain conditions prevail, the plastic deformation at the crack tip is confined to narrow bands whose thickness is of the order of the diameter 2ρ of the crack tip. If $\varepsilon(c)$ is the tensile strain in the specimen adjacent to the tip, the displacement of the crack faces at the crack tip $2V(c)$ is

$$2V(c) = 2\rho\varepsilon(c)$$

If $\varepsilon_f(c)$ is the fracture strain, unstable fracture occurs when

$$\varepsilon(c) = \varepsilon_f(c)$$

$V(c)$ achieves a critical value $V^*(c)$ when

$$V(c) = V^*(c) = \rho\varepsilon_f(c)$$

$V(c)$ can be related to the nominal stress and crack length so that the fracture strength of a cracked plate can be determined if $V^*(c)$ is known.

Fracture Toughness of Spheroidal Graphite Cast Iron

Nanstad, Worzala, and Loper[4] performed fracture toughness tests on spheroidal graphite cast iron of a variety of microstructures. They also described procedures for utilizing the data to predict fail-safe behaviour. Two tests were employed:

(a) Drop weight test.
(b) Plain strain fracture toughness test.

The drop-weight tests were set out in procedures of the Naval Research Laboratory for 16 mm thick plate. The plain strain fracture toughness tests were performed on fatigue pre-cracked compact tensile specimens in accordance with ASTM procedures.[5] The variables studied included nodularity (the geometrical regularity of a spherulite), matrix microstructure, specimen section size, temperature, and strain rate. After failure, the fracture surfaces were observed metallographically and by means of scanning electron microscopy so as to relate microstructure and crack morphology.

Figure 12-2 shows details of two types of specimen used in the investigation. The SEN specimens were fatigue pre-cracked with a Krause bending machine. The CT specimens were fatigued and fractured with an MTS hydraulically closed-loop system. Crack lengths prescribed by the ASTM standard were used. To ensure the presence of the plane strain condition, ASTM procedures[5] require that specimen thicknesses be greater than $2.5(K_1/\sigma_{YS})^2$ where σ_{YS} is the 0.2 per cent offset yield and K_1 is a stress intensity factor for

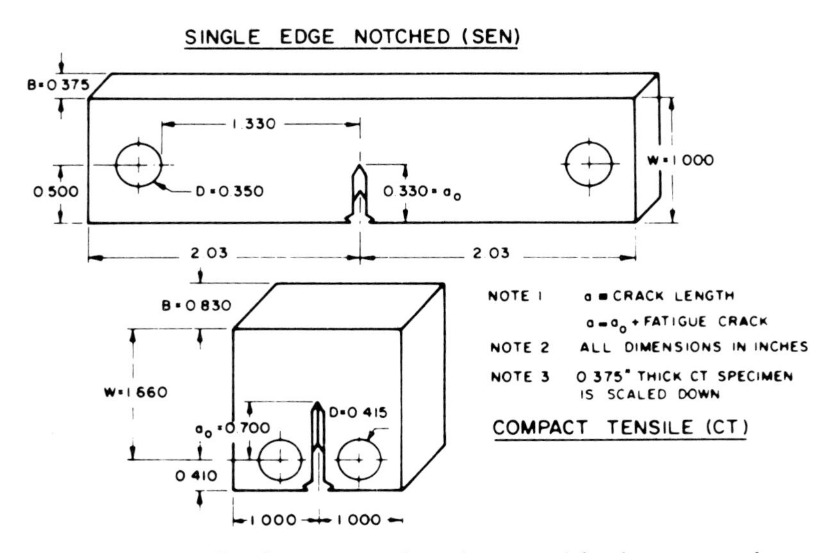

Figure 12-2 Details of two types of specimen used for fracture toughness testing of spheroidal graphite cast iron by Nanstad, Worzala, and Loper[4] (Reproduced by permission of Georgi Publishing Co.)

very tough materials; this requires specimens as thick as 305 mm to achieve valid K_{1c} results at room temperature. In this testing programme, the specimen thickness was 0.5 mm and 21.2 mm as shown in Figure 12-2.

For microstructures which appeared to have failed in a brittle, plane strain manner, calculation methods of ASTM E 399[5] were employed to determine K_{1c}. For microstructures which exhibited a substantial amount of ductile tearing and transverse necking, a fracture toughness value based on gross plastic yielding was determined.

Material Variables

The material tested was a spheroidal graphite cast iron in which microstructural variations were obtained in the as-cast condition by altering the cooling rate and composition. Other microstructures were produced by heat treatment. Matrix variations included as-cast ferritic and pearlitic, annealed ferritic, normalized and tempered microstructures. The nodularity of the graphite was changed and testing was also performed at +25 and −25 °C. The nodule count which varied between 50 and 80 nodules mm^{-2} was not considered to be a significant variable.

With these data a number of conclusions were reported. The value of K_{max} was indicated as being the best for evaluating fracture toughness in the presence of a sharp crack where significant plasticity occurred. If K_{max} was used as a toughness index the matrix structure was shown to play a major role in determining toughness. For example, ferrite rings surrounding the nodules appeared important. These rings failed with a ductile tear even at low temperatures.

The investigation suggested that the shape of the nodule did not significantly affect the toughness. As an example the presence of poorly formed vermicular graphite lowered the toughness values by no more than 10 per cent.

Application of Fracture Toughness Test Results to Design

In general, the use of KK_{1c} values allows the prediction of critical flaw sizes or the calculation of stress levels from given sizes of flaw. In the investigation reported many of the microstructures studied did not yield plane strain data, exhibiting considerable resistance to brittle crack extension. This suggested that they could be classified as elastic–plastic materials and allow design stress up to the yield strength.

The observations of fracture were that the nodules and surrounding ferrite rings absorb strain energy through plastic deformation and ductile tearing. This is the opposite case for steel, where the structure is continuous and in which considerable strain energy can be built up to become available for crack propagation once K_{1c} is reached. The different ductile iron structures and their behaviour under these conditions was part of the subject matter for the investigation.

Crack Opening Displacement and K_{1c} Values

Holdsworth and Jolley[6] used crack opening displacement and K_{1c} values to observe the influence of graphite nodule number and volume fraction on the fracture toughness of ferritic nodular cast iron. The objective was to provide design data in the form of critical defect size for failure under relevant operating conditions. The requirements for this are obtained from either K_{1c} determinations or from crack opening displacement.

Test Pieces

Fracture toughness test pieces having the dimensions 2 mm wide, 10 mm thick, and 50 mm long were used. The stress concentrator (Figure 12-3) consisted of a 0.15 mm slot of length 2 mm cut with a rubber bonded slitting wheel at the root of a milled slot 3.18 mm wide of length 5.70 mm. The authors described the necessity of the type of composite notch as the result of difficulties entailed in machining a 0.15 mm slot of length greater than 2.50 mm. The final portion of the stress concentrator consisted of a fatigue crack 2.30 mm long, initiated at the root of the 0.15 mm slot. The fatigue pre-cracking was carried out on an Amsler Vibraphore at low stresses.

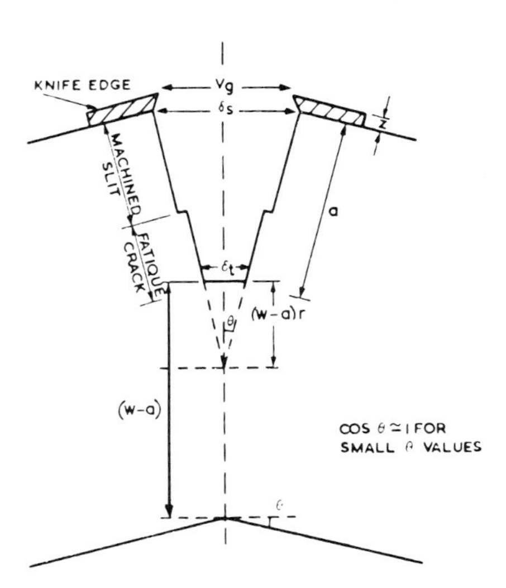

Figure 12-3 Detail of stress concentrator used by Holdsworth and Jolley[6] for determination of crack opening displacement and K_{1c} values in experiments on ferritic spheroidal graphite cast iron (Reproduced by permission of Georgi Publishing Co.)

Cast Iron Examined

The alloys were prepared so that in one series, the mean internodule spacing decreased from 0.139 to 0.075 mm (an increase in nodule number from 37.2 to 106.8 mm^{-2}). The volume fraction was retained constant at about 12.3 per cent. The nodule count was increased by increasing the inoculant addition in the ladle from $\frac{1}{4}$ per cent Fe–Si to $\frac{1}{2}$ per cent Fe–Si, together with bismuth.

In the second series of alloys prepared, the mean spacing between nodules was decreased from 0.191 to 0.094 mm (an increase in nodule number from 20.6 to 73.8 mm^{-2}) as a result of increasing the carbon content from 2.58 to 3.99 per cent (an increase in volume fraction of graphite from 8.58 to 12.62 per cent).

Measurement of COD and K_{1c}

The test pieces were loaded at three-point bending over a loading span of four times the specimen thickness. The loading rate was approximately 50 MN $m^{-3/2}$ min^{-1}. The displacement V_g (Figure 12-3) was measured with a double cantilever beam clip gauge mounted on knife edges attached to the test piece surfaces adjacent to the notch opening.

In order to calculate the COD at the crack tip (δ_t) from the displacement measured above the surface of the test piece, V_g, the value of the rotational factor (r) was first determined for material being investigated. The test pieces were strained to different values of COD between general yielding and maximum load. The opening displacement at the two machined notch roots and the crack tip at mid-thickness were measured by a calibrated graticule in an optical microscope.

Measurement of Rotational Factor

Calibration curves were constructed for each alloy in which the opening displacement was plotted as a function of the position of measurement for increasing values of the applied load. The family of curves were extrapolated so as to cut 'the position of measurement' axis, and these points were defined as the effective points of rotation, assuming that rotation occurred above the centre line of loading. Figure 12-3 shows the definition of rotation factor r in the relationships between $(W - a)$ and $(W - a)r$. The values of r as defined in this figure are plotted against the corresponding off-load clip gauge displacement in Figure 12-4(a). The crack lengths were determined by breaking open the test pieces in liquid N_2 after loading to just past maximum load. Bend tests were performed from -200 to $+30$ °C.

Figure 12-4 (a) The values of the rotational factor, plotted against the knife-edge opening displacement. (b) Test results for six analyses of iron showing the relationship between crack opening displacement and fibrous crack length.[6] (Reproduced by permission of Georgi Publishing Co.)

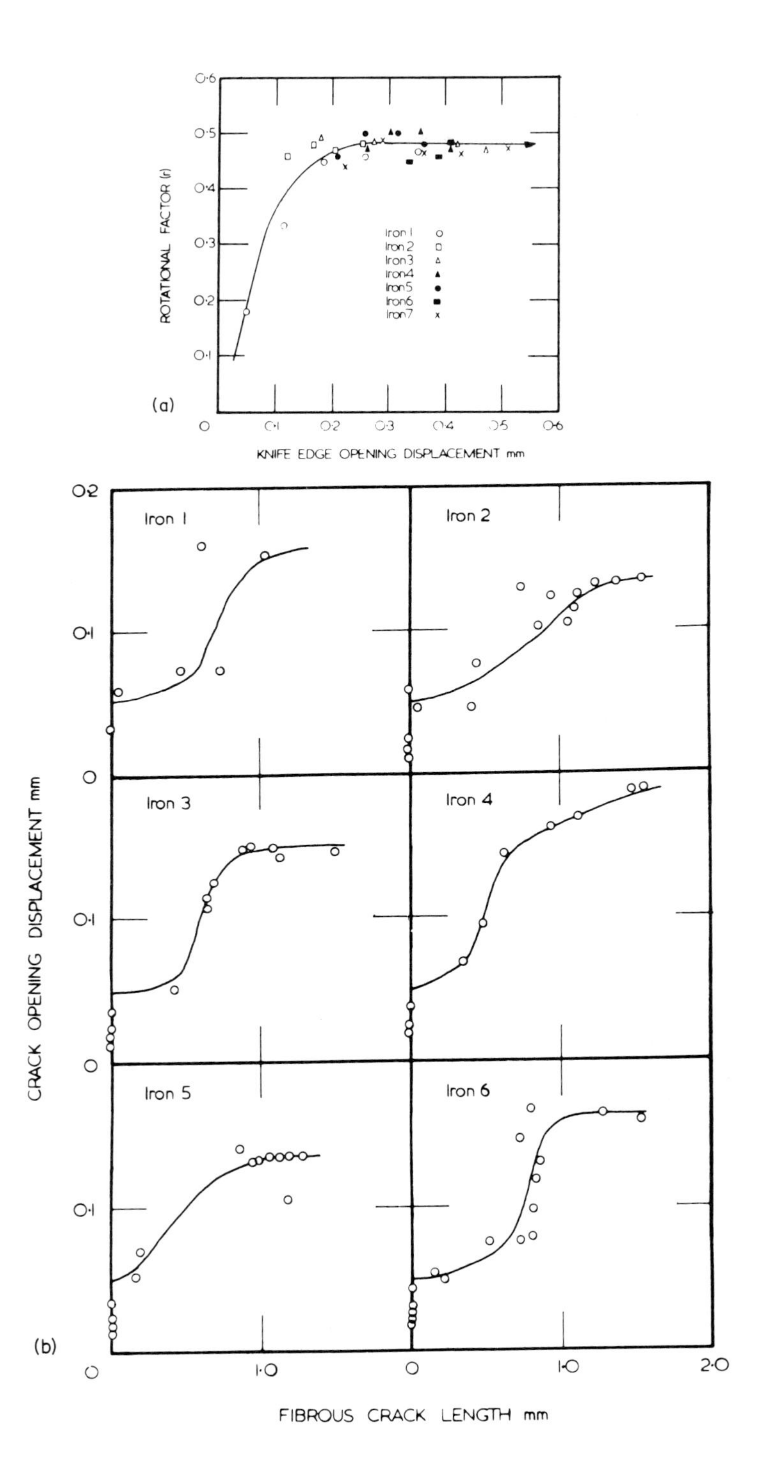

ROTATIONAL FACTOR (r)
KNIFE EDGE OPENING DISPLACEMENT mm
Iron 1
Iron 2
Iron 3
Iron 4
Iron 5
Iron 6
Iron 7
(a)
CRACK OPENING DISPLACEMENT mm
FIBROUS CRACK LENGTH mm
(b)

Measurement of COD

The crack opening displacement (δ_t of Figure 12-3) was derived from V_g, the value of clip gauge displacement, coincident with the onset of maximum loading or instability on the load displacement record. The following relationship was employed

$$\delta_t = \frac{Vg}{1 + (a + z)/[r(W - a)]}$$

In this relationship, z is the knife edge thickness and the other factors are given in Figure 12-3. Fibrous crack length if present was measured on all test pieces with a travelling microscope. Six irons were tested, and the test results of COD in relation to fibrous crack length are shown in Figure 12-4(b). Value of the stress intensity factor K_{1c} were also estimated from the load displacement records by an off-set procedure.[6]

Discussion of Results

Some of the results of the experiments of Holdsworth and Jolley could be interpreted as follows:

(a) From Figure 12-4(b) it is noted that the critical COD below which fibrous crack growth preceded unstable failure was unaffected by nodule spacing.
(b) A decrease of the mean nodular spacing from 0.139 to 0.075 mm by inoculation, while keeping the volume fraction of graphite the same, led to a fall in the transition temperature and COD.
(c) Appreciable effects could be noted when the mean internodule spacing was decreased from 0.191 to 0.094 mm by increasing the volume fraction of graphite. This resulted in a lowering of the transition temperature from −78 to −128 °C and a decrease in the critical COD in the ductile range from 0.185 to 0.130 mm.

Charpy Impact Data

Holdsworth and Jolley[6] referred to observations of the fracture of spheroidal graphite cast iron by Gilbert[7] using Charpy impact testing. The effect of graphite nodule number and spacing on the Charpy impact properties showed similar trends as in the fracture toughness tests. Holdsworth and Jolley suggested that COD and K_{1c} data provide information concerning the critical defect size a component can tolerate in service, while this cannot be provided by Charpy impact data.

In the case of spheroidal graphite cast iron, the flaws to be considered in design would be at the surface and in the interior.

The conclusions were that nodule spacing λ had an effect on the general yielding and linear elastic properties of ferritic spheroidal graphite cast irons. The critical COD in the ductile range is directly proportional to λ. The critical

COD transition temperature is directly related to the mean free path between nodules and K_{1c} is marginally inversely proportional to the mean internodular spacing.

The observation was again made that although conditions of plane strain were present through the temperature range of testing, there was a transition from a fibrous to a cleavage mode of failure.

12-3 Fracture Toughness Testing of White Cast Iron

The fracture toughness of white cast iron has been extensively investigated, for example by Diesburg and Röhrig,[8] and by Burns and Hofer.[9] An investigation of fracture toughness of white cast iron in which the parameters of carbide volume and matrix microstructure were varied, was reported by Zum Gahr and Scholz.[10] The commercial applications of high chromium white cast iron are for mining and milling, and the abrasion resistance is dependent on proper control of the microstructure. Use of these irons, for example in rolling-mill rolls, shows that adequate toughness is possessed by this class of materials. Zum Gahr and Schulz used fracture toughness testing to relate microstructure to toughness and looked at both carbide volume and matrix structure. The irons examined were of the composition given in Table 12-1.

Both static (K_{1c}) and dynamic (K_{1d}) tests were performed. The K_{1c} tests were performed on 12.5 mm thick compact tension specimens machined by grinding and notched by electrical discharge machining (EDM). The specimens were pre-cracked using a fatigue machine. The pre-cracked specimens, with a crack length of at least 2 mm, were fractured under load control with a loading rate of 17.8 kN min^{-1}. The K_{1c} fracture toughness was calculated according to ASTM E 399-78.[5]

The dynamic tests for K_{1d} were performed by instrumented impact methods with an instrumented Charpy tup. The hammer of the testing machine had a velocity of 1.23 m s^{-1} at impact. Testing was performed at 21 °C. Fracture surfaces of selected specimens were examined by scanning electron microscopy and the amount of carbide phase evaluated.

Table 12-1 Chemical composition of white cast irons[10] (Reproduced by permission of Metallurgical Society AIME)

C	Si	Mu	Cr	Mo	Cu	Ni	Carbide volume
1.41	0.58	1.56	11.6	2.39	1.24	0.020	7.1
2.0	0.59	1.54	15.8	2.35	1.14	0.020	14.4
2.58	0.59	1.54	17.6	2.39	1.03	0.020	24.3
2.87	0.59	1.54	20.0	2.36	0.94	0.020	29.1
2.92	0.59	1.54	19.0	2.35	0.94	0.020	28.0
3.50	0.59	1.54	23.4	2.47	0.87	0.020	37.6
3.93	0.59	1.54	24.6	2.45	0.76	0.020	45.4

Microstructure and Fracture Toughness Observations

In the cast irons examined, the as-cast structures were predominantly austenitic. The carbide volume varied between 7 and 45 per cent. The irons were predominantly C–Cr analyses and the structure was varied by variation of the concentration of these elements. The carbide phase at small C + Cr percentages existed as eutectic networks between γ dendrites. As the total C + Cr concentration increased, primary hexagonal carbide rods appeared and radial eutectic carbide structures were noted. To obtain specimens with austenitic matrices, the as-cast material was stress relieved at 200 °C for 2 hours. To obtain martensitic matrices heat treatment between 900 and 1030 °C resulted in secondary carbide precipitation and depletion of the solute elements in the γ phase, which was then transformed to martensite by forced air cooling. Any retained austenite was transformed by refrigerating twice at −78 °C. The hardness values of two sets of structures, martensitic and austenitic, are shown in Figure 12-5(a) as related to carbide volume. The K_{1c} values as a function of carbide volume are shown in Figure 12-5(b). This shows that the values of K_{1c} for either austenitic or martensitic irons remain uniform as the carbide volume increases up to 30 per cent. At this value, K_{1c} becomes similar for both of the structures and falls with further increase of the carbide volume. The K_{1d} values are shown in Figure 12-5(c). The values for austenitic and martensitic irons are separate over the whole range of carbide volume and fall as the present of carbide phase increases.

Further investigation was made of stress relieving and martensite tempering. It was noted that the K_{1c} value of both austenitic and martensitic cast iron could be raised by stress relieving, the martensitic structures tending to have higher K_{1c} values than the austenitic structures.

The carbide volume is seen to have an important influence on K_{1c}, the values of which decrease rapidly, beginning at 30 per cent carbide content. At this figure, fracture was noted to occur in these alloys through the massive carbides. Below 30 per cent carbide, the austenite matrix has a higher K_{1c} than the martensite.

The K_{1d} values show a decrease for both the austenitic and martensitic structures as the carbide volume increases. This was related to the decreasing amount of fracture through the matrix. The difference between K_{1c} and K_{1d} behaviour was related to the influence of high strain rate. This accentuates the better fracture toughness of the as-cast and stress relieved austenite. At the lower strain rates used for K_{1c}, stress relieved austenite and stress relieved martensite have comparable fracture toughness and the carbide matrix determines the fracture.

12-4 Sonic Methods of Testing

Sonic methods of testing cast iron have several applications, two of which are discussed here. The first application is in the measurement of mechanical

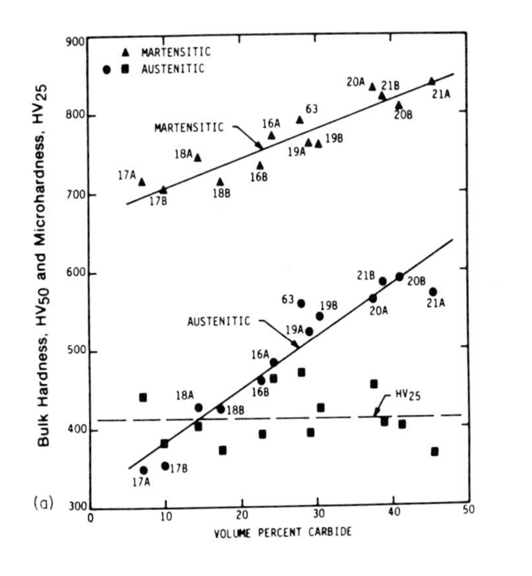

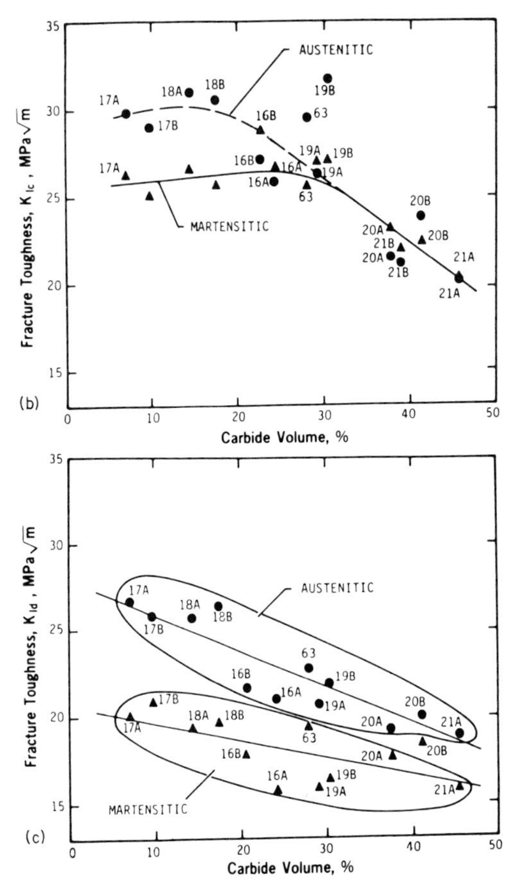

Figure 12-5 (a) Hardness values of martensitic and austenitic white cast irons, related to carbide volume. (b) Fracture toughness, K_{1c}, as a function of carbide volume. (c) Dynamic fracture toughness values as a function of carbide volume[13] (Reproduced by permission of Metallurgical Society of AIME)

properties. The second application to be discussed is as an inspection procedure for graphite shape.

Mechanical Properties

Kovacs and Cole[11] described an experimental procedure for assessing the mechanical properties of spheroidal graphite iron castings. Two basic methods could be employed: (a) pulsed echo and (b) resonance. The pulsed-echo method is based on a distance–time relationship, while the resonance method is based on the relationship:

$$\omega_0 = V/\lambda$$

where ω_0 = resonant frequency and λ = wavelength.

In the pulsed-echo method the velocity of a packet of elastic waves with frequency in the megahertz range, produced by a transducer applied to one end of a sample, is measured by the time taken to pass through the sample.

In the resonance method, a periodic force is applied to the body over a range of frequencies. When the frequency approaches the natural frequency of the system, generally in the kilohertz range, the amplitude of the vibration increases and the system resonates. Additional information can also be obtained by noting the capacity of the material for absorbing a portion of the vibrational energy, which is a measure of the damping capacity.

Both pulsed-echo and resonance methods can allow direct, and dynamic, determination of the elastic constants of the material. The resonant frequency method was given as being more precise. The study made by Kovacs and Cole was on spheroidal graphite iron castings, and related to the effects of graphite volume, morphology, and matrix microstructure on the resonant frequency and damping capacity.

The apparatus employed for resonant frequency determinations is shown in Figure 12-6. This consisted of a drive system, a detection system, and a monitoring apparatus.

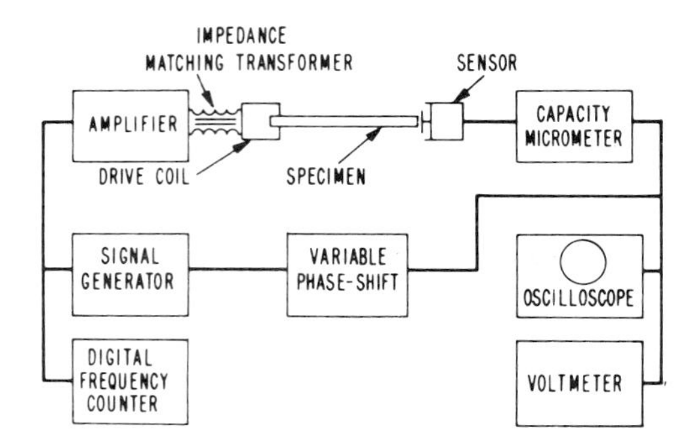

Figure 12-6 Apparatus for resonant frequency and damping determinations.[11] (Reproduced by permission of Georgi Publishing Co.)

The general expression used in the physical determinations for resonant frequency was:

$$\omega_0 = \frac{1}{2l}\sqrt{(E_0/\rho)}$$

This is for the case of continuous longitudinal waves excited parallel to the specimen axis in free harmonic oscillation. Here ω_0 is the natural or resonant frequency of the specimen, l is the length of the test bar, E_0 is the elastic modulus (determined dynamically), and ρ is the density of the material.

The damping capacity δ can be expressed in terms of the logarithmic decrement. Figure 12-7(a) shows an amplitude–frequency curve. Here ω_0 is the resonant frequency and ω_1 and ω_2 are the frequencies at half the maximum amplitude. Then:

$$\delta = \frac{\pi}{\sqrt{3}}\,\frac{\omega_2 - \omega_1}{\omega_0}$$

Figure 12-7(b) shows a graphical representation of results of a regression analysis between the yield strength and tensile strength, and the determined dynamic elastic modulus (DEM). All of the test specimens were machined from a series of 94 castings of unalloyed spheroidal graphite cast iron representing six different compositions, varying mainly in total carbon content. Reasonably good correlation is shown between yield strength and DEM. For both yield strength and tensile strength the correlation diminished somewhat at the higher strengths.

The results demonstrated that for a DEM of 166 GN mm^{-2}, the yield strength is predicted to within ±15 N mm^{-2} and the tensile strength to within 50 N mm^{-2}. A graphical evaluation was also given of the correlation between

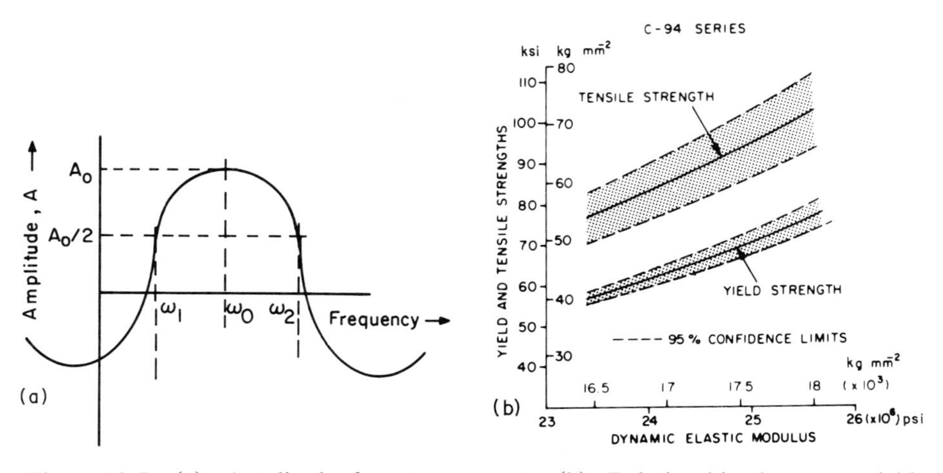

Figure 12-7 (a) Amplitude–frequency curve. (b) Relationship between yield strength and tensile strength, and dynamic elastic modulus. DEM (unalloyed spheroidal graphite cast iron).[11] (Reproduced by permission of Georgi Publishing Co.)

ultrasonic velocities and yield strength. This showed that the DEM determinations were capable of more accurate prediction of yield strength than were the USV determinations.

Further Observations of Kovacs and Cole

Some discussion was given of the use of resonant frequency and its application to prediction of mechanical properties both for gray and spheroidal graphite cast irons. Since resonant frequency is a function of many aspects of microstructure, the spread in correlation between strength and modulus was suggested to be a function of this variation.

Experiments to Determine Structural Influences

Experiments were performed to extend the research and include structural influences. Figure 12-8(a) shows the relationship between dynamic elastic modules (DEM) and carbon content for a series of spheroidal graphite irons cast in a size range of 9, 15, 25, 38, 50, and 75 mm diameter. The carbon contents varied from 2.7 to 4.0 per cent and the carbon equivalents varied from 3.4 to 4.7. The matrix was completely pearlitic.

Figure 12-8(a) shows the large reduction in DEM with increase of casting

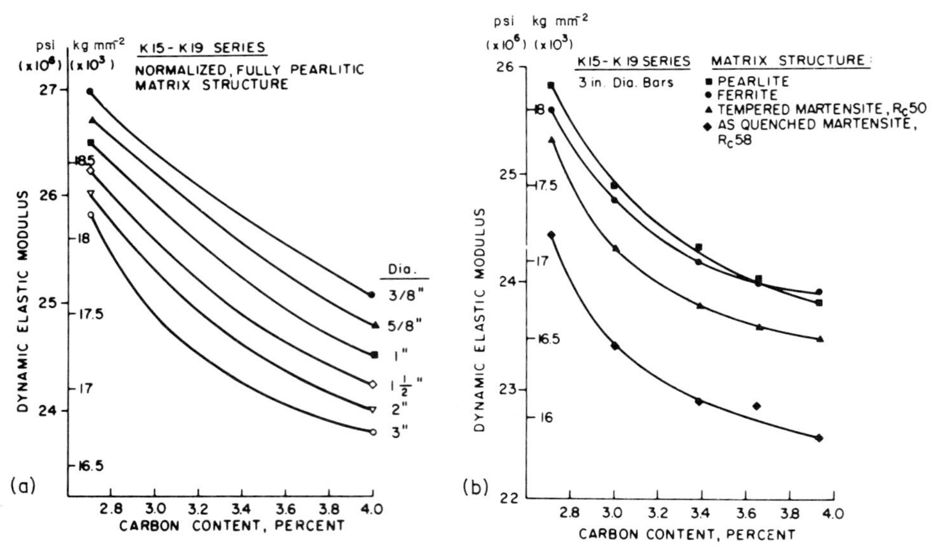

Figure 12-8 (a) Relationship between DEM and carbon content for series of spheroidal graphite irons in different specimen sizes. The results show a large reduction in DEM with increase of casting size, related to an increase in nodule size. (b) Effect of matrix structure on DEM. Results show that DEM is hardly sensitive pearlite–ferrite ratio in the matrix.[11] (Reproduced by permission of Georgi Publishing Co.)

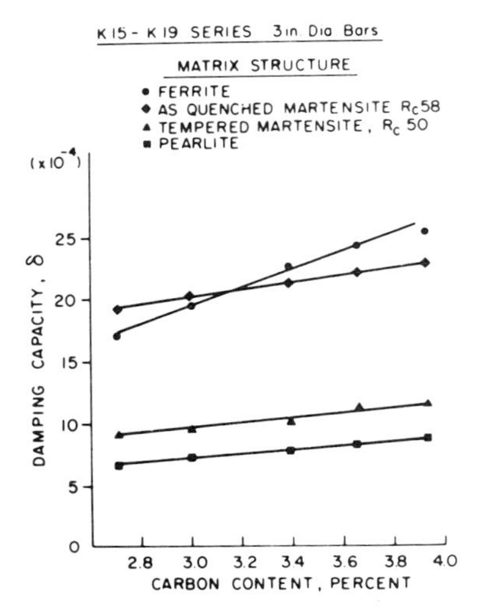

Figure 12-9 Determination of damping capacity on specimens of Figure 12-8(b). Results show that damping capacity is much more sensitive to matrix microstructure than DEM determinations.[11] (Reproduced by permission of Georgi Publishing Co.)

size. This was related to an increase in nodule size. Figure 12-8(b) shows the results of experiments in which the effect of matrix structure on DEM was investigated. Castings of 75 mm diameter were heat treated to give a range of matrix structures, 100 per cent ferritic, 100 per cent pearlitic, 100 per cent martensitic, and tempered martensite. The results show that the dynamic elastic modulus is insensitive to the pearlitic–ferritic ratio in the matrix.

In experiments performed determining the damping capacity on the same set of treated specimens used for DEM, it was shown that this measurement is much more sensitive to the matrix microstructure. The value of δ shows a marked variation as the structure is varied from pearlite to ferrite at the same carbon level (Figure 12-9).

Comparison of DEM and Damping Capacity Data

It is not expected that DEM will change with microstructure at constant graphite volume and shape, since the modulus should be structure insensitive. DEM is more sensitive to carbon content than is the damping. It appears that with so much graphite in the structure, much energy is adsorbed (at the graphite–matrix interface) so that the damping becomes independent of the amount of graphite. Some increase in damping occurs with ferritic microstructures as the graphite volume increases (number of graphite nodules increases).

From the results of Plenard[12] the damping capacity of ferrite is higher than that of pearlite, and hence the sensitivity of damping capacity measurements to this type of microstructural change.

Report by Emerson and Simmons on Graphite Forms and Mechanical Properties of Spheroidal Graphite Cast Iron; Use of Ultrasonic and Sonic Testing

Emerson and Simmons[13] reported on a detailed investigation of sonic methods for testing spheroidal graphite cast iron. The mechanical properties of these irons depend on the perfection or otherwise of the spherulites. Small amounts of imperfect spherulites influence the properties and it is not possible by normal optical methods to adequately ascertain this. Therefore resonant frequency and ultrasonic velocity methods can be resorted to. These measure the modulus of elasticity, and in cast iron this is dependent on graphite form and amount.

While metallography is suggested as applicable for studying the spherulitic form, non-destructive tests using sonic techniques are a better quality control tool for assessing mechanical properties. In particular, the ultrasonic test is applicable to specific locations within castings.

Experimental Procedures

Emerson and Simmons studied ferritic spheroidal graphite cast irons with controlled amounts of non-spherulitic graphite. The longitudinal resonant frequency of the sonic test pieces was measured on a standard manual BCIRA sonic test apparatus to an accuracy of ±0.02 per cent.

The ultrasonic velocity measurements at nominal frequencies of 5, 2.5, and 1.5 MHz were made by measuring the transmission times for ultrasonic waves in the three directions of the ultrasonic test blocks and dividing these times into the measured path lengths. The ultrasonic pulse was obtained from a standard single pulse generator running at a pulse repetition frequency of about 1 kHz

Nodularity Assessment

The percentage nodularity was visually assessed using four experienced metallurgists who worked independently. Using a projection microscope, a count was made in different fields of the number of nodular and non-nodular particles. A particle was considered to be nodular if its aspect ratio (i.e. ratio of axes) was less than 2 to 1.

Further techniques employed were deep etching and examination by a stereo microscope, and use of an image analysing microscope. Evaluation was made by this instrument of the total area of graphite, the sum of the projected heights of the graphite, the number of particles and the number of ends of particles.

Relationship between Mechanical Properties and Nodularity

The mechanical properties measured were UTS, elongation, and yield strength. As the structure deteriorated from nodular form, all these properties decreased. Impact testing was also performed. In notched impact testing, the form of the non-nodular graphite had relatively little effect unless non-spherulitic forms were present in large amounts, and the test was carried out well above the ductil–brittle transformation temperature.

In the unnotched tests, the form of the non-nodular graphite was more significant. When attempting to correlate any of the mechanical properties with nodularity no apparent relationship could be obtained covering the series of irons examined.

Parameters from Image Analysing Microscope

After averaging, the computer was used to calculate the correlation existing between UTS, elongation, yield strength, notched impact, unnotched impact, and modules of elasticity with the following variables of the graphite microstructure: area of graphite, count of graphite particles, projected height of graphite particles, and count of ends of graphite particles. The resultant correlation matrix was examined for relationships which were statistically significant at confidence levels better than 95 per cent. The series of irons examined, having varying magnesium content, showed that few such relationships existed.

Sonic and Ultrasonic Measurements Related to Mechanical Properties

A curve obtained by Emerson and Simmons for the relationship between UTS and ultrasonic velocity is shown in Figure 12-10(a). The results were obtained on two sets of irons and fall on the same line. Figure 12-10(b) for notched impact testing show results for the two sets of irons which fall on two separate curves.

Ultrasonic velocities were measured at frequencies of 1.5, 2.5, and 5.0 MHz. No change in velocity with frequency was observed, showing that the modulus which controls the velocity was independent of frequency in the range employed. The change in velocity measured in the experiments was attributed to variation of the graphite structure.

Note was made of only a fair correlation between ultrasonic velocities measured on cast test blocks and resonant frequencies measured on test bars. the two quantities should be correlated by $(1 - \sigma)/(1 + \sigma)\ (1 - 2\sigma)$, where σ is Poisson's ratio. This was thought to be due to a variation of structure between cast blocks, on which the velocities were measured, and the structure of test bars used for resonant frequency.

Variation of structure occurs from position to position and hence when using these non-destructive methods care must be taken in the selection of location for testing. For calibration, sonic measurements and physical tests

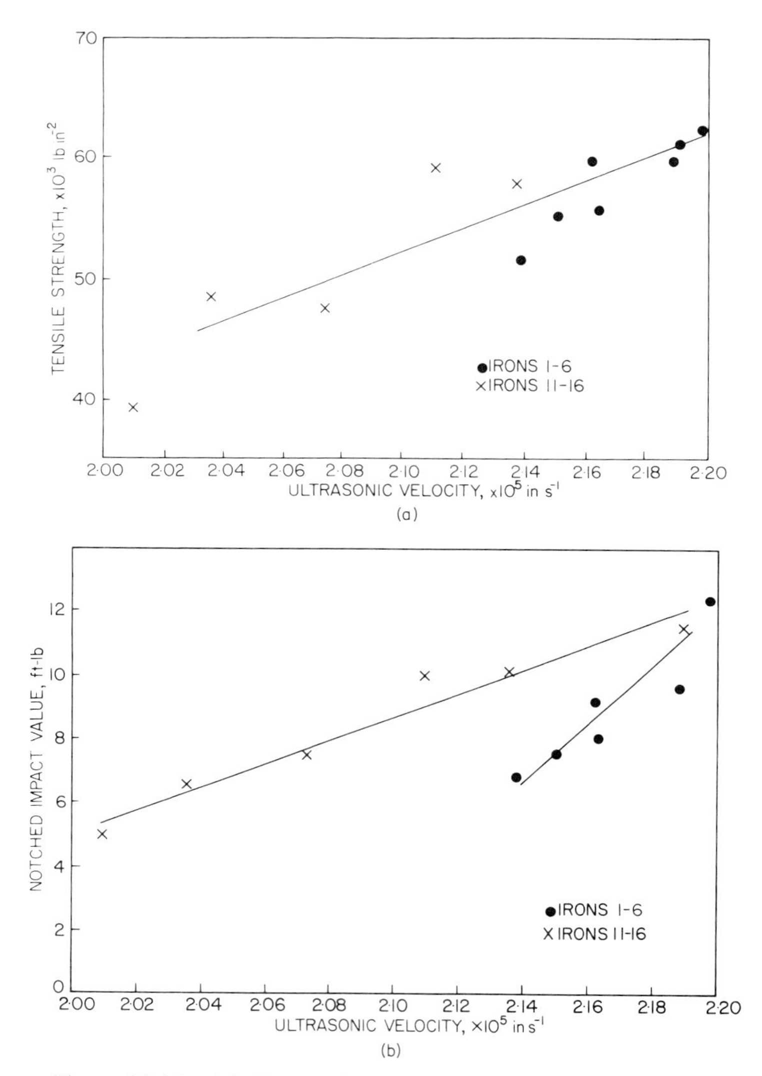

Figure 12-10 (a) Curve obtained by Emerson and Simons[13] showing relationship between UTS and ultrasonic velocity for two sets of irons. The results fall on the same line. (b) Results obtained by Emerson and Simons[13] for relationship between notched impact value and ultrasonic velocity for two sets of irons. The values fall on two separate curves. (Reproduced by permission of American Foundrymen's Society

should be made at the same place. The testing programme showed that non-destructure measurements of resonant frequency or ultrasonic velocity can give a good indication of mechanical properties as influenced by changes in graphite form and amount. It appeared that there was a single relationship applicable to both sets of irons used in the experiment, and this relationship could be extended to include other forms of non-nodular structure. This may obtain provided that the matrix structure remains similar.

References

1. Maranda, R. F., and J. K. Kearns: *Trans. Am. Foundrymen's Soc.*, **82**, 1, 1974.
2. Standke, W.: *Giesserei*, **62(19),** 490, 1975.
3. Ebner, R.: *Giesserei Prax.*, **11**, 183, 1972.
4. Nanstad, R. K., F. J. Worzala, and C. R. Loper: *The Metallurgy of Cast Iron* (Eds. B. Lux, F. Mollard and I. Minkoff), Georgi Publ. Co., Switzerland, 1975.
5. *Test for Plane Strain Fracture Toughness of Metallic Materials*, E-399, Am. Soc. Test. Mater., 1972.
6. Holdsworth, S. R., and G. Jolley.: *The Metallurgy of Cast Iron* (Eds. B. Lux, F. Mollard, and I. Minkoff), Georgi Publ. Co., Switzerland, 1975.
7. Gilbert, G. N. J.: *Iron Steel*, **39**, 432, 1966.
8. Diesburg, D. E., and K. Röhrig: *Giesserei*, **63**, 25, 1976.
9. Burns, D., and K. E. Hofer: *Trans AFS*, **58**, 453, 1977.
10. Zum Gahr, K. H., and W. G. Scholz: *J. of Metals*, **32(10)**, 38, 1980.
11. Kovacs, B. V., and G. S. Cole: *The Metallurgy of Cast Iron* (Eds. B. Lux, F. Mollard, and I. Minkoff), Georgi Publ. Co., Switzerland, 1975.
12. Plenard, E.: *Mod. Cast*, **41**, 144, 1962.
13. Emerson, P. J., and W. Simmons: *Trans. Am. Foundrymen's Soc.*, **84**, 109, 1976.

Bibliography

Roll, F.: *Handbuch der Giesserei-Technik*, Vol. I/I, Springer-Verlag, 1959.
Zener, C.: *Elasticity and Anelasticity of Metals*, Univ. of Chicago Press, 1948.

Index

Acicular Structure, 190, 193
Activity, carbon in liquid iron, 87
  change in ternary system, 90
  coefficient, 87
  coefficient in multicomponent system, 89
  silicon in liquid iron, 87
Adsorption energy, 115
Alloy cast irons, alloy malleable, 228–229
  aluminium, 195–198
  austenitic (manganese), 200–201
  spheroidal graphite (*see* N:-Resist)
  bainitic, spheroidal graphite, 193–195
  copper, 201
  high Cr, high C, white, 175–182
    Mo, white, 183
  low alloy gray, 190
    spheroidal graphite, 190
    white, 183
  martensitic, spheroidal graphite, 193–195
  Ni-Hard, 187–189
  Ni-Resist flake, 189
  Ni-Resist ductile (spheroidal graphite), 190
  silicon, 199–200
  silicon–molybdenum, spheroidal graphite, 200
  tin, 201
Alpha factor, 10
Anisotropic surface free energy, 105
Annealing, 206
Arc furnace melting, 156
Asymmetric coupled zone, 71–74, 78–80
Auger spectroscopy, 51
Austempering, 193–194, 217–218
  and austenitizing temperature, 218
Austenite, carbon content in Fe–Cr–C, 181
  formation, 208
  growth kinetics, 6
  retained, 218
  solidus in Fe–C, 29
  transformation, 185–187, 219

Bainite, 190–192, 211, 213
  crystallography, 214
  metallography, 214
  transformation, 213–214
Blackheart malleable, 228, 235–237
Boron, in Ni–C–B alloys, 162–165
Boundary free energy, 105
Branching of graphite, 63–65
Brittle solid, Orowan model, 248–249
Brinell hardness test, 278
Burton, Cabrera, Frank, growth mechanism, 10–11

C curve derived, 213
Calcium carbide nucleus, 61
Calculated phase diagrams, 28, 30, 179
Carbide phases Fe–Cr–C, 178
Carbide stabilization (table), 180, 237–238
Carbides, hardness, 183
Carbon, activity in liquid iron, 87
  equivalent, 94
  oxygen equilibrium, 135
    in cupola, 146–147
  solubility in multicomponent system, 91–92
Carburizing materials, 149–151
CCT diagrams, 214–217
Cementite, dissolution rate, 238, 241–245
  solubility in austenite, 30
Cerium, interaction with graphite, 103, 115
  treatment of iron, 103
  treatment in intermediate structures, 169–171
Chalmers, constitutional supercooling theory, 15
Charpy impact testing, 286–287
Chill test, 156

Chromium molybdenum irons, 183–184
Chromium white cast irons, 182–183
Chunky graphite (*see* Vermicular graphite)
Coarseness, of graphite in eutectic, 66–68
Coincidence, boundary theory, 42–44
  lattice, 42
Collaud diagram, 264
Compact tensile fracture toughness specimen, 281
Compacted (Vermicular) graphite, 165–171
Competitive growth, 74
Conductivity (*see* Thermal conductivity)
Conical beam camera, 41
Conical helix graphite growth, 109
Constitutional supercooling, 15, 121
Cooperative growth, 26, 74–77
Copper in cast iron, 201
Coral graphite, 123, 161–165
Crack opening displacement, COD, Terminology, 166, 280–284, 286
Critical flaw size, 282
Critical stress intensity factor, K_c, K_{lc}, 280–284
Crystallography of graphite, 9
Cupola, desulphurization, 143–148
Cylindrulite, 106

Damping capacity, 256, 290–291, 293
Degree of saturation, S_c, 93
Dendrite tip growth temperature, 79
Dendritic graphite, bi-crystal, 41
  growth direction, 42
Delta Tee method, 131
Destabilization, 220
Desulphurization in cupola, 143–148
Diesel engine components, 272
Direct reduced iron, 157
Dislocations, edge, 45, 46–47
  growth from screw, 108–109
  impurity interaction, 116–119
  partial, 46
  screw, 12
  twin boundary structure, 45
Distribution coefficient, La in Ni–C, 117
Divorced eutectic growth, 74
Drop weight test, 281
Ductile–brittle failure, 268–271
Ductile iron (*see* Spheroidal graphite cast iron)
Dwell time of atom, 115
Dynamic elastic modulus (DEM) 258, 290–294

Electron beam treatment, 221
Electron microscope microprobe analyser, 51
Electron microscopy of graphite, 48
Electron probe microanalysis, 40, 51
Energy, internal boundaries, 105
Energy theorem of equilibrium form, 103–105
Epitaxial graphite growth, 62
Equilibrium diagrams Fe–A1, 195
  Fe–C, 29, 30, 31
  Fe–C–A1, 195–198
  Fe–C–Cr, 175, 177
  Fe–C–P, 39
  Fe–Cr, 175, 176
  Fe–C–Si, 33–37
  Fe–Ni, 185, 186
  Fe–P, 38
  Fe–Si, 199
Equilibrium form, 104, 105
Eutectic, branching, 63–65
  cell model, 64
  cell frequency, 60
  cell number, 58, 68
  coarseness, 66–67
  competitive growth, 74–77
  cooperative growth, 26, 74–77
  coupled zone, 72, 78
  divorced, 74
  faceted, non-faceted, 23, 25, 80
  fineness, 68
  graphite (gray iron), 63–72
  gray or white, 81–84
  impurity effects, 25
  interface attachment kinetics, 77–78
  interface with liquid, 63
  irregular, 80
  Jackson–Hunt analysis, 23–25
  Kurz–Fisher analysis, 78–81
  lamellar spacing (λ–R) relationships, 66–67
  ledeburite (white iron), 81
  metallic systems, 22–23
  Morrogh–Oldfield model, 64
  nickel–carbon system, 63
  sulphur influence, 67
Eutectoid, area of Fe–C–Si, 33
  transformation, 309–212

Faceted growth, 18
Fatigue, pre-cracked tensile specimen, 283
  thermal, 272–276
Ferrite formation, 97, 98–101
Ferrite–pearlite ratio, 98

Ferrite rings, 282
Ferritic gray cast iron, 94
Ferritic spheroidal graphite iron, 220
Fibrous crack growth, 286
Field ion microscope atom probe mass
 spectometer, 51
First stage graphitization, 229–233
Flake graphite growth, 13
Fracture, ductile brittle, 268–271
 ferrite matrix, 266
 graphite flake, 265
 graphite spherulite, 265–266
 gray iron, 263–268
 mechanics, 263
 pearlite, 267
 toughness testing, 279–283
 toughness white cast iron, 287–288
Furnaces, melting arc, 156
 melting cupola, 143–148
 melting induction, 148

Gamma (γ) phase, *see* Austenite
Gas, effect of dissolved, 155
 hydrogen by mould reaction, 155
 hydrogen effect on eutectic
 undercooling, 69–71
 hydrogen effect on solid state
 graphitization, 246
 hydrogen solubility in liquid alloys,
 151–154
 nitrogen on eutectic undercooling,
 69–71
 nitrogen on mechanical properties,
 137, 149, 155
 nitrogen solubility in liquid alloys,
 137–143
 oxygen, Fe–O–C–Si system, 134–137
 influence on Widmanstätten graphite,
 156
Gibbs–Duhem equation 87
Gibbs–Wulff theorem, 105
Graphite, branching
 chunky (*see* Vermicular graphite)
 63–65
 coarse-fine eutectic, 69
 compacted (*see* Vermicular graphite)
 conical helix form, 109
 conical habit type, 46, 47
 coral, 15, 161–165, 166
 crystal structure, 9
 defect growth, 10
 defect structure (*see* Dislocations)
 dendritic growth, 41
 dislocation density, 11
 eutectic growth, 63–81
 flake (*see* Graphite, lamellar form)
 fracture, 46
 from carbon black, 46
 impurity effects, 15
 instability, 15, 21, 22
 intermediate structures, 169–171
 lamellar form, 13
 morphology, in liquid state growth,
 107–112, 121–123, 171
 in solid state growth, 48, 245–246
 nodule, (*see* Spherulite *and* Spheroidal
 graphite)
 particle spacing and mechanical
 properties, 268
 rate of growth, on (0001), 13
 on $(10\bar{1}0)$, 13
 rotation boundary (twist boundary),
 11–13, 41–44
 solubility, in austenite, 30
 in liquid iron, 30
 spheroid, *see* Spherulite
 vermicular, English terminology
 (table), 166
 growth, 165–171
Graphitization, alloy element influence,
 235–236, 241–245
 Birchenall and Mead model, 233–235
 Burke and Owen model, 229–232
 control by dissolution of cementite,
 238–240
 manganese and sulphur influence,
 236–237
 nucleation, 232–233
 shape of graphite phase, 245–246
 stabilization of carbides, 237
Griffith, fracture criterion, 248–249
Growth, diffusion control, 6–8
 eutectic, 23–25
 faceted crystals, 10–11
 faceted, non-faceted eutectic, 23, 25,
 80
 graphite eutectic, 63–81
 interface control, 8–9
 lamellar form, 13
 ledeburite, 81
 mechanisms, 6
 Zener model, 6
 Zener model, modified Hillert, 7–8

Hardening, surface, 220–226
Hardness, Brinell, 278
Harth diagrams, 223–225
Heat treatment, annealing, 206–207
 austempering, dactile iron,
 217–218

bainitic spheroidal graphite iron, 193–195
CCT diagrams, 214–217
Fe–Cr–C alloys, 182, 219–220
Fe–Cr–Mo alloys, 183–184, 219–220
ferritic spheroidal graphite iron, 220
martensitic spheroidal graphite iron, 193–195
pearlitic spheroidal graphite iron, 220
stress relief, 207–208
TTT diagrams, 211–213
Helical growth mechanisms spherulite, 109
Henry's law, 87
Herring bone structure spherulite, 109
Heterogeneous nucleation, 3–5, 54–58
Hillock growth angle, 22
Holes in graphite crystals, 116–119
Homogeneous nucleation, 2–3
Hopper crystal, 14
Hour glass form, 22

Ice crystals, 112
Image analysis microscopy, 295
Impact toughness tests, 266
Impact transition temperature, 268–271
Imperfect spherulitic forms, 123
Impurity effects from melting process, 155
Induction furnace melting, 148, 156
Inglis, stress distribution, 248
Inoculants, 58–59
Inoculation, 156
model by carbides, 61
model by duplex sulphide oxide, 6
Instability in growth, 15
Chernov model, 18
graphite morphology, 122
graphite types, 22
graphite undercooling, 122
Instrumental techniques, 40
Interaction energy of elements with graphite, 115
Interface, attachment kinetics, 9, 77
coherent, 56
incoherent, 58
semi-coherent, 56
stability (*see also* Instability)
Interfacial energy, in heterogeneous nucleation, 5
in nucleation, 56–58
Intermediate graphite forms, 159–160
Internal stress (*see also* Residual stress), 203–208

Inverse chill, 84
Ion-etched spheroidal graphite sections, 109
Ion-microscope mass analyser, 51
Irregular eutectic, 80
Isothermal transformations, 219

Kinetic attachment coefficient, 19, 22
Kinetic undercooling, 113
by lanthanum, 116
Cabrera theory, 116
calculation, 121
Chernov theory, 114
Kink, atom attachment, 12, 114
spacing, 115

Lamellar spacing (λ–R) relationships, 66–67
Lanthanum, distribution coefficient, 117
interaction with graphite, 115, 118
Laplanche diagram, 95
Lead, influence on graphite, 120
Ledeburite eutectic, 81
Liquid metal preparation, 134
Life time, atom on surface, 115
Low alloy, gray iron, 190
spheroidal graphite iron, 190
Low index faces of crystal, 105

Magnesium, interaction with graphite, 103, 115, 168–171
role in undercooling, 114
Magnesium silicide, hopper growth, 14
Malleable iron, graphitization, 229–246
types, 228–229
Manganese, austenitic cast irons, 200–201
Martensite, 184
hardness, 181
M_s temperature, 218
tempered, 194
Martensitic white cast iron, 187–188
Maurer diagram, 94
Mechanical behaviour, brittle fracture, 248–249
critical flaw size, 282
damping capacity, 291
ductile–brittle failure, 268–271
fracture toughness K_c, K_{Ic}, 280–288
modulus, 257–258
stress-deformation models, 248–255
stress–strain, 255–257
tensile failure, 263–266

thermal fatigue, 272–276
UTS, spheroidal graphite iron, 295–297
Mechanical properties, acicular flake irons (table), 193
aluminium cast irons (table), 195
cast irons, compared (table), 166
pearlitic, flake irons (table), 193
Ni-Hard (table), 189
Mechanical testing, crack opening displacement COD, 283–286
damping capacity, 291, 293
dynamic elastic modulus DEM, 291, 293
elastic modulus, 256
fracture toughness K_c, K_{Ic}, 279–288
hardness, Brinell, 278
impact, Charpy, 286
pulsed echo, 290
resonance, 290–292, 294–297
slow bend, 266
stress–strain, 256
thermal fatigue, 272–275
ultrasonic velocity, 294–296
wedge penetration, 278–279
Meehanite, 58
Melt chemistry and Periodic System, 129
Melting, influence on mechanical properties, 148–149, 156–157
Melting furnaces, arc, 156–157
cupola, 143–148
induction, 148–149
plasma arc, 157
Metastable system, 86
Microprobe analysis, 51–52
Microsegregation, 172–174
Models, mechanical behaviour, Clough and Shank, 253–255
Coffin, 252–253
Fisher, 250–252
Orowan, 248–249
Modulus, calculated, 258–262
dependent on graphite form, 258–259
dependent on graphite quantity, 258
deviation of static and dynamic, 258–259
dynamic, 258–260
influence of particle size, 262
secant, 258
tangent, 257–258
temperature dependence, 262
Molybdenum, effect on austenite transformation, 99
in M_2C carbide, 183
in M_6C carbide, 183
in Ni–Mo irons, 191–195
in Si–Mo cast iron, 200
in White Irons, 183–184
Morphological change, 171
Morphology of graphite, 122
Morrogh and Williams, graphite formation, 102
Morphological stability (*see* Interface Stability), 17
Mullins and Sekerka theory, 17, 18

Network of adsorbed atoms, 116
Nickel, effect on austenite transformation, 185–187
in cast iron, 185–190
–molybdenum bainitic S.G. iron, 193
Ni-Hard, 187–189
ductile types, 190
Ni-Resist, flake graphite types, 189
Nitrogen, influence on structure, 155
influence on undercooling, 71
solubility in Fe–C, 137
solubility in Fe–C–Si, 140–143
solubility effects of O and S, 143
solubility kinetics, 139
Nodule count, 127–128
Nodularity, 127–128
assessment, 294–295
Nomenclature of cast irons, 166
No-partition temperature, 99–101
Nucleated growth on $(10\bar{1}0)$, 12–13
Nucleation, 56
coherent, 58
contaminants, 5
frequency (*see* Nucleation rate)
heterogeneous, 3–5, 54–58
homogeneous, 2–3
rate, 5
two-stage process, 59
Nucleus, carbide, 61
duplex sulphide oxide, 61–63

Orowan, brittle solid, 248
Oxygen, adsorption, 134
and graphite nucleation, 137

Para cementite, 99
Para equilibrium, 99
Para ferrite, 99
Para pearlite, 99
Partition coefficient, La in Ni–C, 117
Partitioning of solute, and γ transformation, 98

Cahn and Hagel analysis, 99
double diagram, 99
Pearlite, deformation and fracture, 267–268
transformation kinetics, 211–212
Periodic system, melt chemistry, 129
Peritectic reactions, Fe–Cr–C, 180
Phase diagrams (*see* Equilibrium diagrams)
Phase transformations, 208–220
Phosphide eutectic, 36
Plane strain fracture toughness tests, 281
Pro-eutectoid ferrite, 209–211
Pulsed echo test, 290
Pyramidal growth, 110
Pyramids, graphite surfaces, 110

Raoult's law, 87
Rare earths, in intermediate structures, 159, 167, 171
Reactive elements, 114, 119
Refining, liquid metal, 157
Residual stress, 204–208
Resonant frequency testing, 290–294
Retained austenite, 218
Rod forms of graphite, 120, 161
Rotation boundaries, 109
Rotational factor, 284
Roughness, degree of, 10

Saturation ratio, S_r, 93
Scanning Auger microscopy, 51
Scanning electron microscopy, 40, 49
Scheil segregation equation, 172
Screw dislocations, 219, 220
Secondary carbides, 220
Segregation, 113, 124, 172–174
Sigma phase, 175, 181
Silicon, activity in liquid iron, 87
cast irons, 199–200
cementite formation and, 98
dioxide and inoculation, 59
effect on γ-pearlite transformation, 99
graphite growth and, 98
graphite nucleation and, 98
–molybdenum S.G. iron, 200
oxygen equilibrium, 135
structural changes and, 97
Single edge notched specimen, 281
Slow bend test, 266
Solidification, in metallurgical systems, 1
spheroidal graphite iron, 102
white influenced by melt chemistry, 83
white or gray, 81–84
Solute, boundary layer and instability, 119
distribution at growing interface, 15
partitioning, 98–101
Sonic testing, 288
BCIRA apparatus, 294
Spacing, between eutectic phases, 66–67
between kinks, 115
Spheroid (*see* Spherulite)
Spheroidal graphite (*see* Spherulite)
Spheroidal graphite cast iron, 102–103 (*see also* Alloy cast irons)
Spherulite, Shubnikov's model, 110
graphite growth types, 50
imperfect forms, 123
Spherulitic growth, control by thermal measurement, 127–129
interface instability, 112
related to undercooling, 113–121
reviews, 124
ribbed structure, 120
screw dislocations, 108
studied by thermal measurements, 124–126
surface energy models, 103, 107
Spinel nucleus, 62–63
Spiral growth, 12
Stability, instability, 15
Stacking faults, 46
Stacking fault ribbons, 46
Step instability, 162–165
Step motion, 10
Steps on crystal surface, 10–12
Strength of cast iron, 248–255
Stress at graphite flake, 204, 257
Stress from martensite transformation, 207
Stress–strain behaviour of cast iron, 255–256
Stress–strain modulus, 257
Stress relief, 206, 208
Strongly adsorbed atoms, 114–115
Structural diagrams, 94–98
Sulphur, effect on eutectic coarsening, 69
effect on graphite surface energy, 107
Supersaturation over faceted surface, 19
Surface alloying, 225
Surface energy, growth models, 107–108
Surface hardening, 220
electron beam, 221
laser beam, 222–226
Symmetrical coupled zone, 71
Synthetic cast iron, 148

Temperature, of eutectic solidification, 59
of undercooling of melt, 59
Tensile failure, 263
Ternary eutectic, 36
Tesselated stresses, 204
Thermal analysis, 124–127
Thermal conductivity values of irons, 167
Thermal fatigue, 272–276
Thermal stress, 272
Thermochemistry of cupola, 144–148
Thermodynamics and structural diagrams, 97
Thermodynamics of Fe–C system, 86
Tilt boundaries, 46
Transformation of γ to ferrite or pearlite, 98
Transmission electron microscopy, 46
Twins in graphite, crystallography, 45
dislocation structure, 45
metallography, 44
Twist boundary (*see* Rotation boundaries)

Ultrasonic velocity testing, 294
and UTS, 295
Undercooling, and composition, 113
and growth forms, 120–122
calculated, 120
kinetic, 116–119, 121
kinetic and constitutional, 119

Vermicular graphite (*also* Chunky and Intermediate graphite), 159–171
coral graphite (England), 166

Wagner interpolation formulae, 89
Wedge penetration test, 278–279
White cast iron, 175
Widmanstätten graphite, 156
Wulff theory, 105

X-ray, data of epitaxy, 62
diffraction techniques, 41
diffraction topography, 46
study of twinning in graphite, 45

Zener, growth model, 6
model modified by Hillert, 7, 8